Uncertainty and Reliability Analysis for Engineering Systems

Uncertainty and Reliability Analysis for Engineering Systems

Guest Editors

Guijie Li

Feng Zhang

MDPI

Basel • Beijing • Wuhan • Barcelona • Belgrade • Novi Sad • Cluj • Manchester

Guest Editors

Guijie Li
School of Mechanics and
Aerospace Engineering
Dalian University of Technology
Dalian
China

Feng Zhang
School of Mechanics, Civil
Engineering and Architecture
Northwestern Polytechnical University
Xi'an
China

Editorial Office
MDPI AG
Grosspeteranlage 5
4052 Basel, Switzerland

This is a reprint of the Special Issue, published open access by the journal *Applied Sciences* (ISSN 2076-3417), freely accessible at: https://www.mdpi.com/journal/applsci/special_issues/7B9VQG5K8J.

For citation purposes, cite each article independently as indicated on the article page online and as indicated below:

Lastname, A.A.; Lastname, B.B. Article Title. *Journal Name* **Year**, *Volume Number*, Page Range.

ISBN 978-3-7258-6986-2 (Hbk)
ISBN 978-3-7258-6987-9 (PDF)
https://doi.org/10.3390/books978-3-7258-6987-9

Contents

About the Editors

Guijie Li

Guijie Li, Associate Professor at Dalian University of Technology, dedicates his research primarily to aircraft mechanisms and structural design, reliability analysis of complex systems, and uncertainty quantification. With a keen focus on reliability design and evaluation of complex structures under the coexistence of aleatory and epistemic uncertainties, as well as systematic importance measure analysis, he has pioneered an innovative framework for uncertainty propagation and importance measure analysis under combined aleatory and fuzzy uncertainties. This framework has been successfully validated and applied to a range of practical engineering scenarios, yielding tangible technical benefits. Recently, targeting the intractable challenges posed by complex high-dimensional nonlinear problems, he has launched cutting-edge research on hierarchical reliability and uncertainty quantification methodologies empowered by advanced artificial intelligence algorithms. Boasting over 40 peer-reviewed academic papers, he also holds key academic positions: serving as an editorial board member of *Equipment Environmental Engineering* and guest editor for special issues of the renowned international journals *Materials* and *Applied Sciences*.

Feng Zhang

Feng Zhang received his PhD in aircraft design from Northwestern Polytechnical University in 2010, and he joined Northwestern Polytechnical University in September 2010 as an assistant professor. In 2014, he became an associate Professor. His research includes aircraft design and reliability engineering.

Preface

As a core performance metric of products, reliability has received increasing attention in recent years. For complex engineering systems, uncertainties in geometric parameters, material characteristics, and applied loads are more prominent. To ensure that engineering systems meet high-reliability standards, these uncertainties should be fully taken into account during design, manufacturing, and testing processes. However, given the complexity of the subject, a host of challenging problems still await in-depth investigation.

This Reprint presents the latest advances in the field of uncertainty and reliability analysis of engineering systems from three perspectives: fundamental innovations (e.g., efficient optimization algorithms and a unified modeling framework), practical applications (including infrastructure protection and structural vibration control), and interdisciplinary integration (utilizing machine learning and Bayesian inference for predictive maintenance and fault analysis). It incorporates research findings from diverse fields, including structural engineering, energy, rail transit, and industrial metrology, and presents the cutting-edge trends of this domain through a variety of case studies. The Reprint is intended to provide high-value resources for researchers, practitioners, and graduate students, and ultimately commits to facilitating academic exchanges and advancing the development of engineering systems characterized by uncertainty quantification and reliability enhancement.

Guijie Li and Feng Zhang

Guest Editors

applied sciences

MDPI

Editorial

Uncertainty and Reliability Analysis of Engineering Systems: Theory, Methods, and Applications

Guijie Li [1,*], Lai Zhang [1], Feng Zhang [2] and Xue Li [2]

[1] School of Mechanics and Aerospace Engineering, Dalian University of Technology, Dalian 116024, China
[2] School of Mechanics and Transportation Engineering, Northwestern Polytechnical University, Xi'an 710129, China
* Correspondence: ligj@dlut.edu.cn

1. Introduction

Uncertainty and reliability analysis constitutes a core pillar in the design, operation, and assessment of engineering systems, directly impacting safety, efficiency, and sustainability. In practical engineering scenarios, uncertain factors—such as material property variations, load fluctuations, environmental disturbances, and human operation differences—are pervasive. Reliability, which is defined as the capacity of a system to stably perform its intended functions under specified conditions for a predetermined period, serves as a fundamental criterion for evaluating the performance of engineering systems. These two concepts are intrinsically linked: uncertainty is the primary source of reliability risks in systems, and scientific reliability analysis must be based on the accurate identification and quantification of uncertainty. Neglecting these critical factors may culminate in catastrophic consequences such as structural failure, equipment malfunction, and decision-making biases. Consequently, uncertainty and reliability analysis has become an essential research priority in the engineering field.

This book presents ten thematic papers centered on "Uncertainty and Reliability Analysis of Engineering Systems". The contributions cover diverse engineering scenarios and cutting-edge analytical methods, providing a comprehensive overview of recent research advancements in the field.

The selected papers exhibit both theoretical depth and practical value, focusing on three core dimensions: firstly, innovations in fundamental theories and methods, including uncertainty-driven reliability assessment, the optimization of efficient algorithms, and the development of unified modeling frameworks; secondly, practical applications in engineering, specifically uncertainty quantification and reliability assurance in structural vibration control, the preservation of historical infrastructure, and industrial measurement; and thirdly, the integration of cutting-edge interdisciplinary technologies, applying machine learning and Bayesian inference to performance prediction, failure prevention, and multi-mode fault analysis. Each achievement contains innovative theoretical insights and technical solutions. We are honored to present this systematic collection, which integrates research achievements on uncertainty and reliability analysis from multiple engineering fields such as structural and energy engineering, rail transit, and industrial measurement. It not only showcases the latest research trends in this field but also provides rich practical application cases. This book serves as a valuable academic reference and technical resource for researchers, frontline engineering and technical personnel, and graduate students specializing in system reliability and uncertainty analysis.

We hope that this book will promote academic discussions and the exchange of achievements in this field, drive the innovative development of theories and methods for uncertainty and reliability analysis, facilitate their practical application in a wider range of engineering scenarios, and provide strong support for improving the safety, reliability, and durability of engineering systems.

2. Overview of Contributions

The following is a brief overview of the ten contributions included in this book, centered on the core themes of uncertainty quantification and reliability analysis in engineering systems, two interconnected concepts critical to ensuring the safety, efficiency, and sustainability of complex engineering infrastructures. Uncertainty, defined as the lack of complete knowledge about system properties, operating conditions, or measurement outcomes, and reliability, the ability of a system to perform its intended function under specified conditions, are addressed through diverse methodological innovations and practical applications across multiple engineering domains.

In [1], the authors tackle the challenge of reliability assessment for complex engineering systems operating under variable conditions that induce component degradation. They propose a two-level framework integrating Reliability Block Diagrams (RBDs), Fault Tree Analysis (FTA), Importance Measures, and Failure Mode and Effects Analysis (FMEA). By incorporating uncertainty through the statistical parameters of component reliability distributions, the framework identifies critical components via Importance Measures and quantifies dominant failure modes using Risk Priority Numbers (RPN). Applied to a fleet of Solid Waste Collection and Compaction Trucks in Colombia, the approach establishes a probabilistic basis for maintenance prioritization, enhancing the accuracy of critical component and failure mode identification.

To address the challenges of limited reliability evaluation efficiency and the inability to ensure platform-independent performance across different multi-core architectures, which arise when Reliability Block Diagrams are applied to the scenario of efficient online evaluation for resource-constrained embedded hardware, the authors of [2] proposed a solution by developing an enhanced version of the open-source librbd library. This library is equipped with optimized evaluation algorithms, restructured computational sequences, cache-aware data structures, and an adaptive parallelization framework. The verification results demonstrate a significant reduction in computational complexity, enabling real-time analysis of larger-scale systems compared with the original implementation.

To improve the reliability and safety in complex energy systems such as wind turbines, the authors of [3] present a unified hybrid modeling framework integrating FTA, RBD, and BowTie methodology. This framework quantifies risk, evaluates safety barrier effectiveness, and supports scenario-based sensitivity analysis using key metrics including availability and failure probability. A simulation-based case study on a wind turbine generator subsystem shows that critical defense points (e.g., protective relays) are identified, and targeted improvements (e.g., enhanced oil analysis, redundant dashboards) reduce consequence frequency by nearly two orders of magnitude. The approach introduces explicit defense-in-depth modeling, improving computational efficiency and providing a practical decision support tool for asset managers. Focusing similarly on wind turbine systems, the authors of [4] address the high time and cost of wind turbine blade fatigue testing by exploring the application of an improved Neural Basis Expansion Analysis (N-BEATS) deep learning model for stiffness degradation prediction. To optimize the model's performance, they expand stiffness data to meet model requirements, modify N-BEATS' basic block structure to treat sequence-to-sequence prediction as a nonlinear multivariate regression problem, and adopt the Pinball Mean Absolute Percentage Error (Pinball-MAPE) loss function to reduce bias. Additionally, two data augmentation techniques (time series combination and random noise injection) mitigate overfitting. Experimental results

https://doi.org/10.3390/app16020957

demonstrate the model's ability to learn underlying stiffness patterns and accurately predict remaining stiffness, offering a cost-effective alternative to traditional fatigue testing.

In [5], the authors address the robust design of Tuned Mass Dampers (TMDs) in linear structural dynamics. For single-degree-of-freedom (SDOF) and multi-degree-of-freedom (MDOF) systems, the authors propose an efficient analytical uncertainty quantification method based on Taylor series expansion, estimating the mean and scatter of key performance indicators without the computational burden of double-loop sampling approaches. An additional mode decomposition technique for MDOF systems with multiple TMDs replaces time-consuming time integration with modal analysis. Formulated as single- or multi-objective optimization tasks, the approach is validated through numerical examples, offering a practical solution for TMD design in structures prone to critical dynamical vibrations.

How to protect historical railway assets amid uncertainties in material performance, structural degradation, and regulatory requirements? The authors of [6] propose a reliability-oriented preservation framework that systematically accounts for legal, material, and procedural uncertainties. Two industrial case studies validate the framework: the successful substitution of timber with certified PVC cladding in a non-listed signal box (improving durability and reducing maintenance needs) and an unsuccessful timber-to-aluminum gable replacement (attributed to planning misalignment and unestimated approval uncertainties). This research provides a structured method for mitigating uncertainties in the asset management of historical infrastructure.

The authors of [7] investigate measurement uncertainty and compliance evaluation in natural gas moisture analysis, a critical factor in preventing pipeline corrosion and ensuring combustion efficiency. The authors compare multiple algorithms for moisture content calculation, estimate and validate measurement uncertainty using the Panametrics PM880 portable hygrometer, and evaluate compliance with Brazilian legislation via guard bands. The results reveal a maximum variation of 1% in moisture content results across the compared algorithms, with an expanded uncertainty of approximately 20%. Notably, this uncertainty value does not compromise risk assessment, as the measured moisture levels remain well within acceptable ranges below the specified limits, which ultimately leads to the establishment of an upper tolerance limit of 58.4 ppmv H_2O.

In [8], the authors focus on uncertainty quantification in shear wave velocity (Vs) prediction, critical for earthquake engineering applications. Geotechnical variability and inherent uncertainties are addressed through a hybrid framework integrating an explainable machine learning (ML) model and Bayesian inference. The Extreme Gradient Boosting (XGBoost) algorithm, coupled with Shapley Additive Explanations (SHAPs) and partial dependency analysis, identifies key geotechnical parameters and delivers high predictive accuracy. A Bayesian Generalized Linear Model (GLM) complements this by providing probabilistic predictions with 95% credible intervals, explicitly quantifying uncertainty. Validated against real case scenarios, the hybrid approach combines ML's predictive power with Bayesian inference's uncertainty quantification, offering an interpretable tool for confident engineering decision-making.

The prevention of situation awareness failure in shearer monitoring operations, an issue that has been exacerbated by the digitization of instrument control systems, is explored from a cognitive function perspective with a focus on the impact of working memory training. In [9], the authors randomly assigned subjects to either a training group or a control group, conducting pre- and post-training assessments that included working memory measurements, simulated shearer monitoring tasks, and recordings of task performance, situational awareness scales, and electroencephalography (EEG) data. The results revealed significant improvements in both monitoring performance and situational awareness scores following training, accompanied by activation in the θ, α_2, and β_1 EEG frequency bands—

https://doi.org/10.3390/app16020957

findings consistent with the established link between these oscillatory bands and cognitive processes underlying working memory and decision-making. The study demonstrates that working memory training enhances operators' rapid reaction and decision-making abilities in complex or emergency scenarios, thereby providing a new direction for mitigating situation awareness failure in high-stakes monitoring environments.

To enhance the efficiency of reliability analysis for nonlinear implicit models of complex structures under multi-failure modes, the authors of [10] propose a surrogate model based on an adaptive AdaBoost algorithm. An adaptive method selects optimal training samples, ensuring even distribution across failure curve boundaries and comprehensive information retention. Leveraging AdaBoost's integration and iterative properties, simple binary classifiers are iteratively combined to build a high-precision alternative model for fault diagnosis, with Monte Carlo simulation used to assess failure probability. Validated through three numerical examples, the method exhibits exceptional accuracy and efficiency, overcoming challenges in multi-failure mode reliability evaluation and facilitating the application of complex mechanical designs.

3. Conclusions

The above overview of the ten papers in this book shows that uncertainty and reliability analysis is a crucial issue in various engineering fields. It can be stated that uncertainty and reliability analysis should be properly considered across all domains, in any structural, energy, rail transit, and industrial measurement application. It is my aspiration that this book integrates contributions from various engineering fields, thereby enriching the research landscape with this pivotal topic.

Author Contributions: Writing—original draft preparation, G.L.; writing—review and editing, L.Z., F.Z. and X.L.; supervision, G.L. All authors have read and agreed to the published version of the manuscript.

Funding: This research was funded by Science Challenge Project (No. TZ2025003) and the National Natural Science Foundation of China (Grant No. NSFC 52275143).

Acknowledgments: The successful publication of this book would not have been possible without the dedicated efforts of the outstanding authors, professional and rigorous reviewers, and the editorial team of *Applied Sciences*. Meanwhile, congratulations are extended to all authors. We would like to take this opportunity to express our sincere gratitude to all reviewers and editors. Finally, we would like to convey our thanks to the editorial team of *Applied Sciences*.

Conflicts of Interest: The authors declare no conflicts of interest.

Abbreviations

The following abbreviations are used in this manuscript:

RBD	Reliability Block Diagrams
FTA	Fault Tree Analysis
FMEA	Monte Carlo Simulation
RPN	Risk Priority Numbers
N-BEATS	Neural Basis Expansion Analysis
Pinball-MAPE	Pinball Mean Absolute Percentage
TMD	Tuned Mass Dampers
SDOF	Single-Degree-of-Freedom
MDOF	Multi-Degree-of-Freedom
Vs	Shear Wave Velocity
ML	Machine Learning
XGBoost	Extreme Gradient Boosting

 https://doi.org/10.3390/app16020957

SHAPSs	Shapley Additive Explanations
GLM	Generalized Linear Model
EEG	Electroencephalography

References

1. Clavijo-Villamizar, M.A.; Guevara Carazas, F.; Arango-Castrillón, J.D.; Patino-Rodriguez, C.E. Uncertainty-Driven Reliability Analysis Using Importance Measures and Risk Priority Numbers. *Appl. Sci.* **2025**, *15*, 11867. [CrossRef]
2. Gori, G.; Papini, M.; Fantechi, A. Efficient Reliability Block Diagram Evaluation Through Improved Algorithms and Parallel Computing. *Appl. Sci.* **2025**, *15*, 11397. [CrossRef]
3. Parnon, M.A.A.; Papadopoulou, K.A.; Sinha, J.K. A Unified Modelling Framework Combining FTA, RBD, and BowTie for Reliability Improvement. *Appl. Sci.* **2025**, *15*, 10902. [CrossRef]
4. Yang, S.; Gao, J.; Yuan, Y.; Zhou, J.; Meng, L. Prediction of Wind Turbine Blade Stiffness Degradation Based on Improved Neural Basis Expansion Analysis. *Appl. Sci.* **2025**, *15*, 1884. [CrossRef]
5. Most, T.; Zabel, V.; Das, R.R.; Khadka, A. An Efficient Uncertainty Quantification Approach for Robust Design of Tuned Mass Dampers in Linear Structural Dynamics. *Appl. Sci.* **2025**, *15*, 9329. [CrossRef]
6. Wailes, T.; Khan, M.K.; He, F. A Reliability-Oriented Framework for the Preservation of Historical Railway Assets Under Regulatory and Material Uncertainty. *Appl. Sci.* **2025**, *15*, 4705. [CrossRef]
7. Moreira, R.M.; Biazon, C.L.; de Oliveira, E.C. Measurement Uncertainty and Compliance Evaluation Applied to Natural Gas Moisture. *Appl. Sci.* **2025**, *15*, 2482. [CrossRef]
8. Chala, A.T.; Ray, R. Uncertainty Quantification in Shear Wave Velocity Predictions: Integrating Explainable Machine Learning and Bayesian Inference. *Appl. Sci.* **2025**, *15*, 1409. [CrossRef]
9. Yuan, X.; Song, R.; Sun, L. Research on the Effect of Working Memory Training on the Prevention of Situation Awareness Failure in Shearer Monitoring Operations. *Appl. Sci.* **2024**, *14*, 11876. [CrossRef]
10. Zhang, F.; Qiao, Z.; Tian, Y.; Wu, M.; Xu, X. Reliability Analysis of Complex Structures Under Multi-Failure Mode Utilizing an Adaptive AdaBoost Algorithm. *Appl. Sci.* **2024**, *14*, 10098. [CrossRef]

https://doi.org/10.3390/app16020957

applied sciences

MDPI

Article

Reliability Analysis of Complex Structures Under Multi-Failure Mode Utilizing an Adaptive AdaBoost Algorithm

Feng Zhang *, Zijie Qiao, Yuxiang Tian, Mingying Wu and Xiayu Xu

School of Mechanics, Civil Engineering and Architecture, Northwestern Polytechnical University, Xi'an 710129, China

* Correspondence: nwpuwindy@nwpu.edu.cn

Abstract: A reliability analysis can become intricate when addressing issues related to nonlinear implicit models of complex structures. To improve the accuracy and efficiency of such reliability analyses, this paper presents a surrogate model based on an adaptive AdaBoost algorithm. This model employs an adaptive method to determine the optimal training sample set, ensuring it is as evenly distributed as possible on both sides of the failure curve and fully contains the information it represents. Subsequently, with the integration and iterative characteristics of the AdaBoost algorithm, a simple binary classifier is iteratively applied to build a high-precision alternative model for complex structural fault diagnosis to cope with multiple failure modes. Then, the Monte Carlo simulation technique is employed to meticulously assess the failure probability. The accuracy and stability of the proposed method's iterative convergence process are validated through three numerical examples. The findings of the study illuminate that the proposed method is not only remarkably precise but also exceptionally efficient, capable of addressing the challenges related to the reliability evaluation of complex structures under multi-failure mode. The method proposed in this paper enhances the application of mechanical structures and facilitates the utilization of complex mechanical designs.

Keywords: AdaBoost algorithm; adaptive iteration; complex structure; multiple-failure mode; failure probability

1. Introduction

The aim of the structural reliability analysis is to meticulously quantify the failure probability that arises from a multitude of uncertain factors, such as the geometric size error, variations in material properties, and external applied load [1,2]. To improve system reliability and simplify the reliability-based design problems widely applied in various engineering practices [3,4], many scholars have studied the reliability of multi-failure mode systems [5–10]. For example, Marozaua et al. [7] conducted a reliability assessment to solve specific safety assessment problems, analyzed the underlying causes of micro-electro-mechanical system device failure, and carried out a failure test using a typical micro-electro-mechanical system device package structure. Wang et al. [8] introduced a probabilistic analysis framework aimed at evaluating the reliability of turbine discs while accounting for the interdependencies among various failure modes. Meng et al. [9] proposed a hybrid reliability-based topology optimization method to deal with cognitive and cognitive uncertainty. Yang et al. [10] proposed an innovative reliability analysis method that integrates Markov chain Monte Carlo method with random forest algorithms. J. Guadalupe Monjardin-Quevedo, Alfredo Reyes-Salazar, et al. [11] explored the properties of steel by evaluating the seismic reliability of deep column smf. The first-order reliability method, response surface method, and an advanced probability scheme were adopted. The efficiency and accuracy of the proposed method were verified by the traditional Monte Carlo simulation method. German Michel Guzman-Acevedo, Juan A. Quintana-Rodriguez, Jose Ramon Gaxiola-Camacho, et al. [12] evaluated the Usumacinta Bridge in Mexico based on probabilistic methods, The Sentinel-1 image was used to define structural reliability. The case for selecting probabilistic methods integrates the reliability of the structure. Displacement theory and probability density function obtained by InSAR technique (pdf).

Common structural reliability analysis methods include the digital simulation method [13,14], approximate analytical method [15,16] and surrogate model methods. The Monte Carlo method [17] is a relatively straightforward digital simulation method. However, the large number of samples required for obtaining an accurate solution makes the method less efficient. Another digital simulation method known as importance sampling [18] utilizes the design point of the limit state equation as its sampling center, thereby enhancing sampling efficiency to a certain extent. However, this design point must be predetermined, which limits the applicability of this method to complex structures. Approximate analytic methods, such as the first-order second-moment method [19], necessitate minimal computational effort but yield large relative errors. Consequently, these methods are only suitable for solving the reliability problem associated with an explicit limit state equation. In view of the nonlinear implicit characteristics of complex mechanical structures, surrogate modeling methods such as the Kriging algorithm [20], neural network algorithm [21], and support vector machines algorithm [22] are frequently utilized to assess the reliability of these intricate mechanical structures. However, these algorithms are simple extensions of the single traditional classification algorithm for specific data, and they inevitably inherit the inherent shortcomings of the original algorithm [23]. Simultaneously, these algorithms rely on relatively idealized assumptions regarding probability distribution and data types when dealing with data uncertainty, which consequently hampers their widespread applicability in practical projects [24].

Traditional reliability analysis methods have accuracy problems, large errors, and low computational efficiency, making it difficult to effectively apply them in practical engineering. To address these challenges, this paper proposes a new reliability analysis method for complex structural systems with multiple failure modes, utilizing an adaptive AdaBoost model that enhances both accuracy and effectiveness in assessing the reliability of complex structures. Adaptive iteration continuously optimizes the sampling center through proactive sampling. During this process, the effectiveness of the sample is quantitatively evaluated by measuring the ratio of the number of samples falling into the ineffective region to the total number of samples. The AdaBoost algorithm is a strong integrated learning method for low-bias prediction tasks, which are not easy to overfit during training [25]. The method employs a training dataset with different weights to construct different weak basis classifiers. Each data sample is then assigned a weight that indicates its importance as a training sample, and all samples have the same weight in the first iteration. In the ensuing iterations, the weight assigned to misclassified samples is elevated (or alternatively, the weight attributed to correctly classified samples is proportionately diminished), causing the new classifier to pay increasing attention to misclassified samples that may gather near the classification edge, thereby ultimately reducing the error rate [26–28]. Many scholars around the world have actively engaged in in-depth research on the AdaBoost algorithm, evaluating its application effects and proposing practical and feasible improvements [29–34]. For example, Liu et al. [31] presented an innovative approach to the naval gun hydraulic system fault diagnosis based on the BP-AdaBoost model. Zhou et al. [32] proposed a new polynomial chaos expansion surrogate modeling method utilizing AdaBoost for uncertainty quantification. Lou [33] artfully integrated the enhanced AdaBoost algorithm with the backpropagation neural network model, employing an evolutionary optimization algorithm to refine the model and proposing a mechanical structure reliability calculation method grounded in this advanced framework. Du et al. [34] proposed a real BP AdaBoost algorithm based on weighted information entropy to solve the problems of low prediction accuracy and poor reliability of the software reliability model of the single neural network. Meng et al. [35–37] proposed a new hybrid adaptive Kriging model and a water-cycle algorithm based on reliability assessment learning and optimization strategies, subsequently applying it to offshore wind turbine monopole and offshore wind power towers. N. Asgarkhani, F. Kazemi, et al. [38] proposed a machine learning (ML) algorithm to provide prediction models for determining the seismic response, seismic performance curve and earthquake failure probability curve of brbf. Farzin Kazemi, Neda Asgarkhani, and Robert Jankowski [39] solved the problem of seismic probability and risk assessment of reinforced concrete shear walls (RCSWs) by introducing a stacked machine learning (stacked ML) model based on Bayesian optimization (BO), genetic algorithm (GA), particle swarm optimization (PSO), and gradient-based optimization (GBO) algorithms. The IDA curves (MIDA) and seismic probability curves of this model have a good curve-fitting ability.

The subsequent sections of this paper will be meticulously organized into three sections. Section 2 will thoroughly explore the distinctive characteristics of complex structures in the context of multiple failure modes, the methodology for adaptive sample selection, and the principle underlying the iterative AdaBoost algorithm. Additionally, it outlines a process for assessing the reliability of mechanical structures based on the AdaBoost surrogate model. The Section 3 validates the effectiveness and precision of the AdaBoost surrogate model in evaluating multi-failure mode complex structures through three illustrative examples and discusses their computational results. The Section 4 summarizes both the advantages and limitations of the AdaBoost surrogate model and establishes the applicable scope of this method.

2. Reliability Analysis Based on the AdaBoost Surrogate Model

2.1. Reliability Modeling of Complex Structures under Multi-Failure Mode

There are two primary reliability models for complex structures exhibiting multi-failure mode: the series system and the parallel system. Nevertheless, in the realm of practical engineering applications, complex structural systems frequently exhibit multiple failure modes, with the overall failure mode typically arising from the interactions among these failures. Particularly within intricate systems, the interrelationships between failure modes become progressively more complex, as each mode engages in reciprocal interactions with others.

Suppose that a problem contains m failure modes, the corresponding limit state equation is denoted by $g^{(k)}(x) = 0(k = 1, 2, \ldots, m)$, and the corresponding failure domain is denoted by $D_f^{(k)} = \left\{g^{(k)}(x) \leq 0\right\}(k = 1, 2, \ldots, m)$, where $x = \{x_1, x_2, \ldots, x_n\}$ represents an n-dimensional random variable.

When m failure modes are present in series, the relationship between the system's failure domain, D_f, and the failure domain, $D_f^{(k)}$, of each individual mode can be expressed as follows:

$$D_f = \bigcup_{k=1}^{m} D_f^{(k)} \tag{1}$$

When m failure modes occur simultaneously, it is a parallel situation, and the relationship between the system's failure domain, D_f, and failure domain, $D_f^{(k)}$, of each individual mode can be expressed as follows:

$$D_f = \bigcap_{k=1}^{m} D_f^{(k)} \tag{2}$$

2.2. Adaptive AdaBoost Algorithm

The adaptive AdaBoost algorithm initially employs an adaptive iteration method to obtain the optimal training sample set, which is as evenly distributed as possible on both sides of the failure curve, fully encompassing the information that describes the failure curve. By using the integration and iteration properties of the AdaBoost algorithm, a high-precision surrogate model for failure discrimination in complex structures under multi-failure modes can be achieved through simple binary classifier iteration (the dichotomy method).

2.2.1. Adaptive Sampling

The idea of adaptive sampling is to determine the optimal design point through the iteration process and pre-sample the sample center in each iteration to obtain an optimal training sample set that is as evenly distributed as possible on both sides of the failure curve and fully contains the information of the failure curve. To swiftly ascertain the optimal design point, this study applied the extended sample variance and adaptive iteration method. The detailed steps of the proposed adaptive sampling method are outlined as follows:

(1) It is presumed that the variables are mutually independent and follow a normal distribution, $x_i \sim N(\mu_i, \sigma_i)(i = 1, 2, \ldots, n)$.
(2) Define the expanded coefficient, f, and let $x_i \sim N(\mu_i, f\sigma_i)(i = 1, 2, \ldots, n)$.
(3) Employ Latin hypercube sampling to derive 200 sample points, and then bring them into the system model for discrimination and calculate failure probability $\eta_j(j = 1, 2, \ldots, m)$. Select the point $\overline{x}^{(j)}$ with the largest joint probability density from among the failure points; this point is taken as the sampling center for subsequent sampling.
(4) Define a sampling efficiency index, $\varepsilon_j = |0.5 - \eta_j|$. The smaller the value of ε_j, the closer the sampling center is to the optimal design point, and the failure probability approaches 50%.
(5) When $\varepsilon_j \leq 0.1$, the loop comes to an end. The sampling center at this time is take as the new sampling center, and resampling is conducted using $x_i \sim N(\mu_i, f\sigma_i)$ $(i = 1, 2, \ldots, n)$.

The detailed flowchart illustrating the expanded sample variance and the adaptive iteration method is presented in Figure 1.

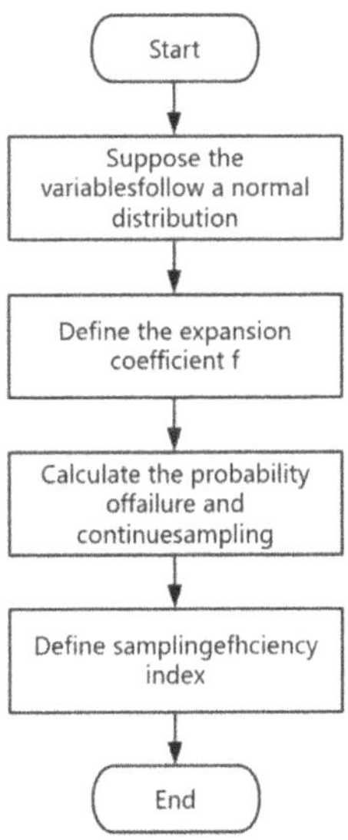

Figure 1. Flowchart of expanded sample variance and adaptive iteration method.

2.2.2. AdaBoost Algorithm

First, the AdaBoost algorithm builds the first simple classifier, h_1, based on the original data distribution. Then, Bootstrap produces T classifiers that are interconnected. Finally, these different classifiers are combined to form a stronger final classifier (strong classifier).

In theory, as long as each weak classifier's classification ability is better than random guessing (a classification accuracy rate greater than 0.5), the error rate of the strong classifier will converge zero as the number of weak classifiers approaches infinity. Within the framework of the AdaBoost algorithm, different training sets are generated by adjusting the weights of each sample. At the beginning, the weight of each sample is the same, and basic classifier, $h_1(x)$, is trained under this sample distribution. The weights of the samples exhibiting poor $h_1(x)$ scores are elevated, while the weights of properly classified samples are decreased to highlight the incorrect samples. Meanwhile, $h_1(x)$ is assigned a weight according to its error, representing its importance; the lower the error rate, the greater the corresponding weight. The basic classifier is trained again using the new sample weight to obtain the basic classifier, $h_2(x)$, and its weight. Therefore, by analogy, after T rounds, T fundamental classifiers and their associated weights can be derived. Finally, T basic classifiers are added according to the previously calculated weights to obtain the final desired strong classifier.

Let us assume that the training set sample is denoted as $T = \{E_{11}, E_{12}, G_{12}, G_{13}, X_c, X_t, Y_c, Y_t, S_{12}, S_{13}, \mathrm{d}\}$, and the output weight of the training set in the *k-th* weak learner is $D(k) = (w_{k1}, w_{k2}, \ldots w_{km}); w_{1i} = \frac{1}{m}; i = 1, 2 \ldots m$.

Let us take a binary classification problem as an example. If the output is $\{0, 1\}$, the weighted error rate of the *k-th* weak classifier, $G_k(x)$, on the training set is $e_k = P(G_k(x_i) \neq y_i) = \sum\limits_{i=1}^{m} w_{ki} I(G_k(x_i) \neq y_i)$.

The weight coefficient of the *k-th* weak classifier,$G_k(x)$, is denoted as $\alpha_k = \frac{1}{2} \log \frac{1-e_k}{e_k}$. An increase in the classification error rate, e_k, is associated with a decrease in the corresponding weak coefficient, α_k, of the classifier weight. Assuming that the sample set-weight coefficient for the *k-th* weak classifier is $D(k) = (w_{k1}, w_{k2}, \ldots w_{km})$, then the sample set-weight coefficient of the corresponding $k + 1$ weak classifier is $w_{k+1,i} = \frac{w_{ki}}{Z_K} \exp(-\alpha_k y_i G_k(x_i))$, where Z_K represents a normalization factor, $Z_K = \sum\limits_{i=1}^{m} w_{ki} \exp(-\alpha_k y_i G_k(x_i))$. It is apparent from the w_{k+1} calculation formula that if the *i-th* sample is classified incorrectly, $y_i G_k(x_i) < 0$, resulting in the increase in the weight of the sample in the $k + 1$ weak classifier. Should the classification be deemed accurate, the weight in the $k + 1$ weak classifier decreases.

The weighted voting method is adopted for AdaBoost classification, and the ultimate strong classifier can be expressed as follows: $f(x) = sign(\sum\limits_{k=1}^{K} \alpha_k G_k(x))$.

2.3. Reliability Analysis Model Based on the Adaptive AdaBoost Algorithm

The basic idea of a reliability analysis based on the adaptive AdaBoost algorithm involves employing an adaptive iterative method to find the optimal training sample set, as elaborated in Section 2.2.1. Then, a high-precision surrogate model for failure discrimination of complex structures under multiple failure modes is obtained through simple binary classifier iteration based on the integration and iteration characteristics to the AdaBoost algorithm, as discussed in Section 2.2.2. Finally, the Monte Carlo method is used to determine the failure probability. The detailed analytical procedures are delineated as follows:

(1) Use the adaptive method to find the optimal sample center.
(2) According to the sample center, use Latin hypercube sampling to generate training samples.
(3) Initialize the weight distribution of the training data. Assign the same weight to each training sample as follows: $\boldsymbol{D_1} = (w_{11}, w_{12}, \ldots, w_{1i}, \ldots, w_{1N}), w_{1i} = \frac{1}{N}$, $i = 1, 2, \ldots, N$.
(4) In the *m-th* iterations, obtain $G_m(x)$, which represents the current m-round iterative classifier; obtain e_m, denoting the present classification error; and obtain α_m, which signifies the cumulative coefficient as elaborated below:

① Obtain the basic classifier $G_m(\boldsymbol{x}) : x_i \rightarrow \{-1, +1\}$ by learning the training dataset with a weight distribution of $\boldsymbol{D_m}$.

② Calculate the classification error rate of e_m for the training dataset:

$$e_m = P(G_m(x_i) \neq y_i) = \sum_{i=1}^{N} w_{mi} \boldsymbol{I}(G_m(x_i) \neq y_i) \tag{3}$$

③ Calculate the coefficient for $G_m(\boldsymbol{x})$, where α_m represents the importance of $G_m(\boldsymbol{x})$ in the final classifier:

$$\alpha_m = \frac{1}{2}\log\frac{1-e_m}{e_m} \tag{4}$$

It is evident from the preceding equation that $e_m \leq \frac{1}{2}$ and $\alpha_m \geq 0$, and that α_m increases as e_m decreases, thereby indicating that basic classifiers with lower classification error rates have a greater impact on the final strong classifier.

④ Update the weight distribution of the training dataset to obtain a new weight distribution for the subsequent interaction:

$$\boldsymbol{D}_{m+1} = (w_{m+1,1}, w_{m+1,2}, \ldots, w_{m+1,i}, \ldots, w_{m+1,N}) \tag{5}$$

$$w_{m+1,i} = \frac{w_{mi}}{Z_m}\exp(-\alpha_m y_i G_m(x_i)), i = 1,2,\ldots,N \tag{6}$$

where $Z_m = \sum_{i=1}^{N} w_{mi}\exp(-\alpha_m y_i G_m(x_i))$.

As a result, the weights of samples misclassified by the fundamental classifier, $G_m(x)$, are augmented, whereas the weights of accurately classified samples are diminished. In this manner, the AdaBoost algorithm can concentrate on samples that are more challenging to differentiate.

(5) Integrate the weak classifiers:

$$f(\boldsymbol{x}) = \sum_{m=1}^{M} \alpha_m G_m(\boldsymbol{x}) \tag{7}$$

The ultimate strong classifier is denoted by

$$G(\boldsymbol{x}) = sign(f(\boldsymbol{x})) = sign(\sum_{m=1}^{M} \alpha_m G_m(\boldsymbol{x})) \tag{8}$$

(6) Use the Monte Carlo method to calculate the failure probability based on the final strong classifier.

Figure 2 illustrates the flowchart of steps 1–6 for the reliability analysis of complex structures utilizing the adaptive AdaBoost algorithm proposed in this paper.

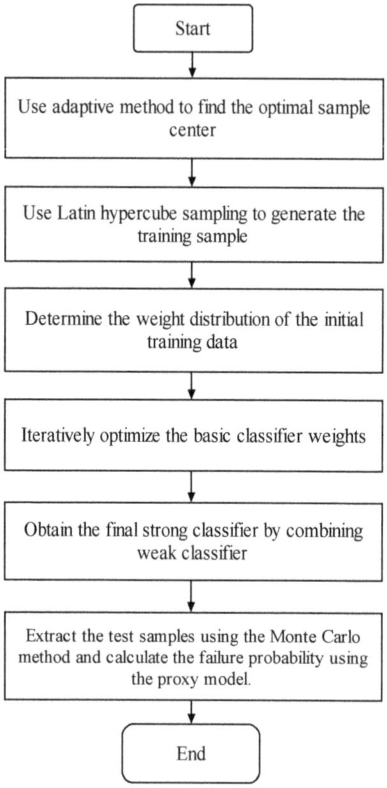

Figure 2. Flowchart of the proposed method for reliability analysis of complex structures using the adaptive AdaBoost algorithm.

3. Examples

To verify the accuracy and efficiency of the proposed adaptive AdaBoost surrogate model, this section provides examples of mechanical structure reliability analysis under different failure functions, with the iteration number set to 200.

3.1. Parallel System

Consider a parallel system distinguished by two distinct failure modes, with the corresponding function articulated as follows.

$$y_1 = x_1^2 - x_2 + 2 \tag{9}$$

$$y_2 = -x_1^2 + x_2 - 6 \tag{10}$$

where x_1 and x_2 denote normally distributed random variables, specifically denoted as $x_1 \sim N(4, 1^2)$ and $x_2 \sim N(3, 0.5^2)$, respectively. The failure conditions of the system are defined by $y_1 < 0$ and $y_2 < 0$.

The proposed adaptive AdaBoost surrogate model is employed to evaluate the failure probability of the parallel system. The variation in error with respect to the number of iterations during the training process (resubstitution loss) is shown in Figure 2.

It can be discerned from Figure 3 that the adaptive AdaBoost surrogate model effectively simulates the failure surface of parallel systems. As the number of iterations escalates, the error of the agent model steadily diminishes and ultimately converges to zero.

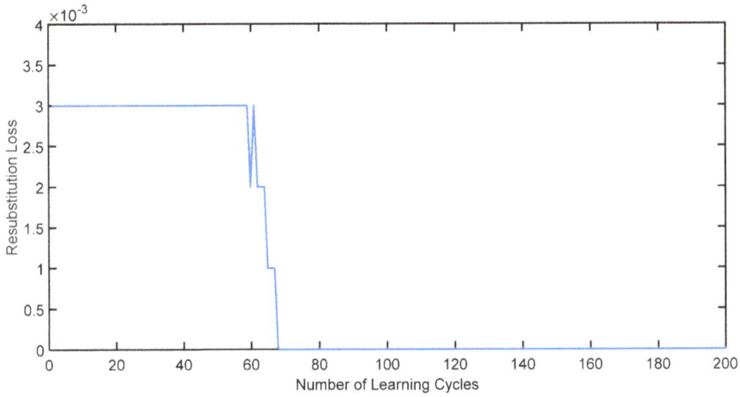

Figure 3. Variation in the parallel system error with the number of iterations during training.

The trained model is utilized to evaluate the failure probability, which is subsequently compared with the result obtained by using the conventional Monte Carlo method. The results are presented in Table 1, where it can be noted that the failure probability obtained derived from the adaptive AdaBoost method is 3.7×10^{-5}, with a relative error of 7.5%, when compared to the conventional Monte Carlo method. This underscores the effectiveness of the proposed adaptive AdaBoost model. It is noteworthy that the conventional Monte Carlo method necessitated millions of samples, whereas the adaptive AdaBoost surrogate model required only 500 samples, leading to a substantial reduction in computation time and a marked enhancement in computational efficiency.

Table 1. Calculated failure probability of the parallel system.

Method	Sampling Frequency	Failure Probability	Relative Error
Monte Carlo method	10^6	4×10^{-5}	/
Adaptive AdaBoost method	500	3.7×10^{-5}	7.5%

3.2. Series System

Consider a series structure system comprising two failure modes, with the corresponding function denoted as follows.

$$y_1 = 2x_1 + 2x_2 - 4.5x_4 - 100 \tag{11}$$

$$y_2 = x_1 + x_2 + 2x_3 - 4.5x_4 + 380 \tag{12}$$

where x_1, x_2, x_3, and x_4 are normal random variables, specifically denoted as $x_1 \sim N(200, 30^2)$, $x_2 \sim N(200, 30^2)$, $x_3 \sim N(200, 30^2)$, and $x_4 \sim N(80, 15^2)$, respectively. The failure conditions of the system are defined by $y_1 < 0$ or $y_2 < 0$.

The adaptive AdaBoost surrogate model is employed to assess the failure probability of the series system. The variation in error concerning the number of iterations throughout the training process (resubstitution loss) is illustrated in Figure 3.

As depicted in Figure 4, it is evident that with an increasing number of iterations, the error of the adaptive AdaBoost surrogate model gradually decreases and converges to zero, indicating its efficacy in failure discrimination for the series system.

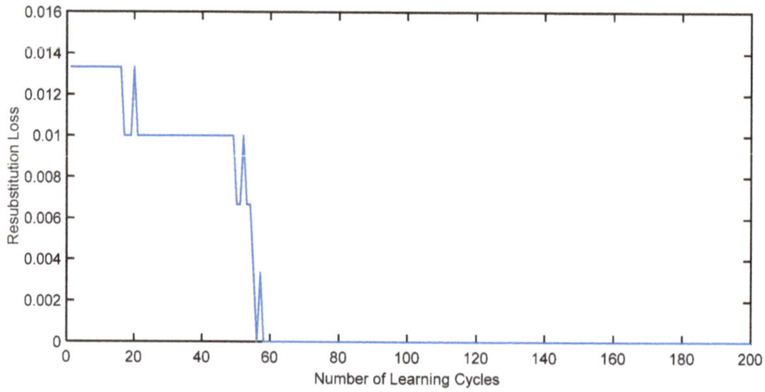

Figure 4. Variation in series system error with the number of iterations during training.

The trained models are utilized to evaluate the failure probability, after which the results are compared with those derived from the conventional Monte Carlo method. The results are presented in Table 2, with a relative error of 5.49% between the result obtained using the adaptive AdaBoost surrogate model and that obtained using conventional Monte Carlo methods. This illustrates the precision and substantiates the efficacy of the proposed method. In contrast to the traditional Monte Carlo method, which necessitated millions of samples, the adaptive AdaBoost surrogate model required only 300 samples, meaning that the sampling efficiency has been significantly improved and the computing time has been greatly shortened.

Table 2. Calculated failure probability of series system.

Method	Sampling Frequency	Failure Probability	Relative Error
Monte Carlo method	10^6	9.1×10^{-4}	/
Adaptive AdaBoost method	300	9.6×10^{-4}	5.49%

3.3. Engineering Example

Figure 5 shows an I-beam system with eight random input variables, $\boldsymbol{X} = \{ d , b , t_w, t_f, L , a , P , S \}$ [40]. The specific distribution is presented in Table 3. The response function of the I-beam system is given by

$$Y = g(\boldsymbol{X}) = S - \sigma_{max} = S - \frac{Pa(L-a)d}{2LI} \tag{13}$$

where S is the strength, σ_{max} is the maximum stress, and $I = \frac{bd^3 - (b-t_w)(d-2t_f)^3}{12}$.

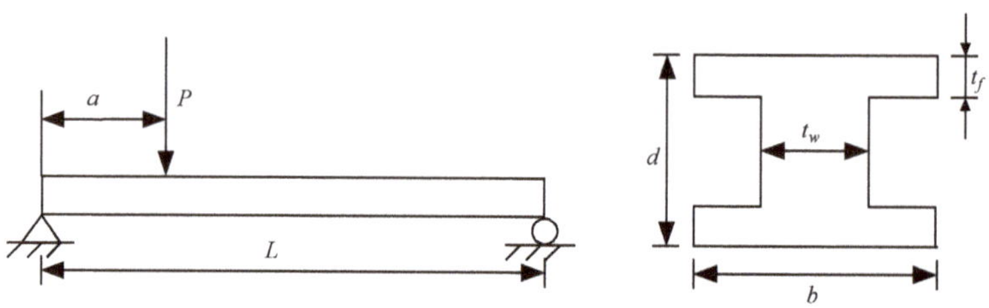

Figure 5. Schematic diagram of the I-beam system.

Table 3. Distribution of random variables of the I-beam system.

Variable	Mean Value	Standard Deviation	Distribution Type
d/in	2.3	1/24	Normal
b/in	2.3	1/24	Normal
t_w/in	0.16	1/24	Normal
t_f/in	0.26	1/24	Normal
L/in	120	6	Normal
a/in	72	6	Normal
P/N	6070	200	Normal
S/kPa	170,000	4760	Normal

The adaptive AdaBoost surrogate model is employed to obtain the failure probability of the I-beam system. The variation in the error with respect to the number of iterations during the training process (resubstitution loss) is shown in Figure 6.

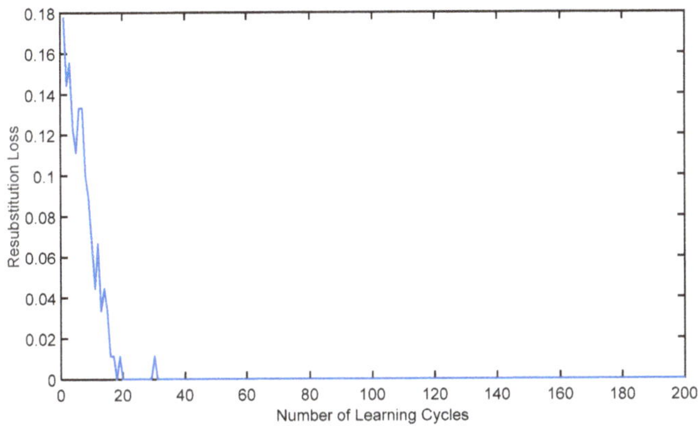

Figure 6. Variation in system error with the number of iterations in the training process.

As illustrated in Figure 5, it is evident that with an increasing number of iterations, the error of the adaptive AdaBoost surrogate model gradually decreases and converges to zero; this suggests that the surrogate model possesses a robust fitting capability.

The trained model is employed to ascertain the failure probability, which is subsequently compared with that determined using the conventional Monte Carlo method. The results are presented in Table 4, in which the relative error between the results obtained from the adaptive AdaBoost surrogate model and those obtained from the conventional Monte Carlo method is 1.59%, satisfying accuracy requirements and demonstrating the effectiveness of the adaptive AdaBoost surrogate model. In comparison to the traditional Monte Carlo method, which necessitated millions of samples, only 70 samples are required by the adaptive AdaBoost surrogate model, indicating considerably improved sampling efficiency and a reduction in calculational time.

Table 4. Calculated failure probability of the I-beam system.

Method	Sampling Frequency	Failure Probability	Relative Error
Monte Carlo method	10^6	0.2074	/
Adaptive AdaBoost method	70	0.2041	1.59%

3.4. Analysis and Discussion

The previously mentioned examples illustrate that as the number of iterations escalates, the calculation error under the adaptive AdaBoost surrogate model converges to a minimum value for both parallel and series systems. This suggests that the AdaBoost surrogate model demonstrates an exceptional fitting performance and remarkable versatility in both parallel and series systems. Furthermore, when comparing the adaptive AdaBoost method with the Monte Carlo method, it is evident that the relative error between the failure probability derived from the adaptive AdaBoost alternative model method and that computed using the Monte Carlo method is smaller, aligning more closely with Monte Carlo simulation results, while significantly enhancing computational efficiency compared to traditional methods. Therefore, the accuracy of structural reliability calculated by the method proposed in this paper is higher than that of the traditional algorithm model, and calculation efficiency has been significantly enhanced, rendering it suitable for real engineering structures.

4. Conclusions

An adaptive AdaBoost algorithm is proposed to evaluate the reliability of multi-failure mode structures. This method shows excellent effectiveness in complex mechanical structures where explicit functions are difficult to establish. The optimal training sample set is obtained by the adaptive method, and then the AdaBoost algorithm is trained on this refined sample set. The generated proxy model not only meets the accuracy requirements, but also can better distinguish whether the target structure is invalid under the given parameters.

(1) Compared with the traditional Monte Carlo method, this model significantly improves the computational efficiency and can accurately calculate the failure probability in a shorter time.
(2) The method has good universality. Compared with the general alternative model, the adaptive AdaBoost algorithm proposed in this paper has the advantages of strong applicability, low dependence on the operator's engineering experience, and high precision.
(3) However, we must be aware that the studies conducted to date have some limitations. The adaptive AdaBoost algorithm proposed by us has not been applied to a multi-failure mode reliability analysis of more complex structures. More and more complex research scenarios put forward higher requirements and challenges for our proposed algorithms.

However, the method proposed in this paper is of great significance to the exploration of complex mechanical structures, which is helpful to improve the application level of mechanical structures and promote their wide application.

Author Contributions: Conceptualization, F.Z. and Z.Q.; Methodology, F.Z., Y.T., M.W. and X.X.; Software, Z.Q.; Formal analysis, F.Z., Z.Q., Y.T., M.W. and X.X.; Writing—original draft, Z.Q., Y.T. and M.W.; Funding acquisition, F.Z. All authors have read and agreed to the published version of the manuscript.

Funding: This work was supported in part by the Fundamental Research Funds for the Central Universities (NWPU-310202006zy007).

Institutional Review Board Statement: Not applicable.

Informed Consent Statement: Not applicable.

Data Availability Statement: The raw data supporting the conclusions of this article will be made available by the authors upon request.

Conflicts of Interest: The authors declare no conflict of interest.

References

1. Yazdi, M.; Golilarz, N.A.; Nedjati, A.; Adesina, K.A. An improved lasso regression model for evaluating the efficiency of intervention actions in a system reliability analysis. *Neural Comput. Appl.* **2021**, *33*, 7913–7928. [CrossRef]
2. Yan, W.; Deng, L.; Zhang, F.; Li, T.; Li, S. Probabilistic machine learning approach to bridge fatigue failure analysis due to vehicular overloading. *Eng. Struct.* **2019**, *193*, 91–99. [CrossRef]
3. Yang, J.; Chen, C.; Ma, A. A Fast Product of Conditional Reduction Method for System Failure Probability Sensitivity Evaluation. *CMES-Comput. Model. Eng. Sci.* **2020**, *125*, 1159–1171. [CrossRef]
4. Zhu, S.-P.; Keshtegar, B.; Trung, N.-T.; Yaseen, Z.M.; Bui, D.T. Reliability-based structural design optimization: Hybridized conjugate mean value approach. *Eng. Comput.* **2021**, *37*, 381–394. [CrossRef]
5. Xu, J.-G.; Cai, Z.-K.; Feng, D.-C. Life-cycle seismic performance assessment of aging RC bridges considering multi-failure modes of bridge columns. *Eng. Struct.* **2021**, *244*, 112818. [CrossRef]
6. Zhang, F.; Xu, X.; Cheng, L.; Tan, S.; Wang, W.; Wu, M. Mechanism reliability and sensitivity analysis method using truncated and correlated normal variables. *Saf. Sci.* **2020**, *125*, 104615. [CrossRef]
7. Marozaua, I.; Auchlina, M.; Pejchal, V.; Souchon, F.; Vogel, D.; Lahti, M.; Saillen, N.; Sereda, O. Reliability assessment and failure mode analysis of MEMS accelerometers for space applications. *Microelectron. Reliab.* **2018**, *88–90*, 846–854. [CrossRef]
8. Wang, R.; Liu, X.; Hu, D.; Mao, J. Reliability assessment for system-level turbine disc structure using LRPIM-based surrogate model considering multi-failure modes correlation. *Aerosp. Sci. Technol.* **2019**, *95*, 105422. [CrossRef]
9. Meng, Z.; Pang, Y.; Pu, Y.; Wang, X. New hybrid reliability-based topology optimization method combining fuzzy and probabilistic models for handling epistemic and aleatory uncertainties. *Comput. Methods Appl. Mech. Eng.* **2020**, *363*, 112886. [CrossRef]
10. Yang, F.; Ren, J. Reliability Analysis Based on Optimization Random Forest Model and MCMC. *CMES-Comput. Model. Eng. Sci.* **2020**, *125*, 801–814. [CrossRef]
11. Monjardin-Quevedo, J.G.; Reyes-Salazar, A.; Tolentino, D.; Gaxiola-Camacho, O.D.; Vazquez-Becerra, G.E.; Gaxiola-Camacho, J.R. Seismic reliability of steel SMFs with deep columns based on PBSD philosophy. *Structures* **2022**, *42*, 1–15. [CrossRef]
12. Guzman-Acevedo, G.M.; Quintana-Rodriguez, J.A.; Gaxiola-Camacho, J.R.; Vazquez-Becerra, G.E.; Torres-Moreno, V.; Monjardin-Quevedo, J.G. The Structural Reliability of the Usumacinta Bridge Using InSAR Time Series of Semi-Static Displacements. *Infrastructures* **2023**, *8*, 173. [CrossRef]
13. Wang, Z. Markov chain Monte Carlo sampling using are servoirmethod. *Comput. Stat. Data Anal.* **2019**, *139*, 64–74. [CrossRef]
14. Liu, X.-X.; Elishakoff, I. A combined Importance Sampling and active learning Kriging reliability method for small failure probability with random and correlated interval variables. *Struct. Saf.* **2020**, *82*, 101875. [CrossRef]
15. Lu, H.; Cao, S.; Zhu, Z.; Zhang, Y. An improved high order moment-based saddlepoint approximation method for reliability analysis. *Appl. Math. Model.* **2020**, *82*, 836–847. [CrossRef]
16. Meng, Z.; Li, G.; Yang, D.; Zhan, L. A new directional stability transformation method of chaos control for first order reliability analysis. *Struct. Multidiscip. Optim.* **2017**, *55*, 601–612. [CrossRef]
17. Chen, H.-N.; Mao, Z.-L. Study on the Failure Probability of Occupant Evacuation with the Method of Monte Carlo Sampling. *Procedia Eng.* **2018**, *211*, 55–62. [CrossRef]
18. Xiao, S.; Oladyshkin, S.; Nowak, W. Reliability analysis with stratified importance sampling based on adaptive Kriging. *Reliab. Eng. Syst. Saf.* **2020**, *197*, 106852. [CrossRef]
19. Yang, Z.; Ching, J. A novel reliability-based design method based on quantile-based first-order second-moment. *Appl. Math. Model.* **2020**, *88*, 461–473. [CrossRef]
20. Zhang, X.; Pandey, M.D.; Yu, R.; Wu, Z. HALK: A hybrid active-learning Kriging approach and its applications for structural reliability analysis. *Eng. Comput.* **2022**, *38*, 3039–3055. [CrossRef]
21. Nezhad, H.B.; Miri, M.; Ghasemi, M.R. New neural network-based response surface method for reliability analysis of structures. *Neural Comput. Appl.* **2019**, *31*, 777–791. [CrossRef]
22. Ni, T.; Zhai, J. A matrix-free smoothing algorithm for large-scale support vector machines. *Inf. Sci.* **2016**, *358–359*, 29–43. [CrossRef]
23. Kotsiantis, S.B. Supervised Machine Learning: A Review of Classification Techniques. *Informatica* **2007**, *31*, 249–268.
24. Liu, H.; Zhang, X.; Zhang, X. PwAdaBoost: Possible world based AdaBoost algorithm for classifying uncertain data. *Knowl.-Based Syst.* **2019**, *186*, 104930. [CrossRef]
25. Huang, X.; Li, Z.; Jin, Y.; Zhang, W. Fair-AdaBoost: Extending AdaBoost method to achieve fair classification. *Expert Syst. Appl.* **2022**, *202*, 117240. [CrossRef]
26. Ravikumar, S.; Sekar, S.; Jeyalakshmi, S.; Narayanan, S.; Vivekanandan, G.; Sundarakannan, N. An optimized AdaBoost Multi-class support vector machine for driver behavior monitoring in the advanced driver assistance systems. *Expert Syst. Appl.* **2023**, *212*, 118618.
27. Jiang, H.; Zheng, W.; Luo, L.; Dong, Y. A two-stage minimax concave penalty based method in pruned AdaBoost ensemble. *Appl. Soft Comput. J.* **2019**, *83*, 105674. [CrossRef]
28. Zhou, Y.; Mazzuchi, T.A.; Sarkani, S. M-AdaBoost-A based ensemble system for network intrusion detection. *Expert Syst. Appl.* **2020**, *162*, 113864. [CrossRef]
29. Yu, Q.; Zhou, Y. Traffic safety analysis on mixed traffic flows at signalized intersection based on Haar-AdaBoost algorithm and machine learning. *Saf. Sci.* **2019**, *120*, 248–543. [CrossRef]

30. Wang, W.; Sun, D. The improved AdaBoost algorithms for imbalanced data classification. *Inf. Sci.* **2021**, *563*, 358–374. [CrossRef]
31. Liu, X.; Hu, Y.; Xu, Z.; Ren, Y.; Gao, T. Fault diagnosis for hydraulic system of naval gun based on BP-AdaBoost model. In Proceedings of the 2017 Second International Conference on Reliability Systems Engineering (ICRSE), Beijing, China, 10–12 July 2017.
32. Zhou, Y.; Lu, Z.; Cheng, K. AdaBoost-based ensemble of polynomial chaos expansion with adaptive sampling. *Comput. Methods Appl. Mech. Eng.* **2022**, *388*, 114238. [CrossRef]
33. Luo, P. Reliability analysis of mechanical structure based on improved BP-AdaBoost algorithm. *Intern. Combust. Engine Parts* **2019**, *15*, 41–42.
34. Du, R.C.; Hua, J.X.; Zhai, X.Y.; Li, Z.P. Research on Software Reliability Prediction Based on Improved Real AdaBoost. *J. Air Force Eng. Univ. (Nat. Sci. Ed.)* **2018**, *19*, 91–96.
35. Meng, D.; Yang, S.; De Jesus, A.M.P.; Fazeres-Ferradosa, T.; Zhu, S.-P. A novel hybrid adaptive Kriging and water cycle algorithm for reliability-based design and optimization strategy: Application in offshore wind turbine monopole. *Comput. Methods Appl. Mech. Eng.* **2023**, *412*, 116083. [CrossRef]
36. Meng, D.; Yang, S.; de Jesus, A.M.P.; Zhu, S.-P. A novel Kriging-model-assisted reliability-based multidisciplinary design optimization strategy and its application in the offshore wind turbine tower. *Renew. Energy* **2023**, *203*, 407–420. [CrossRef]
37. Yang, S.; Meng, D.; Wang, H.; Yang, C. A novel learning function for adaptive surrogate-model-based reliability evaluation. *Philos. Trans. R. Soc. A-Math. Phys. Eng. Sci.* **2024**, *382*, 20220395. [CrossRef]
38. Asgarkhani, N.; Kazemi, F.; Jakubczyk-Gałczyńska, A.; Mohebi, B.; Jankowski, R. Seismic response and performance prediction of steel buckling-restrained braced frames using machine-learning methods. *Eng. Appl. Artif. Intell.* **2024**, *128*, 107388. [CrossRef]
39. Kazemi, F.; Asgarkhani, N.; Jankowski, R. Optimization-based stacked machine-learning method for seismic probability and risk assessment of reinforced concrete shear walls. *Expert Syst. Appl.* **2024**, *255 Pt D*, 124897. [CrossRef]
40. Li, G.; Lu, Z.; Song, S. A new method to measure the importance of fundamental variables to failure probability. *Mech. Eng.* **2010**, *32*, 71–75.

applied sciences

MDPI

Article

Uncertainty-Driven Reliability Analysis Using Importance Measures and Risk Priority Numbers

Maria Valentina Clavijo [1], Fernando Guevara Carazas [2], Juan David Arango Castrillón [2] and Carmen Elena Patino-Rodriguez [3,*]

1 Energy Department, Politecnico di Milano, 20156 Milan, Italy; mariavalentina.clavijo@polimi.it
2 Mechanical Engineering, Universidad Nacional de Colombia, Medellín 050034, Colombia; fjguevarac@unal.edu.co (F.G.C.); judarangoca@unal.edu.co (J.D.A.C.)
3 Industrial Engineering, Universidad de Antioquia, Medellín 050010, Colombia
* Correspondence: elena.patino@udea.edu.co

Featured Application

The framework was applied to Solid Waste Collection and Compaction Trucks, using real failure data to identify critical components and improve maintenance planning under uncertainty. Beyond this case, it can be extended to other complex systems such as heavy vehicles, industrial machinery, and energy infrastructure.

Abstract

Uncertainty is a key factor in the reliability assessment of complex engineering systems, especially when they operate under variable conditions that affect component degradation. This study presents a framework for the systematic and uncertainty-based prioritization of critical components and failure modes. The method combines Reliability Block Diagrams, Fault Tree Analysis, and Importance Measures with Failure Mode and Effects Analysis. This two-level approach links component failures with their effect on system reliability. Uncertainty is introduced through the statistical parameters of component reliability distributions and the resulting impact on system behavior is examined. Components with the highest importance are then examined through Failure Mode and Effects Analysis to identify main failure modes and calculate their Risk Priority Numbers. The framework is applied to a fleet of Solid Waste Collection and Compaction Trucks used by a waste management company in a Colombian city. This system operates under high-load variability, mechanical shocks, and environmental stress. The combined Importance Measures and Risk Priority Number analysis provides a probabilistic basis for identifying critical components and their dominant failure modes, linking reliability uncertainty with maintenance prioritization. The results show that combining Importance Measures and Risk Priority Number improves the identification of critical components and dominant failure modes, supporting maintenance prioritization based on reliability impact. The framework offers a practical approach for reliability assessment and maintenance planning under uncertainty, linking component-level uncertainty with system performance to guide decision-making in complex systems.

Keywords: reliability analysis; importance measures (IM); risk priority number (NPR); uncertainty quantification; system criticality; waste collection trucks

1. Introduction

Reliability analysis is fundamental to the design, operation, and maintenance of complex engineered systems [1–3]. Across energy production, transportation, and healthcare sectors, anticipating failures and quantifying risks are essential to ensure safety, performance, and sustainability [4–10]. However, traditional reliability approaches often assume deterministic component behavior, overlooking the pervasive influence of uncertainty [11,12]. Uncertainty may stem from incomplete knowledge, stochastic variability, or operational complexity, and inadequate treatment can compromise decision quality [9,13]. Consequently, reliability engineering is shifting toward probabilistic frameworks that explicitly quantify and propagate uncertainty in both component and system analyses [3,12,14,15].

Among reliability tools, combining Failure Mode and Effects Analysis (FMEA) with Fault Tree Analysis (FTA) or Reliability Block Diagrams (RBD) has proven effective. FMEA identifies and prioritizes potential failure modes, while FTA and RBD introduce probabilistic modeling at the system level, allowing interactions and dependencies to be represented more realistically.

A persistent challenge is the accurate assessment of component criticality, which directly influences maintenance planning, safety protocols, and resource allocation [3,12,16–18]. Critical analysis forms the bridge between reliability modeling and operational decision-making. Achieving informative rankings under deterministic and uncertain conditions requires methods that are analytically rigorous yet operationally practical [19,20]. Within this paradigm, two established instruments are Importance Measures (IMs) and the Risk Priority Number (RPN).

IMs quantify how changes in component reliability affect overall system performance [11,21–23]. They provide a mathematically grounded way to identify critical components, support redundancy allocation, and analyze uncertainty propagation across a system [13,15,24,25]. By contrast, the RPN, derived from FMEA, offers a semi-quantitative and highly interpretable index for prioritizing risk in industrial contexts. Its simplicity has made it widely used for decades, though it often lacks the probabilistic depth needed to represent uncertainty [26–28].

A comparison of these two strands highlights complementary strengths and limitations. IMs offer analytical rigor but can be challenging for practitioners to interpret, while RPNs are intuitive but rely on heuristic scoring. Although conceptual links exist—such as multi-factor RPN refinements resembling composite IMs or simulation-based IMs aligning with probabilistic FMEA—comprehensive frameworks that unify both perspectives remain uncommon [3,12,29]. This gap means that few approaches jointly address uncertainty quantification, interpretability and practical implementation in reliability-based prioritization.

Several recent studies illustrate the practical importance of integrating structured reliability tools into real-world systems. Applications in sectors such as medical waste management, transport logistics, and power generation demonstrate how probabilistic modeling can support maintenance and safety planning under uncertainty [30]. However, the literature remains fragmented, with IMs advancing in academic rigor and RPN remaining dominant in industrial practice, without sufficient integration between the two [4,22,23,31,32].

Despite these advances, most studies rely on deterministic parameter assumptions. They treat time-to-failure (TTF) parameters as fixed, disregarding the statistical uncertainty inherent in estimation. This simplification can lead to overconfident reliability predictions and unstable component rankings, limiting decision value in data-limited settings. Few works explicitly quantify how parameter-estimation uncertainty affects component prioritization metrics.

This study presents a statistics-based framework integrating parameter inference with uncertainty-informed prioritization, validated using operational data. The framework incorporates parametric uncertainty based on maximum-likelihood estimation and the observed information matrix, followed by Monte-Carlo propagation of uncertainties through FTA or RBD formulations. It defines IMs under parametric uncertainty, providing probabilistic intervals for classical reliability indices, and introduces a rank stability summary that quantifies the consistency of component prioritization under uncertainty. These tools are integrated into a single workflow linking parameter inference, uncertainty propagation and decision relevance. Specifically, the work advances reliability assessment by (i) combining inductive and deductive reliability perspectives to represent uncertainty propagation from components to the system; (ii) interpreting classical IMs as sensitivity indicators under parametric uncertainty; and (iii) incorporating these results into an RPN-based prioritization structure to support decision-making under variable conditions.

The approach is validated using field failure data from Solid Waste Collection and Compaction Trucks (SWCCTs). These systems operate under variable conditions where mechanical degradation and environmental stress amplify uncertainty effects, providing a relevant setting for evaluation. Results show that combining the probabilistic structure of IMs with the practical interpretability of RPN improves the identification of critical components and supports reliability-based maintenance decisions.

This paper is organized as follows: Section 2 reviews the theoretical background. Section 3 presents the methodological framework for parametric uncertainty modeling and propagation. Section 4 describes the case study, including data preparation and model construction. Section 5 reports the main results. Section 6 discusses implications for reliability and maintenance planning and concludes with directions for future research.

2. Theoretical Background

2.1. Importance Measures in Reliability Analysis

IMs have been explored in diverse contexts, with applications expanding from structural analyses to data-driven and cost-oriented frameworks [13,33–36]. Early work focused on binary systems, using IMs as sensitivity indices to quantify how component reliability affects system performance. Several authors advanced this line by applying Monte-Carlo methods to estimate differential IMs in nuclear reactor protection systems, demonstrating the value of stochastic simulation in capturing uncertainty. Differential importance has also been used for comparative statics beyond reliability (e.g., inventory settings). For multi-state systems, composite IMs address the limits of binary formulations and integrated generalizations have been used in aviation applications. Simulation-based analyses in offshore oil and gas have introduced measures tailored to repairable systems [11,15,24,37].

IMs are commonly classified as structural, reliability, lifetime, or cost-based [24]. The appropriate category depends on the available information and data quality [38]. For instance, structural IMs are used when only the system design is known, whereas reliability IMs are used when failure probability data is available. Lifetime IMs consider operational factors, such as age, usage, and environment. Cost-based IMs account for the financial impact of maintenance or improvements. Some of these measures also integrate uncertainty directly into the criticality assessment. However, the present study concentrates on Structural IMs and Reliability IMs, as they are the most widely used in reliability-centered decision-making. We use Birnbaum ($I_i^B(t)$), Risk Reduction Worth—RRW ($I_i^{RRW}(t)$), Risk Achievement Worth—RAW ($I_i^{RAW}(t)$), Fussell–Vesely ($I_i^{FV}(t)$), and Criticality ($I_i^{CR}(t)$), which together span structural sensitivity, failure contribution, and operational consequence.

Theoretical IMs clarify how reliability propagates through engineering systems (Table 1). Each measure captures a specific aspect of system behavior: $I_i^B(t)$ defines structural dependency, $I_i^{RRW}(t)$ and $I_i^{RAW}(t)$ quantify system improvement or degradation limits, $I_i^{FV}(t)$ measures the contribution of each component to overall risk, and $I_i^{CR}(t)$ combines structural sensitivity with operational reliability. Together, these indices form a diagnostic that helps engineers understand how and why system reliability changes under different configurations and operational conditions.

Table 1. Mathematical definitions of selected IMs.

Classification	IM	Used for Identify	Equation	Characteristics	
Structural	Birnbaum	Components whose marginal reliability variation exerts disproportionate influence on overall system reliability.	$I_i^B(t) = \frac{\partial R_s(t)}{\partial R_i(t)}$	Quantifies the sensitivity of system reliability to changes in the reliability of an individual component	[21,22,39,40]
Structural	RRW	Best candidates for reliability improvement	$I_i^{RRW}(t) = \frac{F_s(t)}{[F_s(t)\|F_i(t)=0]}$	Measures system unreliability change when a component is assumed perfect	[22,38,39]
Structural	RAW	Components where abrupt failure-probability escalation maximally degrades system performance	$I_i^{RAW}(t) = \frac{[F_s(t)\|F_i(t)=1]}{F_s(t)}$	Identifies the system components that are essential to rendering the system unreliable	[22,38,39]
Reliability	Fussell-Vesely	Allowed performance degradation	$I_i^{FV}(t) = \frac{F_i(t)}{F_s(t)}$ $I_i^{FV}(t) = F_s(t) - \frac{[F_s(t)\|F_i(t)=0]}{F_s(t)}$	A low probability of failure is not always the most suitable candidate for change	[41,42]
Reliability	Criticality	Relative contribution of component reliability to overall system reliability.	$I_i^{CR}(t) = \left[\frac{\partial R_s(t)}{\partial R_i(t)}\right] \cdot \left[\frac{R_i(t)}{R_s(t)}\right]$ $I_i^{CR}(t) = I_i^B(t) \cdot \left[\frac{R_i(t)}{R_s(t)}\right]$	Assesses component suitability beyond just failure probability	[24]

$I_i^B(t)$ quantifies the sensitivity of system reliability to variations in a component's reliability. It identifies components for which small reliability changes produce larger effects on overall performance. In practice, it informs maintenance prioritization, inspection schedules and redundancy allocation. Components with high $I_i^B(t)$ are those whose failure leads to a significant loss of functionality.

$I_i^{RRW}(t)$ and $I_i^{RAW}(t)$ complement Birnbaum's measure by expressing practical limits on improvement and degradation. $I_i^{RRW}(t)$ quantifies the gain if a component were perfect; $I_i^{RAW}(t)$ quantifies the loss if it were failed. High $I_i^{RRW}(t)$ indicated where upgrades yield the largest benefit, while high $I_i^{RAW}(t)$ flag elements that warrant close monitoring or redundancy. These measures are used in nuclear, aerospace, and industrial applications to support maintenance planning, redundancy design, and resource allocation. Taken together, they show how extreme component conditions influence overall system behavior, guiding risk-informed decisions.

$I_i^{CR}(t)$ evaluates a component's contribution relative to system reliability. It is defined as the product of $I_i^B(t)$ and the ratio between component and system reliability, linking structural sensitivity with probabilistic significance. Whereas $I_i^B(t)$ measures absolute influence, $I_i^{CR}(t)$ expresses influence in proportional terms, indicating where a unit increase in component reliability most effectively increases system reliability. It is useful when resources are constrained, prioritizing components that are both influential and at reliability levels where improvements are most effective. It is often discussed alongside $I_i^{FV}(t)$ because both describe proportional system effects rather than purely structural dependence.

In applied reliability analysis, the measures provide complementary views: $I_i^B(t)$ identifies where the system is most structurally sensitive; $I_i^{RRW}(t)$ and $I_i^{RAW}(t)$ quantify the best and worst operational cases; $I_i^{FV}(t)$ shows current unreliability contributions; and $I_i^{CR}(t)$ highlights where proportional reliability improvements are most effective. Using them together allows engineers to balance structural significance, practical improvement potential, and probabilistic impact, providing a complete view of system reliability. This integrated approach supports maintenance prioritization, redundancy planning, and design optimization in complex engineering systems.

2.2. Uncertainty Modeling

Uncertainty in reliability modeling arises from the limited information available to estimate the statistical parameters of component lifetime models. For a component i, let the random variable $t_i = (t_{i1}, t_{i2}, \ldots, t_{in},)$ represent its TTF, characterized by a parametric probability density function (PDF) $f_i(t \mid \theta_i)$ and corresponding reliability function. For component i, let θ_i denote the parameters of the selected lifetime model [2,31]. The parameters θ_i are estimated from observed failure data t_i using Maximum Likelihood Estimation (MLE):

$$\hat{\theta}_i = \arg\max_{\theta_i} \prod_{j=1}^{n_i} f_i(t_{ij} \mid \theta_i)$$

Since these estimates are subject to sampling variability, they are treated as random variables. Under standard regularity conditions, the MLE is asymptotically normal,

$$\theta_i \sim \mathcal{N}(\hat{\theta}_i, \hat{V}_i)$$

where $\hat{V}_i = \boldsymbol{I}_i^{-1}(\hat{\theta}_i)$ is the inverse observed information matrix and $I_i(\hat{\theta}_i)$ is the observed Fisher information. This provides a probabilistic representation of epistemic uncertainty in the parameter estimates.

$$I_i(\hat{\theta}_i) = -\left.\frac{\partial^2 \log L_i(\hat{\theta}_i)}{\partial\theta_i \partial\theta_i^T}\right|_{\theta_i=\hat{\theta}_i}$$

This formulation enables the propagation of parameter uncertainty through the reliability structure function $\phi(\cdot)$, which defines the system reliability as follows:

$$R_{sys}(t) = \phi(R_1(t \mid \theta_1), R_2(t \mid \theta_2), \ldots, R_n(t \mid \theta_i))$$

As a result, $R_{sys}(t)$ becomes a random variable, and its distribution captures the uncertainty propagation from component-level parameters to the system level.

This parametric uncertainty affects the estimation of component reliability and, consequently, the derived IMs. By treating parameters as random variables, the variability in the data is reflected in the estimated reliability functions $R_i(t)$, the system reliability $R_{sys}(t)$, and the associated measures [2]. Accordingly, the analysis reports interval estimates of reliability and importance rather than single deterministic values, offering a more realistic basis for decision-making.

2.3. Risk Priority Numbers in Failure Mode and Effect Analysis

The RPNs a classical metric derived from Failure Mode and Effects Analysis (FMEA), used to identify and prioritize risk in engineering systems [41]. It guides maintenance decisions by ranking failure modes that need preventive or corrective actions and supports the definition of maintenance priorities [27,43,44]. The RPN is computed from three factors: occurrence (O), severity (S), and detectability (D). Which reflect how often a failure occurs, the impact of its effects and how easily it can be detected before it occurs [19,29,41,45]. The product of these three values provides a single numerical indicator that allows maintenance teams to focus on the most critical issues. Its simplicity and clarity make it widely used in industrial environments, where fast decisions are often required. Table 2 summarizes common RPN formulations, including the traditional multiplicative model and several adapted versions that consider different relations between failure causes, effects, and modes.

Table 2. Formulations of RPN.

Expression	Description	
$RPN_j^i = S_j^i \cdot D_j^i \cdot O_j^i$	Classical RPN definition for the j-th failure mode of the i-th component	[19,41,45]
$RPN_j^i = \left(D_{jk}^i \cdot O_{jk}^i\right) \cdot \left(S_j^i\right)$	Adjusted RPN when severity is assigned to jth failure mode, and occurrence/detectability are assigned to kth causes.	[29]

Although the RPN is widely used in sectors such as automotive manufacturing, process industries, and aerospace, it has known limitations [17,26,43]. Because it uses ordinal scales, it depends strongly on expert judgment and assumes linear relationships between factors that may not be independent [19,29]. The lack of a probabilistic basis limits the description of how uncertainty propagates across related failure modes. To address these issues, studies propose variants that incorporate probabilistic reasoning, fuzzy logic [29,46], or cost-based weighting, aiming to improve consistency while retaining simplicity for practical use [27,46–48].

From a practical point of view, the main challenge is to keep the method easy to apply while improving its accuracy and reliability. Maintenance engineers often prefer tools that are simple, transparent, and adaptable to real data limitations. Probabilistic versions of the RPN can include confidence intervals or uncertainty ranges, making the results more consistent with the actual operating conditions. When used together with Importance Measures, the RPN supports not only the ranking of failure modes but also the evaluation of how stable those rankings remain when data or parameters vary. This connection helps link reliability analysis with maintenance planning, providing clear guidance for decisions in complex systems.

Despite its limitations, the RPN is a valuable tool for identifying and prioritizing risks early in the design phase and throughout the asset lifecycle. Its persistence in industrial practice underscores the importance of frameworks that build upon its strengths rather than discarding them.

The theoretical framework presented the basis for identifying critical components using IMs and RPN. The approach does not propose new reliability indices but extends classical IMs to a probabilistic context that considers parameter uncertainty. Using reliability estimates obtained from FT or RBD models allows the evaluation of how uncertainty in each component affects the overall system reliability. This helps to describe reliability sensitivities under changing operating conditions. The combination of IMs with the FMEA structure component-reliability assessment and failure-mode prioritization. IMs indicate how much each component contributes to system reliability, while RPN identifies the most critical failure modes by combining severity, occurrence, and detection. Comparing both results provides a complete view of system behavior and supports maintenance planning.

From a practical point of view, this theoretical foundation helps maintenance teams to focus their resources on components and failure modes that have the greatest effect on system reliability. Considering uncertainty in IMs reflects the natural variability of real operating conditions and the limitations of available data. The methodology serves as a tool to reduce maintenance costs and improve planning by concentrating efforts on the failures that most affect system performance. The integration of IMs and RPN converts reliability analysis into useful guidance for maintenance planning and decision-making under real operational uncertainty.

3. Methodology

The objective of this methodology is to enable a systematic prioritization of critical components and failure modes in complex engineering systems that account for uncertainty. The framework integrates quantitative reliability techniques, such as RBD, FTA, and IMs, with qualitative failure analysis via FMEA. This two-level approach evaluates failure mode impact and assesses each component's effects on system reliability. The method traces failure propagation through critical components and identifies the failure modes with the highest impact on system performance. By linking the identification of critical failures with their quantified effects on reliability, the framework indicates which components should be prioritized, while facilitating the development of actionable mitigations.

Integrating these techniques provides a stronger basis for maintenance planning and risk management by accounting for epistemic uncertainty. The methodology follows a conditional workflow that adapts based on the level of available system knowledge. Figure 1 presents a workflow that combines deterministic, probabilistic and qualitative reliability tools for decision-making, showing the progression from structural definition to decisions under uncertainty in the reliability assessment process.

Numerically, the workflow after qualitative assessment proceeds as follows: estimate lifetime parameters for each component $(i \in n)$ $\hat{\theta}_i \rightarrow$ compute reliabilities $R_i(t)$ $\forall i \in n \rightarrow$ evaluate system reliability $R_{sys}(t) = \phi(R_1(t), R_2(t), \ldots, R_n(t))$ via FTA or RBD$\rightarrow$ compute IMs $\{I_i^B, I_i^{RRW}, I_i^{RAW}, I_i^{FV}, I_i^C\} \forall i \in n \rightarrow$ synthesize with FMEA by pairing each prioritized component with its top failure modes and RPNs.

The integration of these four analytical blocks establishes a closed-loop reliability framework in which each stage reinforces the subsequent stage. System Characterization defines the structural basis that constrains probabilistic modeling; System Reliability quantifies these relationships and reveals component dependencies; Uncertainty Modeling evaluates the robustness of these dependencies under parameter variability; and FMEA validates and complements quantitative findings by linking them to observable failure behavior and performable mitigations. Their interaction enables consistency between structural logic, probabilistic evidence, and operational experience, improving the traceability and reliability of maintenance decisions.

3.1. System Characterization

The first methodological block establishes the functional analysis through information from technical manuals, the literature, and expert knowledge. This stage is not merely descriptive; it serves as the source of conditions for the subsequent quantitative modeling. Defining system configuration and performing functional analysis enables a clear mapping between physical subsystems and their logical representation in reliability models. Based on this description, the method takes a systematic approach to identifying failure modes (j) for each component (i) of the system (f_i^j). A failure mode is the distinct mechanism by which a failure presents itself, resulting in a functional failure [49–51].

3.2. System Reliability

The second block converts the qualitative system representation into a quantitative structure using FTA or RBD. At this point, it is used to model system behavior and compute component and system failure probabilities [13,51,52]. This stage shifts the analysis from descriptive modeling to data-based evaluation. The estimation of TTF and failure rates connects historical records and manufacturer data with the probabilistic behavior of each component.

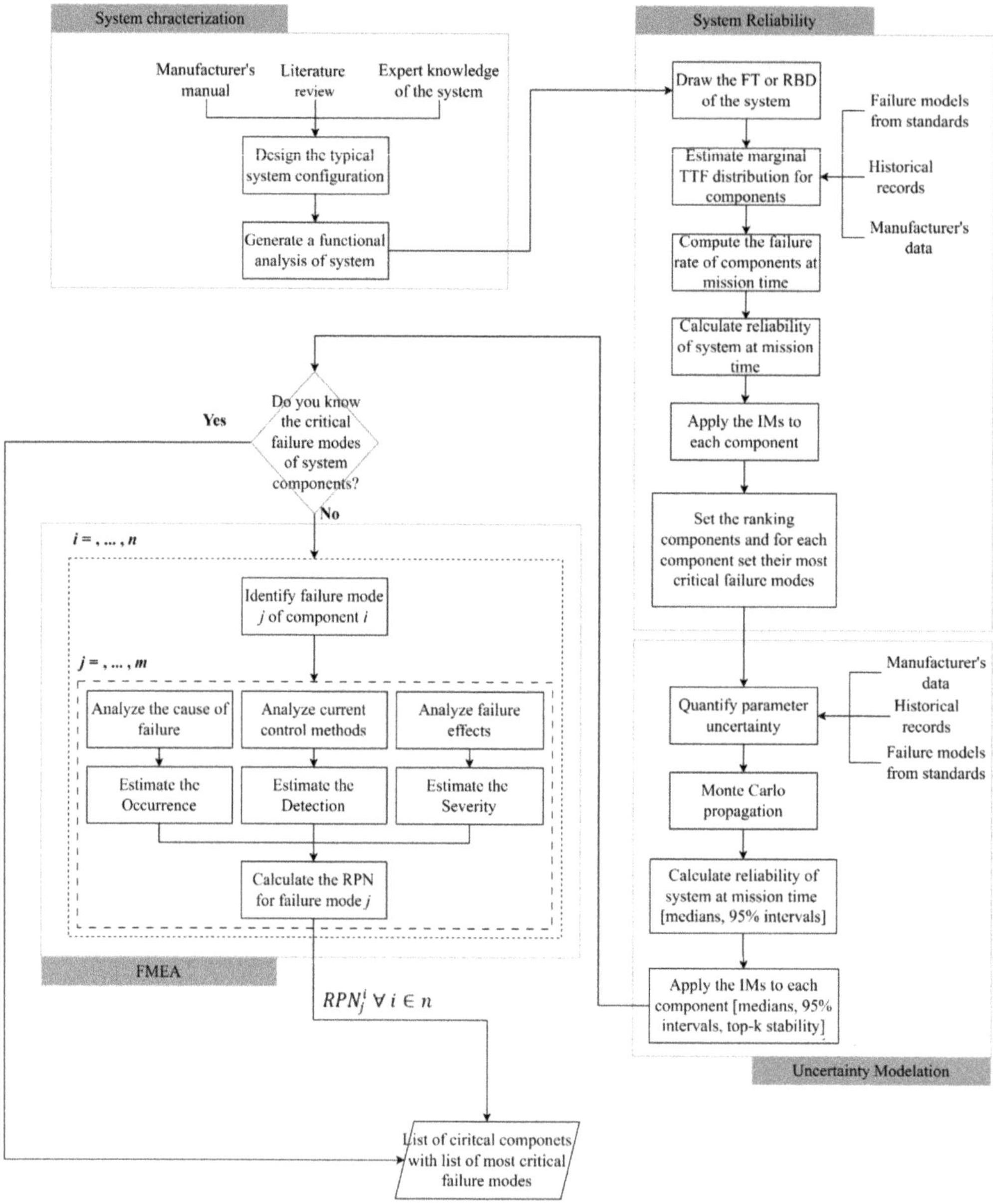

Figure 1. Proposed Methodology.

Let $\mathcal{F} = \{\mathcal{F}_p(\cdot|\theta_{i,p})\}$ denote a candidate set of lifetime distributions (e.g., Exponential, Weibull, Gamma). Given observed TTF data for component i, $TTF^i = \{t_1^i, t_2^i, \ldots, t_l^i\}$, the parameters $\theta_{i,p}$ are estimated by maximum likelihood.

$$\hat{\theta}_{i,p} = arg \max_{\theta} \mathcal{L}\left(TTF^i \middle| \mathcal{F}_p, \theta\right)$$

The preferred model for the i-th component is selected using goodness-of-fit diagnostics (e.g., Kolmogorov–Smirnov) and information criteria (the Akaike Information Criterion—AIC or the Bayesian Information Criterion—BIC). The resulting reliability function $R_i(t)$ follows from the chosen family, for instance,

$$R_i(t) = \begin{cases} exp(-\theta_i t) \; exponential \\ exp\left[-\left(\frac{t}{\theta_{i,1}}\right)^{\theta_{i,2}}\right] \; weibull \end{cases}$$

The procedure for estimating the parameters of these distributions and using them to support reliability assessment and asset management follows the approach previously proposed [53].

If TTF data are unavailable, parameter estimates may be elicited from expert judgment, sectoral standards or analogous systems, ensuring methodological adaptability.

Subsequently, once $R_i(t)$ has been specified for $i = 1, \ldots, n$, the overall system reliability is obtained from the structure function $\phi(\cdot)$ defined by the FT or RBD:

$$R_{sys}(t) = \phi(R_1(t), R_2(t), \ldots, R_n(t))$$

where series, parallel, or k-out-of-n relationships capture how component failure propagates within the system.

Once system reliability $R_{sys}(t)$ is obtained, the application of IMs extends the evaluation beyond failure probability to include the structural contribution of each element. Through this process, the influence of each component on the overall system reliability is quantified. The resulting hierarchy of component criticality provides a clear basis for prioritizing maintenance and risk mitigation actions.

Consistent with the concepts reviewed in Section 2.2 and the definitions in Table 1, we quantify component criticality from the system reliability function $R_{sys}(t)$ obtained via FTA/RBD. The goal is to translate system structure and behavior into operational indices that rank the n components by their marginal influence on overall performance, while retaining the structural vs. reliability-based categories introduced earlier.

For each component i, IMs are evaluated by re-computing $R_{sys}(t)$ under targeted perturbations of $R_i(t)$, or under ideal/worst assumptions. Evaluating these metrics over the mission time of interest yields a quantitative ranking of components. In the synthesis step, this IM-based ranking is linked to the qualitative input of FMEA: for each prioritized component i, we report its IM values and rank together with its top failure modes and RPNs. This preserves the complementary roles of methodologies: interpretability of failure modes from FMEA, and probabilistic rigor plus systemic perspective from FTA/RBD and IMs, supporting risk-informed maintenance planning and resource allocation.

3.3. Propagation of Uncertainty

Acknowledging that, in practice, it is impractical to assume fixed parameters for the TTF of each component, we explicitly treat parametric uncertainty in the TTF models introduced in Section 3.2 and propagate its effect to the system reliability and IMs of Section 3.3, following our previous approach [7].

For component i, let θ_i denote the parameters of the selected lifetime model (e.g., θ for Exponential, (θ_1, θ_2) for Weibull).

Uncertainty propagation is performed via Monte-Carlo. For $b = 1, \ldots, B$ (e.g., $B = 10,000$) and all components $i = 1, \ldots, n$.

1. Draw $\theta_i^b \sim \mathcal{N}(\hat{\theta}_i, \hat{V}_i)$;

2. Compute $R_i^b(t)$ from the chosen lifetime model;
3. Evaluate $R_{sys}^b(t) = \phi\left(R_1^b(t), R_2^b(t), \ldots, R_n^b(t)\right)$ via FTA or RBD;
4. Compute IMs $\left\{I_i^{B(b)}, I_i^{RRW(b)}, I_i^{RAW(b)}, I_i^{FV(b)}, I_i^{C(b)}\right\}$.

After all iterations, we summarize $R_{sys}\ (t)$, subsystems failure probabilities and IMs by medians and 95% uncertainty intervals (Monte-Carlo percentiles). We also report rank stability, i.e., the proportion of iterations in which each element appears at the top-k by a given IM, to indicate prioritization robustness.

This procedure quantifies how uncertainty in lifetime-model parameters propagates to system-level reliability and IMs, providing interval-based evidence rather than point estimates alone.

General assumptions: Unless stated otherwise, component failures are treated as statistically independent (dependences are represented by the FTA/RBD structure function); lifetime parameters are time-homogeneous over the analysis horizon; operating conditions are stationary within the mission window; repairs are not modeled (focus on reliability rather than availability). Candidates' lifetimes, families, and parameter estimation follow Section 3.2; IM are computed as shown in Section 3.3; and parametric uncertainty is quantified and propagated as shown in Section 3.4. Dataset-specific assumptions (e.g., mission time, data sources) are detailed in the case study.

3.4. Closing the Loop: IMs, FMEA and Prioritization

The fourth methodological block introduces a conditional mechanism for knowledge completion. If the critical failure modes of system components are unknown, the process transitions toward the qualitative domain. Here, FMEA acts as an inference tool that decomposes uncertainty about specific failure mechanisms into measurable indices: occurrence, detection, and severity. The derived RPNs close the inferential loop by generating a criticality list that complements and validates the ranking obtained through IMs.

FMEA provides a structured list of components, their potential failure modes, and the corresponding risk rankings ($RPN_j^i\ \forall\, i \in n$). These qualitative outputs are retained to support the interpretation of quantitative results at a later stage.

Each component $i \in \{1, 2, \ldots, n\}$ is decomposed into potential failure modes f_i^j with $j = 1, 2, \ldots, m_i$. A failure mode is the distinct mechanism by which a failure presents itself, resulting in a functional failure.

For each mode, three evaluation criteria are defined: occurrence O_i^j, severity S_i^j, and detection D_i^j; the RPN is computed as shown in Table 2.

Conceptually, this block closes the inferential loop: uncertainty about the system's behavior is reduced by aligning deductive reliability results with inductive field evidence. The resulting criticality ranking not only identifies where failures are most probable but also where they are most consequential, guiding maintenance decisions that balance analytical rigor and empirical knowledge. The following sections demonstrate the practical value of this framework by presenting its application to SWCCT within a solid waste management company.

4. Case Study

For empirical validation, the proposed methodology was implemented in a real-world case study involving a fleet of SWCCT operated by a solid waste management company in a Colombian city, thereby providing a relevant testbed to assess its applicability and effectiveness under operational uncertainty.

In this section, the case study is introduced by presenting its operational context, the functional understanding of the SWCCT, and the scope of the analyses conducted at both qualitative and quantitative levels. Rather than focusing on results, which are developed in Section 5, the emphasis here is placed on explaining the system configuration, outlining the operational challenges faced by the fleet, and clarifying the procedures followed in each stage of the analysis. This structure ensures that the reader gains a clear overview of the system and its environment before addressing the findings of the methodology's application.

The company operates a fleet that serves 248 urban routes dedicated to domestic and recyclable waste collection, exposing the trucks to highly demanding operational conditions such as steep terrain, long-distance travel to landfills, continuous loading and unloading cycles, and frequent stops. These challenges make SWCCT an appropriate testbed for assessing how reliability strategies that account for uncertainty can support engineering decision-making, enhance resource allocation, and improve system resilience in real-world environments.

4.1. Functional Analysis of System

SWCCT is the core unit of the vehicle, enabling reception, compaction, and temporary storage of solid waste prior to its final transport. From a methodological perspective, the decomposition of the system into hierarchical levels (system, subsystems, and components) is a prerequisite for implementing reliability-based assessment frameworks. Structurally, the SWCCT integrates two major subsystems: (i) the compaction system (CS) associated with electric and control units responsible for waste compression and operational control; and (ii) the heavy-duty truck system (HDTS), which collectively ensure vehicular mobility, load-bearing capacity and reliable operation under demanding service

The structured breakdown not only allows the identification of functional interdependencies but also facilitates the quantification of failure probabilities at different aggregation levels, which is fundamental for integrating uncertainty into risk evaluation. In summary, the primary function of the SWCCT: the collection and transportation of municipal solid waste to treatment facilities.

Building upon this system-level perspective, the subdivision into the compaction-box structure (CBS) and the hydraulic subsystem (HS) provides a more detailed view for analyzing reliability and risk propagation. CBS directly governs the mechanical integrity of waste containment, while the HS determines the performance and operational continuity of compaction (Figure 2).

Uncertainty in these subsystems arises from heterogeneous loading conditions, hydraulic pressure fluctuations, and wear mechanisms, all of which may trigger cascading effects across the system. Therefore, emphasizing subsystems in the reliability framework ensures a more detailed understanding of failure propagation, enabling a probabilistic treatment that accounts for both epistemic and aleatory uncertainties inherent in waste collection operations.

Extending this system-level integration, the subdivision of the HDTS into critical subsystems—such as the engine, brakes and pneumatic subsystem, suspension, steering, power transmission, and electric and control—introduces a finer resolution for analyzing risk propagation (Figure 3).

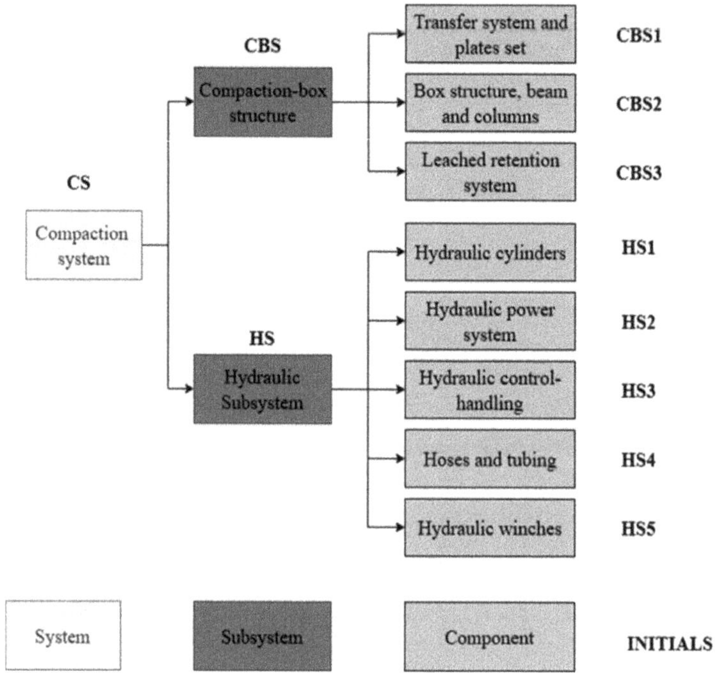

Figure 2. Compaction system of waste collection vehicles.

These subsystems represent diverse physical domains, and their interactions are inherently nonlinear, making them key determinants of overall reliability. From a reliability perspective, this interconnectedness highlights the importance of capturing both aleatory uncertainties (stemming from variable loads, environmental stresses, and wear) and epistemic uncertainties (derived from limited knowledge of degradation mechanisms).

4.2. Critical Components Analysis of SWCCT

Using failure records from December 2015 to January 2020 (hours), Exponential, Weibull, and Lognormal TTF models were fitted using maximum likelihood for each component, selected the best family by AIC (BIC as tiebreaker), and assessed absolute fit with the Cramér–von Mises statistic (KS bootstrap p-values in Appendix A). Figure 4 summarizes both the empirical support (number of records) and the selected marginal family—panel (a) for the compaction system and panel (b) for the heavy-duty truck system. In line with mechanical wear mechanisms, Weibull predominates, while Exponential and Lognormal appear where memoryless or multiplicative effects are plausible. Full parameter estimates (with SEs) and diagnostics are reported in Appendix A. These fitted marginals feed directly into the system-level computation described next.

Given the complexity of the SWCCT and the uncertainty regarding which subsystems contribute most critically to overall reliability, the quantitative stage of the case study was structured around an FTA (Figure 5). This approach was chosen because it provides a direct representation of how specific component or subsystem failures propagate to total system loss, aligning with the operational realities of the fleet. The FTA, when applied to the failure data collected from the trucks, revealed that the HS5 showed no record failures; the compaction system and truck-level subsystems—including power transmission, brakes, and hydraulics—were the main contributors to system risks.

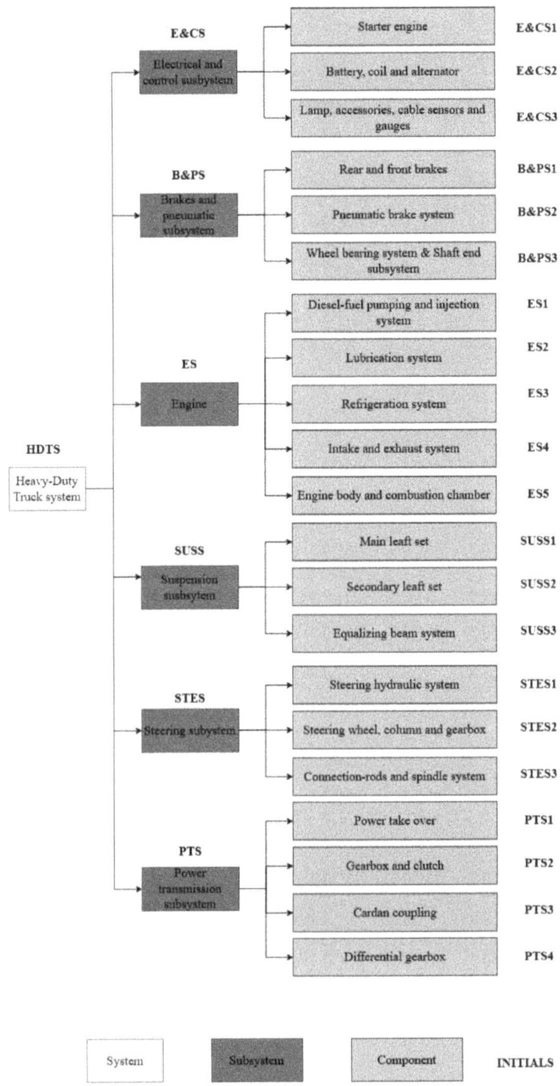

Figure 3. Heavy-Duty Truck System of waste collection vehicles.

By mapping these interdependencies, the FTA demonstrated that the overall reliability of the SWCCT is particularly vulnerable to failures in the compaction box and its hydraulics, or in any major truck subsystem such as the engine, electrical, or control systems. The detailed examination of the power transmission subsystem illustrated how single-component failures can escalate into total subsystem collapse, underscoring the systemic fragility created by tightly coupled elements. This practical insight provided by the FTA enabled the identification of high-impact failure pathways, supporting targeted maintenance actions and risk-informed decision-making for the fleet's operation. The estimated reliability of the SWCCT was derived from the FTA and component failure rates. The FTA included 28 basic events, resulting in 28 minimum cut sets due to the use of OR gates. Based on these quantifications and failure rates, IMs were applied, and the results are discussed in the next section.

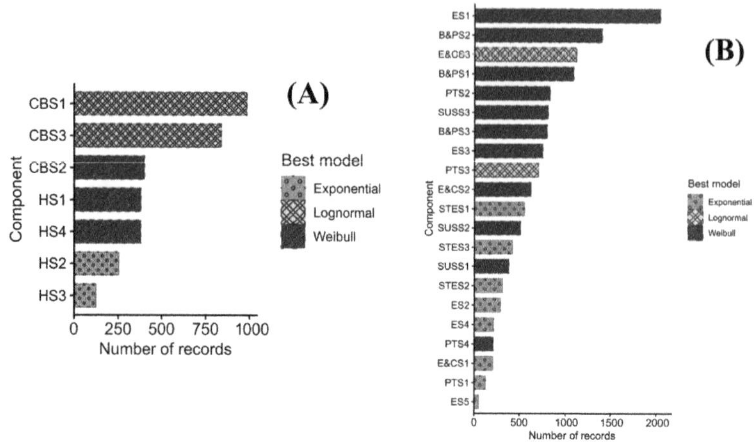

Figure 4. Selected marginal TTF model by component: (**A**) compaction system; (**B**) Heavy-Duty Truck System.

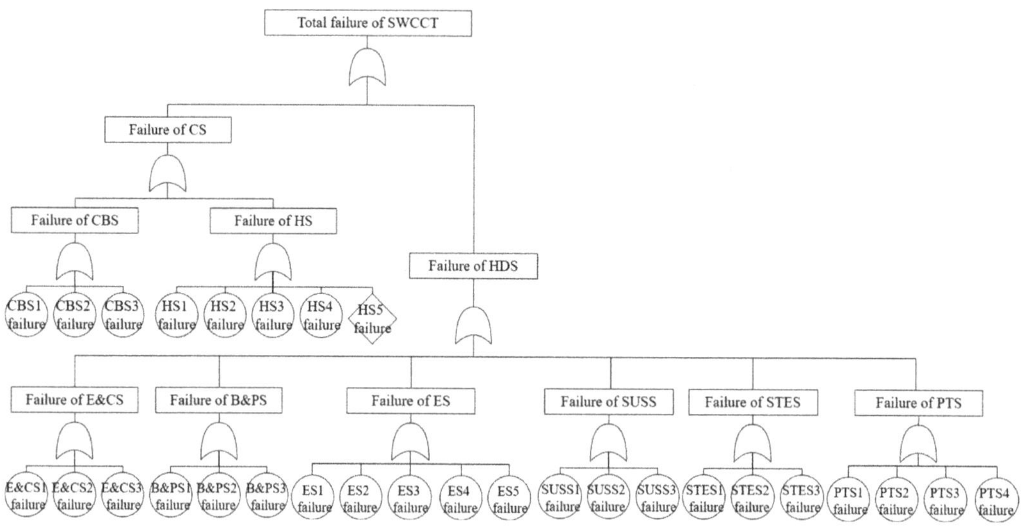

Figure 5. Fault Tree Analysis of waste collection vehicles.

4.3. Uncertainty Quantification of SWCCT

In accordance with what is described in Section 3.3, parametric uncertainty was quantified under the marginal lifetime models and propagated to the SWCCT reliability and IMs. For each basic event (failure of component), the fitted distribution parameters (Appendix A) were treated as random: Exponential rates, Weibull and Lognormal were sampled from their asymptotic MLE normal centered at with standard errors taken from Appendix A. Positivity-constrained (rates, shapes, scales) were sampled from truncated normal to enforce physical bounds. Then, for mission time, each draw yields the failure probability of the component, i.e., the probability of a basic event. These probabilities are then mapped through the FTA to obtain the top-event probability (complete failure of the SWCCT) and, by complement, system reliability.

A Monte-Carlo simulation was run with parameter drawings. For each draw, we computed component failure probabilities, propagated them through the FTA, and then evaluated the IMs. We summarize medians and 90% uncertainty intervals (5th and 95th percentiles) for all IMs across the iterations. To support maintenance planning, we also provide an uncertainty-informed ranking (top five of SWCCT components per IM) and discuss their failure modes identified in the qualitative analysis.

4.4. Critical Modes Failures of SWCCT

Once critical components were identified, the subsequent FMEA revealed 322 potential failure modes across systems and subsystems. This evaluation integrated both static sources, such as manufacturer catalogs, troubleshooting guides, operations manuals, and dynamic sources, including failure records and maintenance work orders collected over a four-year period. Severity and occurrence rankings were assigned using standardized scales from the literature [54,55], ensuring methodological consistency and comparability. However, given the specific complexity of the SWCCT architecture, control design mechanisms and maintenance execution dynamics for the SWCCT fleet, it was necessary to develop tailored detectability criteria to better reflect diagnostic challenges faced in practice. These criteria account not only for the technical difficulty of identifying a failure but also for the expertise and personnel required to perform effective monitoring. As a result, the detectability index captured dimensions not addressed by conventional scales, thereby unifying the ranking procedure across all subsystems. This adaptation guaranteed a consistent and system-specific evaluation of failure modes, as summarized in Table 3.

Table 3. FMEA detectability criteria.

Detection	Likelihood of Detection	Personal (Mechanic/Engineer)	Ranking
Absolute Uncertainty	Design control (DC) cannot or will not detect a potential cause, mechanism, or failure mode	NA	10
Very Remote	Extremely unlikely that DC will detect a potential cause, mechanism, or failure mode	NA	9
Remote	Low likelihood that DC will detect a potential cause, mechanism, or failure mode	Diagnosis of the cause/mechanism requires disassembling multiple components	8
Very Low	Very small chance that DC will detect a potential cause, mechanism, or failure mode	Diagnosis of the cause/mechanism requires disassembling the component.	7
Low	Limited chance that DC will detect a potential cause, mechanism, or failure mode	Diagnosis of the cause/mechanism requires disassembling the sub-system	6
Moderate	Reasonable chance that DC will detect a potential cause, mechanism, or failure mode	Diagnosis of the cause/mechanism is possible without disassembling the sub-system	5
Moderately high	Fairly good chance that DC will detect a potential cause, mechanism, or failure mode	The operator can diagnose the cause/mechanism by disassembling the component	4
High	High likelihood that DC will detect a potential cause, mechanism, or failure mode.	Available sensors allow diagnosis of the cause/mechanism	3
Very High	Very strong chance that DC will detect a potential cause, mechanism, or failure mode.	An external observer can diagnose the cause/mechanism	2
Almost Certain	DC will almost certainly detect a potential cause, mechanism, or failure mode	The subsystem/component can self-diagnose the cause/mechanism	1

In practice, detectability was further refined to reflect the real-world constraints of maintenance operations, including the level of expertise and personnel required to diagnose specific failures. This final adjustment ensured that subsystem-specific design variations and operational realities were fully captured, reinforcing the robustness and applicability of the evaluation framework.

5. Results

The critical evaluation of the components in the SWCCT is presented, using the discussed methodology. Building on the functional analysis overview provided in Figures 2 and 3, critical components of the SWCCT were identified by evaluating each item using importance measures. For each item, measures of importance such as $I_i^B(t)$, $I_i^{RRW}(t)$, $I_i^{RAW}(t)$, $I_i^{FV}(t)$, and $I_i^{CR}(t)$ were applied (see Figure 6).

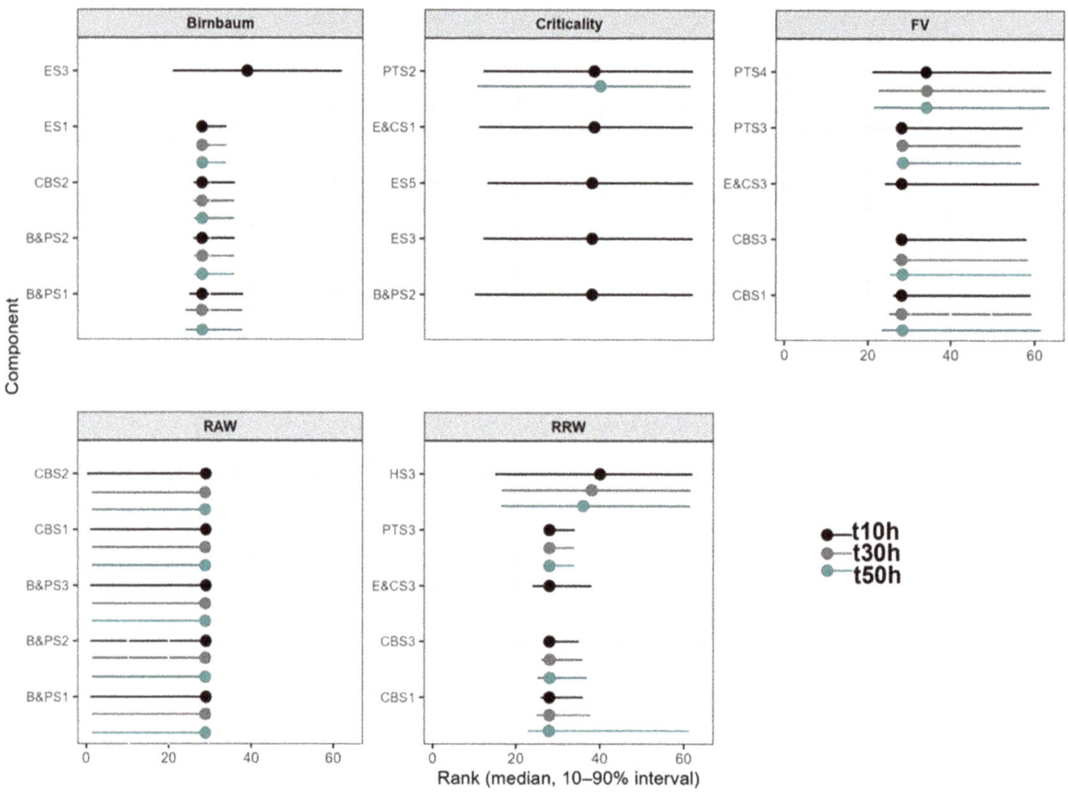

Figure 6. Top 5 Components—Rank Stability for critical components for waste collection vehicles.

Consequently, Figure 6 presents the computation of the four IMs obtained from the FTA of SWCCT. The Heavy-Duty Truck subsystems, particularly the Brakes and Pneumatic System (B&PS), the Electrical and Control Subsystem (E&CS), and the Power Transmission System (PTS), show the highest Importance Measure (IM) values, indicating the strongest influence on the overall reliability of the SWCCT. In contrast, the Compaction Box Structure (CBS) and Hydraulic System (HS) exhibit moderate IM values, suggesting that although they are relevant for operational performance, they are not the dominant contributors to system reliability.

Higher values of $I_i^{(B)}$ indicate components whose reliability variations produce significant effects on the system-level reliability. In this case, B&PS1, E&CS3, and B&PS2

exhibit the largest $I_i^{(B)}$ values, confirming that they belong to minimal cut sets directly associated with the top event. Components in SUSS and STES show intermediate influence, whereas HS and ES components display low marginal contributions, indicating a degree of redundancy in those subsystems.

$I_i^{(FV)}$ values lie between 0.03 and 0.08, maintaining a consistent ranking pattern with $I_i^{(B)}$. The dominance of the B&PS and E&CS subsystems indicates that they act as failure-propagation paths. The reduced dispersion across components within HS and ES confirms that these subsystems act as reliability buffers under stochastic load conditions.

$I_i^{(RRW)}$ represents the relative improvement in system reliability if components B&PS1, E&CS3, and B&PS2 were made perfectly reliable, indicating the largest potential gain from maintenance or design improvement. Conversely, components from HS and ES present $I_i^{(RRW)} \approx 1.00$, confirming minimal potential for reliability enhancement through local interventions.

For $I_i^{CR}(t)$, it is observed that all components contribute proportionally to the overall system reliability. The values range between 0.93 and 0.99, indicating a narrow dispersion and suggesting that reliability improvements would benefit the entire system in a nearly uniform manner. This pattern implies that proportional enhancements in component reliability translate directly into global reliability gains. Such behavior is consistent with the series-dominant configuration of the SWCCT, where overall performance depends on the correct functioning of every subsystem.

For this case, $I_i^{CR}(t)$ and $I_i^{RAW}(t)$ are less meaningful because they become identical when the unreliability of any component reaches one, indicating that system unreliability is maximal.

In systems with series topology, the system reliability R_{sys} can be approximated as the product of individual component reliability. Under this structure, even small variations in the reliability of any component have a noticeable influence on R_{sys}, which explains the relatively uniform criticality (I_i^{CR}) across components. Consequently, maintenance or design actions that target general reliability improvement—rather than focusing solely on isolated elements—will yield system-level benefits.

Figure 7 complements the deterministic (point) estimates of SWCCT reliability reported in Table 4 by displaying the median trajectory and the 90% uncertainty bands, obtained by propagating parametric uncertainty through the FTA.

Regarding the reliability of the vehicles, Figure 7 presents the results obtained when quantifying the main systems of the SWCCT for different periods of operation (10, 20, 30, 40, and 50 h), in accordance with the operators' interest in evaluating fleet reliability at the beginning and at the end of the week (the SWCCT analyzed operates 10.44 h/day).

Table 4. Failure and Reliability under uncertainty of SWCCT.

	Operational Time (Hours)				
	10	20	30	40	50
Reliability of SWCCT (5th percentile)	89.54%	80.01%	71.26%	63.34%	56.20%
Reliability of SWCCT (median)	90.74%	82.10%	74.06%	66.64%	59.83%
Reliability of SWCCT (95th percentile)	91.70%	83.79%	76.37%	69.42%	62.96%
Failure of SWCCT (5th percentile)	8.30%	16.21%	23.63%	30.58%	37.04%
Failure of SWCCT (median)	9.26%	17.90%	25.94%	33.36%	40.17%
Failure of SWCCT (95th percentile)	10.46%	19.99%	28.74%	36.66%	43.80%

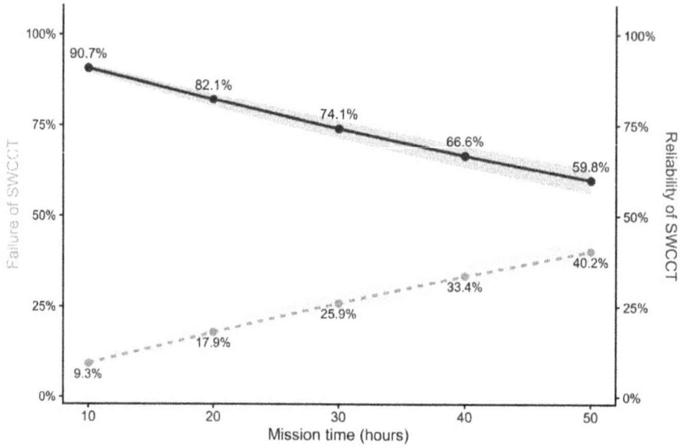

Figure 7. Uncertainty Analysis.

As outlined during methodology description (Section 3.3), once a baseline for system reliability was established, continuing with uncertainty analysis provides a more realistic view of prediction confidence over time. Figure 7 complements the deterministic estimates of SWCCT reliability reported in Table 4 by displaying the median trajectory and the 90% uncertainty bands, obtained by propagating parametric uncertainty through the FTA.

The reliability profile decreases monotonically with time, and the uncertainty bands widen at longer horizons, indicating the accumulation of parameter uncertainty in the system-level prediction. Relative to the deterministic curve, the uncertainty-aware median is slightly lower at all horizons, which signals that point estimates are mildly optimistic when parameter variability is acknowledged.

The IMs derived from the Fault Tree Analysis indicated that the Brakes and Pneumatic System (B&PS), Electrical and Control Subsystem (E&CS), and Power Transmission System (PTS) are the most critical contributors to the overall reliability of the SWCCT. Components within these subsystems exhibited the highest Birnbaum and Fussell–Vesely importance values, confirming that small variations in their reliability have significant effects at the system level. Conversely, the Compaction Box Structure (CBS) and Hydraulic System (HS) showed moderate IM values, while the Electrical Supply (ES) and Structural Subsystem (SUSS) presented the lowest ones, indicating lower marginal effects on the top event probability.

Maintenance prioritization was further supported by failure modes ranked using Risk Priority Numbers (RPNs), as described in the methodology (Figure 1). RPNs were evaluated across components and grouped by subsystem, reinforcing the relevance of B&PS, E&CS, and PTS—already identified as critical through IMs. Figure 8 presents NPR values, highlighting failure modes concentrated in these subsystems.

The pneumatic braking subsystems require attention because two of their components show the highest RPN values and have safety implications. Brake drum overheating is a significant failure mode for braking subsystems, potentially caused by excessive braking without engine assistance. These results align with the IM analysis, confirming that the B&PS subsystem represents the main reliability bottleneck of the vehicle. The high NPR concentration in a limited set of modes implies that corrective actions focused on these faults can significantly reduce the overall operational risk.

Actions include inspection of braking assemblies, adjustment and replacement of linings and drums, verification of air compressor operation, and calibration of pressure regulators and governors. The intermediate group of modes, mainly related to compressor cycling, pedal response, and minor leaks, can be addressed through scheduled preventive maintenance. Low-ranked modes require only periodic monitoring, since their individual contribution to total risk is limited.

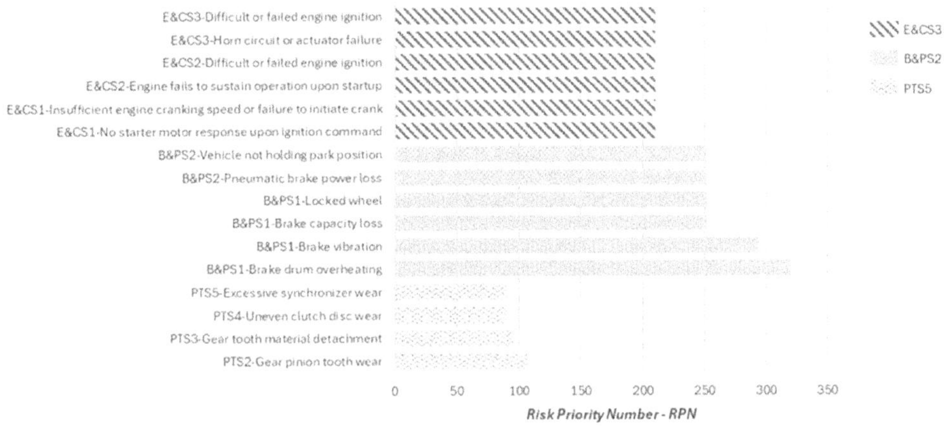

Figure 8. RPN summary for critical components for Systems according to IM prioritization.

The distribution of NPR values supports a tiered maintenance strategy. The highest-ranked modes, which represent roughly 20% of the total number of modes, account for most of the aggregated NPR. These should receive immediate preventive and corrective attention. Actions include inspection of braking assemblies, adjustment and replacement of linings and drums, verification of air compressor operation, and calibration of pressure regulators and governors. The intermediate group of modes, mainly related to compressor cycling, pedal response, and minor leaks, can be addressed through scheduled preventive maintenance. Low-ranked modes require only periodic monitoring, since their individual contribution to total risk is limited.

The following section presents the discussion of findings, examining their technical implications, the relationship between system architecture and observed reliability patterns, and the practical relevance of the proposed framework for maintenance management and design improvement.

6. Discussion

The consistency between IM's FMEA results demonstrates the internal validity of the proposed reliability framework. Both methods identified the same dominant subsystems within SWCCT, particularly the Brakes and Pneumatic System, Electrical and Control Subsystem, and Power Transmission System. This alignment confirms that the IM-based structural reliability model and the FMEA-based operational assessment converge in identifying the most critical components. The result validates that the methodology consistently links system-level reliability with component-level failure behavior.

The IMs provide a quantitative basis for determining which components have the highest structural influence on system reliability. In contrast, the NPR index obtained from the FMEA specifies which failure mechanisms should be prioritized within those components. When analyzed together, the IMs define where design or maintenance interventions produce the greatest effect, while the FMEA clarifies what failure modes must be addressed. This complementarity ensures that system-level reliability analysis is translated into concrete maintenance actions. Discrepancies between IM and NPR rankings arise mainly because IMs reflect structural sensitivity, whereas NPR combines severity, occurrence, and detection—factors that describe the operational impact of failures. The partial correlation observed between both approaches confirms that they are not redundant but complementary in nature.

The uncertainty analysis confirms the stability of the framework. Variations in component reliability parameters produced measurable changes in the system-level reliability indices, confirming that uncertainty propagation has a direct effect on prioritization outcomes. Subsystems that contain a higher number of components in series exhibit greater sensitivity and therefore account for a higher proportion of overall system failures. This structural characteristic explains why the heavy-duty truck section contributes to approximately 80% of total failures. Conversely, subsystems with partial redundancy, such as the Hydraulic System and Electrical System, show reduced sensitivity and act as reliability buffers under stochastic operating conditions.

The findings discussed above confirm that the integration of Importance Measures and FMEA provides a consistent and operationally relevant framework for reliability assessment and maintenance planning of SWCCTs. The combined approach ensures that structural reliability analysis and failure mode prioritization converge toward practical decision-making criteria. By incorporating uncertainty into both system-level and component-level evaluations, the methodology establishes a quantitative basis for prioritization that reflects real operating conditions. The following section presents the main conclusions derived from this study, emphasizing the methodological contributions, the implications for maintenance strategy design, and the potential extensions for future research.

7. Conclusions

This paper presents a methodology for identifying critical components in complex systems by integrating FMEA with IMs. This approach combines the qualitative risk assessment provided by FMEA, which evaluates potential failure modes based on their effects, causes and occurrence, with IMs, which use historical reliability data to identify critical components. This dual approach adds structure to the analysis and provides an understanding of system vulnerabilities, supporting the prioritization of maintenance efforts.

The application to SWCCT demonstrated that combining IMs and FMEA clarifies the interpretation of reliability behavior. IMs determine which components most affect system reliability, while FMEA identifies the specific failure modes that require attention. Together, they enable more accurate maintenance prioritization and resource allocation under uncertain operating conditions. The results obtained provide a consistent and quantifiable overview of the reliability behavior of the SWCCT. The integration of IMs and FMEA allowed the identification of subsystems and specific failure modes with the highest impact on overall system performance, as well as establishing clear priorities for maintenance planning and resource allocation. The coherence between both analytical approaches supports the consistency of the proposed methodology for diagnosing reliability structures in complex mechanical systems.

From a maintenance perspective, the integration of IMs and FMEA provides a systematic approach for reliability-based decision-making. The IMs identify the most reliability-sensitive components, while the NPR refines maintenance priorities by ranking failure modes within those components. The resulting prioritization map supports maintenance scheduling, spare-part management, and inspection planning. Furthermore, by quantifying uncertainty, the framework allows preventive maintenance intervals to be set that are both risk-informed and operationally feasible. Overall, the integrated analysis transforms system reliability results into a decision-support tool for maintenance planning, ensuring that efforts and resources are directed toward the components and failure mechanisms with the highest impact on SWCCT reliability.

Future work should focus on extending the methodology toward dynamic reliability updating using stochastic or Bayesian inference, integrating real-time condition monitoring, and applying data-driven models to improve predictive maintenance and decision support in uncertain operational environments.

Author Contributions: Conceptualization, M.V.C., J.D.A.C. and C.E.P.-R.; Methodology, F.G.C. and C.E.P.-R.; Formal analysis, J.D.A.C. and C.E.P.-R.; Investigation, M.V.C.; Writing—original draft, M.V.C.; Writing—review & editing, F.G.C. and C.E.P.-R.; Supervision, F.G.C. and C.E.P.-R.; Project administration, F.G.C. All authors have read and agreed to the published version of the manuscript.

Funding: This research received no external funding. And The APC was funded by Universidad Nacional de Colombia.

Institutional Review Board Statement: Not applicable.

Informed Consent Statement: Not applicable.

Data Availability Statement: The original contributions presented in this study are included in the article. Data is unavailable due to privacy. Further inquiries can be directed to the corresponding author.

Conflicts of Interest: The authors declare no conflict of interest.

Abbreviations

B&PS	Brakes and Pneumatic Subsystem
CDF	Cumulative Distribution Function
CBS	Compaction-Box Structure
CS	Compaction System
C&BS	Chassis and Bodywork
ES	Engine
E&CS	Electric and Control Subsystem
FMEA	Failure Modes and Effects Analysis
FTA	Fault Tree Analysis
HDTS	Heavy-Duty Truck System
HS H	Hydraulic Subsystem
IM I	Importance Measures
PTS	Power Transmission Subsystem
RAW	Risk-Achievement Worth
RBD	Reliability Block Diagrams
RPN	Risk Priority Number
RRW	Risk-Reduction Worth
STES	Steering Subsystem
SUSS	Suspension Subsystem
SWCT	T Solid Waste Collection and Compaction Truck
SWM	Solid Waste Management

Appendix A. Marginal Distribution Fit of SWCCT Components

This appendix reports the marginal TTF models fitted per component using the failure records collected from December 2015 to January 2020. All TTF values are recorded in hours. For each component, we fitted Exponential, Weibull and Lognormal distributions by maximum likelihood and selected the best by AIC (with BIC as tiebreaker). Table A1 reports, for each component: sample size, selected model, parameter estimates with standard errors (SE), Kolmogorov p-value, AIC, and the Cramér-von Mises. Across components, the number of failure records ranges from 49 (Engine—Body and combustion chamber) to 2043 (Engine—Diesel-fuel pumping and injection system), providing robust samples for distribution fitting.

Table A1. Results of marginal fitting for SWCCT's components.

Component	Amount of TTF Records	Best Model	p-Value KS	AIC	Cramér-Von Mises Statistic	Parameter 1	Parameter SE 1	Parameter 2	Parameter SE 2
B&PS1	1090	Weibull	0.042	10,925.75	0.082	0.992	0.022	1450.69	1.828
B&PS2	1406	Weibull	0.067	13,254.34	0.198	1.021	0.019	1064.24	1.199
B&PS3	798	Weibull	0.717	8422.58	0.034	1.023	0.025	2086.00	2.795
CBS1	984	Lognormal	0.083	10,054.58	0.258	7.200	0.045	1.27	0.032
CBS2	400	Weibull	0.003	4898.79	0.314	1.010	0.034	1035.07	9.540
CBS3	838	Lognormal	0.001	8741.77	0.534	7.459	0.047	1.29	0.033
E&CS1	209	Exponential	0.125	2736.78	0.180	1.15×10^{-4}	0.000		
E&CS2	625	Weibull	0.099	6855.31	0.065	1.002	0.028	2373.38	3.955
E&CS3	1120	Lognormal	0.051	10,791.35	0.684	6.635	0.041	1.27	0.028
ES1	2043	Weibull	0.067	17,979.67	6.34×10^{-9}	1.029	0.018	775.26	0.644
ES2	292	Exponential	0.555	3668.60	0.080	1.86×10^{-4}	0.000		
ES3	753	Weibull	0.536	8142.38	0.068	0.977	0.024	2261.67	3.432
ES4	215	Exponential	0.553	2809.04	0.053	1.41×10^{-4}	0.000		
ES5	49	Exponential	0.338	700.24	0.132	8.39×10^{-5}	0.000		
HS1	379	Weibull	0.010	4505.43	0.159	1.007	0.033	5979.51	8.244
HS2	254	Exponential	0.139	3210.05	0.184	1.19×10^{-4}	0.000		
HS3	124	Exponential	0.479	1725.24	0.128	8.61×10^{-5}	0.000		
HS4	378	Weibull	0.111	4490.65	0.087	1.003	0.032	4562.83	8.835
PTS1	125	Exponential	0.331	1741.62	0.102	7.96×10^{-5}	0.000		
PTS2	831	Weibull	0.098	8594.84	0.153	1.023	0.022	1957.76	2.622
PTS3	705	Lognormal	0.032	7552.40	0.243	8.049	0.050	1.28	0.035
PTS4	211	Weibull	0.616	2719.82	0.045	1.017	0.047	7600.38	18.137
STES1	554	Exponential	0.050	6186.12	0.220	1.83×10^{-4}	0.000		
STES2	311	Exponential	0.270	3873.64	0.073	3×10^{-4}	0.000		
STES3	421	Exponential	0.882	4939.82	0.049	3.60×10^{-4}	0.000		
SUSS1	381	Weibull	0.116	4549.82	0.086	1.018	0.045	4109.83	7.233
SUSS2	510	Weibull	0.006	5828.77	0.294	1.009	0.029	4857.29	5.689
SUSS3	810	Weibull	0.022	8538.50	0.189	1.014	0.023	2449.14	2.881

Parameters conventions: Exponential uses rate λ; Weibull uses shape k and scale λ; lognormal uses meanlog μ and sdlog σ. Standard errors (SEs) are the asymptotic MLE SEs. The Kolmogorov p-value reported is obtained via parametric bootstrap for the selected model.

References

1. Zio, E. Reliability Engineering: Old Problems and New Challenges. *Reliab. Eng. Syst. Saf.* **2009**, *94*, 125–141. [CrossRef]
2. Clavijo Mesa, M.V.; Patino-Rodriguez, C.E.; Guevara Carazas, F.J.; Gunawan, I.; Droguett, E.L. Asset management strategies using reliability, availability, and maintainability (RAM) analysis. *J. Braz. Soc. Mech. Sci. Eng.* **2021**, *43*, 495. [CrossRef]
3. Zio, E. The Future of Risk Assessment. *Reliab. Eng. Syst. Saf.* **2018**, *177*, 176–190. [CrossRef]
4. Patel, R.; Patel, P.B. A Review on Mechanical System Reliability and Maintenance Strategies for Maximizing Equipment Lifespan. *ESP J. Eng. Technol. Adv.* **2022**, *2*, 173–179.
5. Brown, J.I.; Colbourn, C.J.; Cox, D.; Graves, C.; Mol, L. Network Reliability: Heading Out on the Highway. *Networks* **2021**, *77*, 146–160. [CrossRef]
6. Chaal, M.; Ren, X.; Bahootoroody, A.; Basnet, S.; Bolbot, V.; Banda, O.A.V.; Van Gelder, P. Research on Risk, Safety, and Reliability of Autonomous Ships: A Bibliometric Review. *Saf. Sci.* **2023**, *167*, 106256. [CrossRef]
7. Clavijo, M.V.; Schleder, A.M.; Droguett, E.L.; Martins, M.R. RAM analysis of dynamic positioning system: An approach taking into account uncertainties and criticality equipment ratings. *Proc. Inst. Mech. Eng. Part O J. Risk Reliab.* **2022**, *236*, 1104–1134. [CrossRef]
8. Sardi, A.; Rizzi, A.; Sorano, E.; Guerrieri, A. Cyber Risk in Health Facilities: A Systematic Literature Review. *Sustainability* **2020**, *12*, 7002. [CrossRef]
9. Muriel-Villegas, J.E.; Alvarez-Uribe, K.C.; Patiño-Rodríguez, C.E.; Villegas, J.G. Analysis of Transportation Networks Subject to Natural Hazards—Insights from A Colombian Case. *Reliab. Eng. Syst. Saf.* **2016**, *152*, 151–165. [CrossRef]

10. Jamshidi, A.; Rahimi, S.A.; Ait-Kadi, D.; Ruiz, A. A Comprehensive Fuzzy Risk-Based Maintenance Framework for Prioritization of Medical Devices. *Appl. Soft Comput.* **2015**, *32*, 322–334. [CrossRef]
11. Borgonovo, E.; Apostolakis, G.E. A New Importance Measure for Risk-Informed Decision Making. *Reliab. Eng. Syst. Saf.* **2001**, *72*, 193–212. [CrossRef]
12. Flage, R.; Aven, T.; Zio, E.; Baraldi, P. Concerns, Challenges, and Directions of Development for the Issue of Representing Uncertainty in Risk Assessment. *Risk Anal.* **2014**, *34*, 1196–1207. [CrossRef]
13. Azarkhail, M.; Modarres, M. A Study of Implications of Using Importance Measures in Risk-Informed Decisions. In Proceedings of the Probabilistic Safety Assessment and Management: PSAM 7—ESREL'04, Berlin, Germany, 14–18 June 2004; Springer: Berlin/Heidelberg, Germany, 2004; Volume 6, pp. 456–461.
14. Selech, J.; Rogula-Kozłowska, W.; Piątek, P.; Walczak, A.; Pieniak, D.; Bondaronok, P.; Marcinkiewicz, J. Failure and Reliability Analysis of Heavy Firefighting and Rescue Vehicles: A Case Study. *Eksploat. Niezawodn.* **2024**, *26*, 175505. [CrossRef]
15. Ramirez-Marquez, J.E.; Coit, D.W. Composite Importance Measures for Multi-State Systems with Multi-State Components. *IEEE Trans. Reliab.* **2005**, *54*, 517–529. [CrossRef]
16. Lindbom, H.; Tehler, H.; Eriksson, K.; Aven, T. the Capability Concept—On How to Define and Describe Capability in Relation to Risk, Vulnerability and Resilience. *Reliab. Eng. Syst. Saf.* **2015**, *135*, 45–54. [CrossRef]
17. Carazas, F.G.; Souza, G.F.M.D. Risk-Based Decision Making Method for Maintenance Policy Selection of Thermal Power Plant Equipment. *Energy* **2010**, *35*, 964–975. [CrossRef]
18. Rodriguez, C.P.; Souza, G.F.M. Decision-Making Model for Offshore Offloading Operations Based on Probabilistic Risk Assessment. In *Vulnerability, Uncertainty, and Risk: Analysis, Modeling, and Management*; American Society of Civil Engineers: Reston, VA, USA, 2011; pp. 385–393.
19. Chang, K.-H. Evaluate the Orderings of Risk for Failure Problems Using A More General RPN Methodology. *Microelectron. Reliab.* **2009**, *49*, 1586–1596. [CrossRef]
20. Fang, H.; Li, J.; Song, W. Failure Mode and Effects Analysis: An Integrated Approach Based on Rough Set Theory and Prospect Theory. *Soft Comput.-A Fusion Found. Methodol. Appl.* **2020**, *24*, 6673. [CrossRef]
21. Birnbaum, Z.W. On the Importance of Different Components in A Multicomponent System. 1968. Available online: https://apps.dtic.mil/sti/html/tr/AD0670563/ (accessed on 2 November 2025).
22. Chacko, V.M. On Birnbaum Type Joint Importance Measures for Multistate Reliability Systems. *Commun. Stat.—Theory Methods* **2023**, *52*, 2799–2818. [CrossRef]
23. Huseby, A.B.; Kalinowska, M.; Abrahamsen, T. Birnbaum Criticality and Importance Measures for Multistate Systems with Repairable Components. *Prob. Eng. Inf. Sci.* **2022**, *36*, 66–86. [CrossRef]
24. Dui, H.; Chen, L.; Wu, S. Generalized Integrated Importance Measure for System Performance Evaluation: Application to A Propeller Plane System. Maintenance and Reliability. *Eksploatacja* **2017**, *19*, 279–286. [CrossRef]
25. Natvig, B. *Multistate Systems Reliability Theory with Applications*; John Wiley and Sons: Hoboken, NJ, USA, 2010.
26. Lipol, L.S.; Haq, J. Risk Analysis Method: FMEA/FMECA in the Organizations. *Int. J. Basic Appl. Sci.* **2011**, *11*, 74–82.
27. Liu, H.-C.; Liu, L.; Liu, N. Risk Evaluation Approaches in Failure Mode and Effects Analysis: A Literature Review. *Expert Syst. Appl.* **2013**, *40*, 828–838. [CrossRef]
28. Braglia, M. MAFMA: Multi-Attribute Failure Mode Analysis. *Int. J. Qual. Reliab. Manag.* **2000**, *17*, 1017–1033. [CrossRef]
29. Zammori, F.; Gabbrielli, R. ANP/RPN: A Multi Criteria Evaluation of the Risk Priority Number. *Qual. Reliab. Eng. Int.* **2012**, *28*, 85–104. [CrossRef]
30. Makajic-Nikolic, D.; Petrovic, N.; Belic, A.; Rokvic, M.; Radakovic, J.A.; Tubic, V. The Fault Tree Analysis of Infectious Medical Waste Management. *J. Clean. Prod.* **2016**, *113*, 365–373. [CrossRef]
31. Rentong, C.; Shaoping, W.; Hongyan, D.U.I.; Yang, L.I. Component Uncertainty Importance Measure in Complex Multi-State System Considering Epistemic Uncertainties. *Chin. J. Aeronaut.* **2024**, *37*, 31–54. [CrossRef]
32. Cao, Y.; Lu, C.; Dong, W. Importance Measures for Multi-State Systems with Multiple Components Under Hierarchical Dependences. *Reliab. Eng. Syst. Saf.* **2024**, *248*, 110142. [CrossRef]
33. Modarres, M.; Groth, K. *Reliability and Risk Analysis*; CRC Press: Boca Raton, FL, USA, 2023.
34. Hwang, S.K.; Kim, D.-H.; Kim, S.-C. Analysis of Risk Priority Number of FMEA and Surprise Index for Components of 7 Kw Electric Vehicle Charger. *J. Loss Prev. Process Ind.* **2024**, *91*, 105375. [CrossRef]
35. Mohsen, O.; Fereshteh, N. An Extended VIKOR Method Based on Entropy Measure for the Failure Modes Risk Assessment—A Case Study of the Geothermal Power Plant (GPP). *Saf. Sci.* **2017**, *92*, 160–172. [CrossRef]
36. Stamatis, D.H. *Failure Mode and Effect Analysis*; Quality Press: Seattle, WA, USA, 2003.
37. Meng Tay, K.; Peng Lim, C. Fuzzy FMEA with A Guided Rules Reduction System for Prioritization of Failures. *Int. J. Qual. Reliab. Manag.* **2006**, *23*, 1047–1066. [CrossRef]

38. Tarum, C.D. FMERA—Failure Modes, Effects, and (Financial) Risk Analysis. 2001-01-0375. 2001. Available online: https://legacy.sae.org/publications/technical-papers/content/2001-01-0375/ (accessed on 22 September 2025).
39. Kmenta, S.; Ishii, K. Scenario-Based FMEA: A Life Cycle Cost Perspective. In Proceedings of the ASME 2000 International Design Engineering Technical Conferences and Computers and Information in Engineering Conference, Baltimore, MA, USA, 10–13 September 2000; American Society of Mechanical Engineers: New York, NY, USA, 2000; Volume 5.
40. Salazar, J.C.; Nejjari, F.; Sarrate, R.; Weber, P.; Theilliol, D. Reliability Importance Measures for A Health-Aware Control of Drinking Water Networks. In Proceedings of the 2016 3rd Conference on Control and Fault-Tolerant Systems (SysTol), Barcelona, Spain, 7–9 September 2016; pp. 572–578. [CrossRef]
41. Abreu, D.T.M.P.; Barraza, J.E.F.; Clavijo, M.V.; Maturana, M.C.; Santos, J.R.; Schleder, A.M.; Martins, M.R. Research in System Reliability and Risk Analysis Under Development at the Naval Architecture and Ocean Engineering Department from University of São Paulo. *IFAC-PapersOnLine* **2022**, *55*, 193–198. [CrossRef]
42. Chang, Y.-R.; Amari, S.V.; Kuo, S.-Y. Computing System Failure Frequencies and Reliability Importance Measures Using OBDD. *IEEE Trans Comput.* **2004**, *53*, 54–68. [CrossRef]
43. Randall, R.B.; Antoni, J.; Chobsaard, S. The Relationship Between Spectral Correlation and Envelope Analysis in the Diagnostics of Bearing Faults and other Cyclostationary Machine Signals. *Mech. Syst. Signal Process.* **2001**, *15*, 945–962. [CrossRef]
44. Natvig, B.; Eide, K.A.; Gåsemyr, J.; Huseby, A.B.; Isaksen, S.L. Simulation Based Analysis and An Application to An Offshore Oil and Gas Production System of the Natvig Measures of Component Importance in Repairable Systems. *Reliab. Eng. Syst. Saf.* **2009**, *94*, 1629–1638. [CrossRef]
45. Van Der Borst, M.; Schoonakker, H. An Overview of PSA Importance Measures. *Reliab. Eng. Syst. Saf.* **2001**, *72*, 241–245. [CrossRef]
46. Miziula, P.; Navarro, J. Birnbaum Importance Measure for Reliability Systems with Dependent Components. *IEEE Trans. Rel.* **2019**, *68*, 439–450. [CrossRef]
47. Fussell, J.B. How to Hand-Calculate System Reliability and Safety Characteristics. *IEEE Trans. Reliab.* **1975**, *R-24*, 169–174. [CrossRef]
48. Rausand, M. Reliability Quantification. In *Reliability of Safety-Critical Systems*; John Wiley & Sons: Hoboken, NJ, USA, 2014; pp. 91–164. [CrossRef]
49. *ISO 14224*; Petroleum, Petrochemical and Natural Gas Industries—Collection and Exchange of Reliability and Maintenance Data for Equipment. ISO: Geneva, Switzerland, 2006.
50. Yu, Y.; Ireri, E.; Schmidt, B.; Swope, V. A Component FMECA Development Methodology to Support the DO-254 Compliance. In Proceedings of the 2023 Annual Reliability and Maintainability Symposium (RAMS), Orlando, FL, USA, 23–26 January 2023; pp. 1–5.
51. Zhang, C.; Zhang, Y.; Dui, H.; Wang, S.; Tomovic, M. Component Maintenance Strategies and Risk Analysis for Random Shock Effects Considering Maintenance Costs. *Eksploat. I Niezawodn.—Maint. Reliab.* **2023**, *25*, 162011. [CrossRef]
52. *SAE ARP5580*; Recommended Failure Modes and Effects Analysis (FMEA) Practices for Non-Automobile Applications. Society of Automotive Engineers: Warrendale, PA, USA, 2001.
53. Carlson, C.S. *Understanding and Applying the Fundamentals of FMEAs*; Reliasoft Corporation: Tucson, AZ, USA, 2024.
54. Hansen, M.J. FMEA Scales for Severity, Occurrence and Detection. Diunduh Dari. 2011. Available online: https://statstuff.com/ssfiles/tools/FMEAScalesGuide.pdf (accessed on 2 November 2025).
55. Distefano, S.; Puliafito, A. Dynamic Reliability Block Diagrams vs Dynamic Fault Trees. In Proceedings of the 2007 Annual Reliability and Maintainability Symposium, Orlando, FL, USA, 22–25 January 2007; pp. 71–76.

applied sciences

MDPI

Article

Uncertainty Quantification in Shear Wave Velocity Predictions: Integrating Explainable Machine Learning and Bayesian Inference

Ayele Tesema Chala * and Richard Ray

Structural and Geotechnical Engineering Department, Faculty of Architecture, Civil and Transport Sciences, Széchenyi Istvan University, Egyetem Ter 1, H-9026 Gyor, Hungary; ray@sze.hu
* Correspondence: chala.ayele.tesema@hallgato.sze.hu

Abstract: The accurate prediction of shear wave velocity (Vs) is critical for earthquake engineering applications. However, the prediction is inevitably influenced by geotechnical variability and various sources of uncertainty. This paper investigates the effectiveness of integrating explainable machine learning (ML) model and Bayesian generalized linear model (GLM) to enhance both predictive accuracy and uncertainty quantification in Vs prediction. The study utilizes an Extreme Gradient Boosting (XGBoost) algorithm coupled with Shapley Additive Explanations (SHAPs) and partial dependency analysis to identify key geotechnical parameters influencing Vs predictions. Additionally, a Bayesian GLM is developed to explicitly account for uncertainties arising from geotechnical variability. The effectiveness and predictive performance of the proposed models were validated through comparison with real case scenarios. The results highlight the unique advantages of each model. The XGBoost model demonstrates good predictive performance, achieving high coefficient of determination (R^2), index of agreement (IA), Kling–Gupta efficiency (KGE) values, and low error values while effectively explaining the impact of input parameters on Vs. In contrast, the Bayesian GLM provides probabilistic predictions with 95% credible intervals, capturing the uncertainty associated with the predictions. The integration of these two approaches creates a comprehensive framework that combines the strengths of high-accuracy ML predictions with the uncertainty quantification of Bayesian inference. This hybrid methodology offers a powerful and interpretable tool for Vs prediction, providing engineers with the confidence to make informed decisions.

Keywords: shear wave velocity prediction; geotechnical variability; uncertainty quantification; machine learning; extreme gradient boosting; generalized linear model

1. Introduction

Shear wave velocity (Vs) is a critical parameter in geotechnical and earthquake engineering. It has the greatest effect on the determination of dynamic behavior of soil and rock materials during seismic events [1–3]. Vs can be measured with either invasive (e.g., seismic cone penetration test, SCPT, cross-hole test, CHT, downhole test, DHT) or noninvasive (e.g., multichannel analysis of surface waves, MASW) techniques [4–8]. Integrating these tests with laboratory tests enables site characterization and the identification of potentially liquefiable lithological units, which is essential for hazard assessment [9,10]. While these methods offer the benefit of testing the soil in its natural environment, they tend to be time-consuming and economically infeasible for routine geotechnical exploration programs. Consequently, researchers have developed empirical correlations to estimate Vs using

CPT parameters (cone tip resistance qc and sleeve friction fs). Many notable examples of such correlations can be found in the literature (e.g., see the works of Robertson and others [11–13]). However, these empirical models often exhibit limitations in generalization, as they are based on site-specific data and are unable to capture the complex, non-linear relationships inherent in the soil properties influencing Vs. Regardless of the techniques, there is unavoidable uncertainty in the final Vs profile, which contributes to uncertainty in site response predictions [14].

Accounting for Vs uncertainty is of paramount importance in engineering applications. While there has been a great deal of research on the significance of incorporating uncertainty of Vs into engineering practices, no universally accepted methods exist for realistically addressing this critical issue [14]. Matasovic and Hashash [15] highlighted the use of bounding-type profiles, which is developing upper and lower bounds (e.g., $\pm$20%) around a base Vs profile to represent potential variability. Additionally, statistical approaches, such as Monte Carlo simulations and the Toro [16] Vs randomization model, have been utilized to generate a range of Vs profiles that reflect the inherent uncertainty in subsurface conditions.

With the increasing availability of high-quality, freely accessible data (e.g., CPT data), the use of machine learning (ML) algorithms for soil characterization has become increasingly popular in recent years [17,18]. Research related to Vs prediction using various algorithms shows promising results. Notable examples include predictions of Vs from CPT parameters (e.g., see authors' previous research and many others) [19–22]. ML algorithms excel at learning complex relationships within data without relying on the prior knowledge of the underlying mathematical models. This capability makes them highly effective for analyzing highly variable geotechnical data. However, the development of ML models must be approached with caution, as their performances heavily depend on several factors, including data preprocessing, feature selection, and hyperparameter tuning.

With the growing interest in predictive modeling for soil characterization, Bayesian generalized linear models (GLMs), have also emerged as a focus of study. Bayesian GLMs can be considered valuable tools alongside ML algorithms, particularly in geotechnical applications where data variability and uncertainty are prevalent. To the best of the authors' knowledge, no previous study has reported on assessment of Vs predictions by integrating ML and Bayesian GLM approaches. The aim of this study is to develop Bayesian GLM and ML models to predict Vs using CPT parameters. For the ML model, the Extreme Gradient Boosting (XGBoost) algorithm coupled with Shapley Additive Explanations (SHAP) analysis was chosen due to its high performance in predictive models. Following the development of Bayesian GLM and XGBoost models, a comparative analysis was conducted against real SCPT data from different locations to validate the predictive capabilities of the proposed models.

2. Data-Driven Shear Wave Velocity (Vs) Prediction Models

In the following subsections, we will discuss Bayesian GLM and XGBoost algorithms for the development of Vs prediction models.

2.1. Bayesian Generalized Linear Model

The Bayesian framework has been utilized in geotechnical parameter estimations for some time now. This approach offers a powerful paradigm to estimate soil parameters in a probabilistic manner. However, studies on Vs prediction using the Bayesian framework are limited. Bayesian GLM provides a robust framework for integrating CPT data to enhance the accuracy and reliability of Vs predictions. Bayesian GLM are readily available through rstanarm [23] package and can be implemented through R programming language [23]. In this study, we formulate a Bayesian approach to model the relationship between Vs and CPT parameters using a GLM framework. The general form of Bayesian GLM can be expressed as follows [24]:

Let Y represent Vs and X represent CPT parameters (e.g., cone tip resistance, qc; sleeve friction, fs; friction ratio, Rf). Then, the relationship between Y and predictor variables X can be expressed as

$$Y = M(X) + E \tag{1}$$

where Y is predicted Vs, $M(X)$ is the linear predictor function based on the predictor variables and E is an error term.

The linear predictor can be defined as

$$M(X) = \beta_0 + \beta_1 qc + \beta_2 fs + \ldots + \beta_p X_p \tag{2}$$

where β_0 is the intercept and β_i $(i = 1, 2, \ldots, p)$ is the coefficient of X_i.

The Bayesian GLM requires specifying prior distributions for the intercept and coefficients. The values of these parameters are equally likely to be positive or negative, but they are highly unlikely to be far from zero. These beliefs can be represented using normal distributions with a mean of zero and a small standard deviation. The default priors in rstanarm package perform well for most cases. Prior distributions for the coefficients can be defined as

$$\beta_i \sim Normal\left(0, \sigma^2_{\beta_i}\right) \tag{3}$$

where $\sigma^2_{\beta_i}$ is variance of prior distribution of the coefficients. Similarly, the prior distribution for the intercept can be defined as

$$\beta_0 \sim Normal\left(0, \sigma^2_{\beta_0}\right) \tag{4}$$

where $\sigma^2_{\beta_0}$ is variance of prior distribution of the intercept.

The error term is assumed to follow normal distribution with a mean of 0 and variance σ^2).

$$E \sim Normal\left(0,\ \sigma^2\right) \tag{5}$$

Common choices for the prior distribution of the variance of error include the exponential, inverse gamma, and half-normal distributions. In this study, the exponential prior distribution (the default for GLM in the rstanarm package, with rate parameter $\lambda = 1$) was considered. It is worth noting that the priors are internally adjusted in rstanarm to account for data scaling and centering.

$$\sigma \sim Exponential(\lambda) \tag{6}$$

Assuming normally distributed errors, the likelihood function can be expressed as

$$L\left(Y \middle| X, \beta, \sigma^2\right) = \prod_{i=1}^{n} \frac{1}{\sqrt{2\pi\sigma^2}} exp\left(-\frac{(Y_i - M(X_i))^2}{2\sigma^2}\right) \tag{7}$$

By using Bayes' theorem, the posterior distribution of the parameters can be derived as

$$P(\beta | Y,\ X) \propto L\left(Y \middle| X, \beta, \sigma^2\right) P(\beta) \tag{8}$$

The actual values of posterior distributions of the parameters are rarely obtainable through analytical methods. Therefore, Markov Chain Monte Carlo (MCMC) simulation is commonly used to explore the posterior target distribution through random sampling [25,26]. In MCMC simulations, the Markov chains should converge to stationary distributions. To ensure that the chains mix well within parameter space, visual inspections of the chains and statistical tests, such as the Gelman and Rubin Rhat statistics [27], should be conducted. At convergence, Rhat is expected to be below 1.2, which indicates that the between-chain and within-chain variances for the model parameters are in good agreement [28–30]. This threshold serves as a guideline to confirm that the Markov chains have mixed well and that the posterior estimates are reliable and representative of the target distribution.

2.2. Extreme Gradient Boosting (XGBoost) Algorithm

XGBoost algorithm is a powerful machine learning algorithm known for its efficiency, scalability, and ability to handle large datasets with high-dimensional feature spaces and non-linear relationships [31]. In this study, XGBoost was selected to develop a Vs predictive model using seismic cone penetration tests (SCPTs) and SCPTs with a pore water pressure measurement (SCPTu) dataset. Each observation in the dataset represents a measurement from SCPT or SCPTu (e.g., cone resistance (qc), sleeve friction (fs), and depth, among others). These observations (features) are used as an input to the XGBoost model for the prediction of output (in this case the Vs). The XGBoost algorithm has the capability to discover patterns and relationships in the data without requiring pre-defined mathematical models. For instance, the relationships between Vs and geotechnical parameters (e.g., qc and fs) may involve non-linearities and are difficult to express mathematically. XGBoost overcomes this limitation by learning directly from the data, making it particularly suited for SCPT-based Vs predictions where the physical or mathematical relationships may be unknown or too complex to model explicitly. The model uses an ensemble of decision trees, built iteratively to minimize the prediction error by optimizing a loss function. Specifically, XGBoost combines weak learners (individual trees) in a sequential manner, where each tree corrects the errors of its predecessors.

In XGBoost, a regularization parameter was incorporated to reduce model complexity and prevent overfitting. The objective function of XGBoost is defined as follows [31]:

$$\mathcal{L}(\phi) = \sum_{i=1}^{N} l(y_i, \hat{y}_i) + \sum_{i=1}^{T} \Omega(f_t) \tag{9}$$

where $\mathcal{L}(\phi)$ is objective function, $l(.)$ is loss function that measures difference between actual value y_i and predicted value $\hat{y}_i$, N is number of observations, T is number of estimators, and $\Omega(f_t)$ is regularization parameter defined as

$$\Omega(f_t) = \gamma T + \frac{1}{2}\lambda\omega_t \tag{10}$$

where T is the number of nodes, and ω_t weight of the leaf nodes, and γ and λ are regularization coefficients that control the model's complexity.

The predictive performance and generalization capability of XGBoost depends on the values of its hyperparameter. Therefore, the optimal hyperparameter values must be determined through hyperparameters tuning techniques. A next-generation optimization framework, Optuna, is widely used [32]. To prevent overfitting during the tuning process, k-fold cross validation techniques are commonly employed. This process helps the identification of the optimal parameter configuration that maximizes the model's coefficient of determination (R^2) while avoiding overfitting.

To analyze the influence of input features on the optimized XGBoost model, partial dependence plots (PDPs) and SHAPs were utilized. SHAP, introduced by Lundberg [33], offers a unified framework for interpreting feature importance in the model. SHAP assigns a unique Shapley value (ϕ_j) to each feature based on its contribution to the model output. For a prediction $f(x)$, SHAP is expressed as [33]:

$$f(x) = \phi_0 + \sum_{j=1}^{N} \phi_j \tag{11}$$

where ϕ_0 is base value (average model output for the dataset), ϕ_j is SHAP value for feature j and it is defined as

$$\phi_j = \sum_{S \subseteq N \setminus \{j\}} \frac{|S|!(|N| - |S| - 1)!}{|N|!} [f(S \cup \{j\}) - f(S)] \tag{12}$$

where N is total number of input features, S is a subset of features excluding j, $f(S)$ is the model prediction for the subset S, $|S|$ is the number of features in subset S, and $S \subseteq N \setminus \{j\}$ means that S is a subset of the set of features N excluding feature j. The feature with the highest absolute SHAP value is considered the most influential in the model's decision-making process.

PDPs can be used to visualize the interactions between the input features and the target variable in machine learning models [34]. In this study, PDPs were employed to analyze the relationship between input features and the target variable, Vs. By marginalizing over all other features, PDPs enable a clear depiction of the isolated effect of a single feature on Vs. The PDPs can reveal whether the relationship between Vs and input features is linear, monotonic, or non-linear. From the PDP analysis, it is possible to identify key features that contribute significantly to the predictions of Vs.

3. Methodology

This study focuses on developing XGBoost and Bayesian GLMs for predicting Vs. Figure 1 presents the methodology used for the development and evaluation of these predictive models. The process begins with the preprocessing of SCPT and SCPTu datasets. The preprocessed dataset was used for the development of the XGBoost model, conducting partial dependence analysis to identify impactful features, and fitting the Bayesian GLM. Model evaluations include PDPs and SHAP analysis for the XGBoost model and a comparison of the predictive performance of both models. The following sub-sections provide a detailed description of the research data, performance metrics, and the assessment of the models' predictive capabilities.

3.1. Training and Testing Dataset

The data used in this study were provided by Graz University of Technology [35] (Graz, Austria). The complete dataset consists of 1339 cone penetration tests (CPT, CPTu, SCPT, and SCPTu) conducted in Austria and Germany by Premstaller Geotechnik ZT GmbH [35] (Hallein, Austria). For this study, we utilized only the SCPT (46 tests in total) and SCPTu (50 tests in total) datasets. The SCPT and SCPTu can measure Vs at various intervals (e.g., at every 1 m) along the penetration depth, in addition to cone penetration data (i.e., cone tip resistance, qc and sleeve friction, fs). Before training the ML algorithm, the dataset underwent preprocessing. Raw data are rarely error-free or perfectly structured for ML training. Data often arrive in raw, messy, or incomplete forms, rendering them unsuitable for direct use in ML models. Initially, the dataset was filtered to include only SCPT and SCPTu tests. Unnecessary columns, and non-relevant measurements, were removed. Rows with zero or negative values for critical parameters were excluded. The differences in sampling intervals between CPT and Vs measurements were reconciled as per procedures described in [36]. This involved computing interval-based statistics (mean and coefficient of variation, cv) for relevant input features within the depth intervals where Vs data were recorded. Initial input features considered include mean cone tip resistance (qc_mean) and its coefficient of variation (qc_cv); mean sleeve friction (fs_mean) and its coefficient of variation (fs_cv); mean friction ratio (Rf_mean) and its coefficient of variation (Rf_cv); mean of soil behavior type index (Ic_mean); total overburden stress (σ,v); and depth of soil while Vs served as the target variable. Outliers in Vs were identified and removed using the interquartile range (IQR) method, with a threshold of 2.5 times the IQR above the third quartile. The cleaned and processed dataset was then saved for further modeling and analysis. Table 1 presents a statistical summary of the dataset after preprocessing, while Figure 2 displays a pair plot of the input features and the target variable, Vs. The diagonal of the matrix contains histograms that illustrate the distribution of each individual feature. The off-diagonal scatter plots depict the pairwise relationships between features. Additionally, the upper triangle includes linear correlation coefficients, highlighting the strength and direction of these relationships.

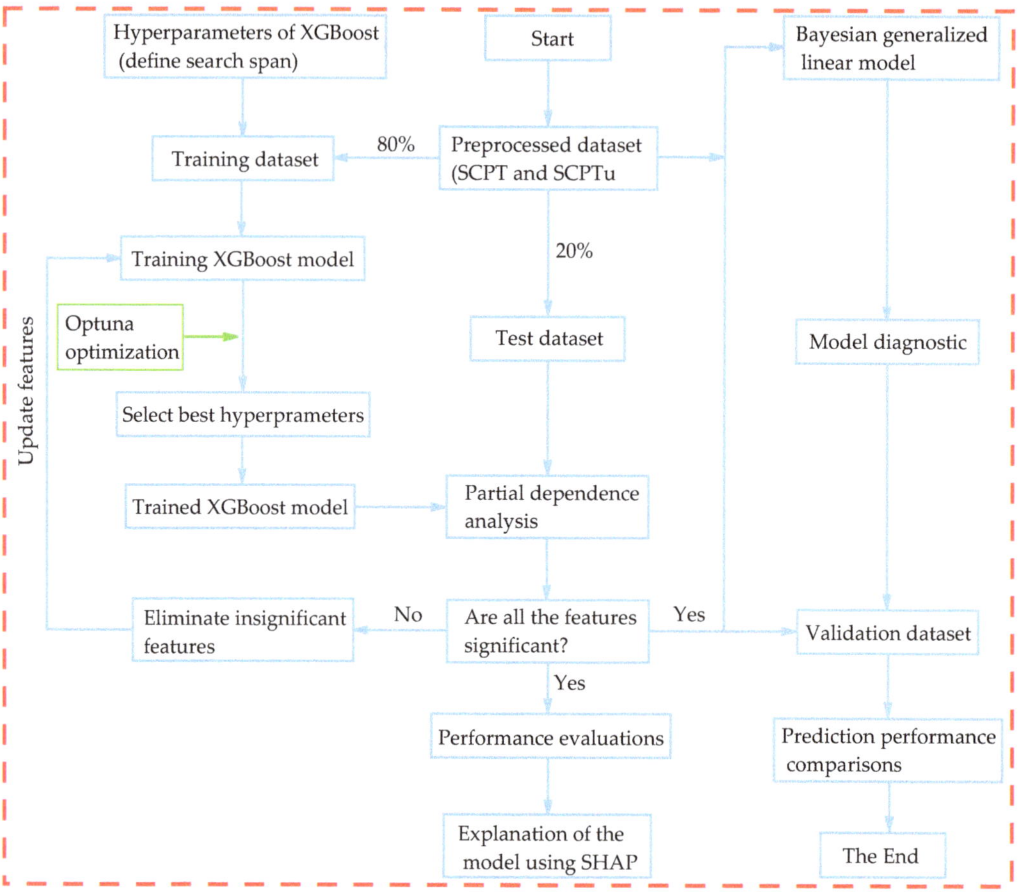

Figure 1. Flow diagram illustrating the methodology for developing and evaluating XGBoost and Bayesian GLMs.

Table 1. Statistical summary of input features.

Metrics	qc_mean (MPa)	qc_cv (-)	fs_mean (kPa)	fs_cv (-)	Rf_mean (%)	Ic_mean (-)	Depth (m)	σ,v (kPa)	Vs (m/s)
Mean	4.24	0.24	50.53	0.26	2.29	2.74	13.38	254	237.6
STD	7.12	0.25	56.57	0.21	3.16	0.65	9.01	171	91.1
Minimum	0.02	0	0.35	0	0.09	0	0.5	9.5	22
Maximum	74.75	2.59	820.25	1.79	112.95	4.06	49.5	941	547
Count	3600	3600	3600	3600	3600	3600	3600	3600	3600

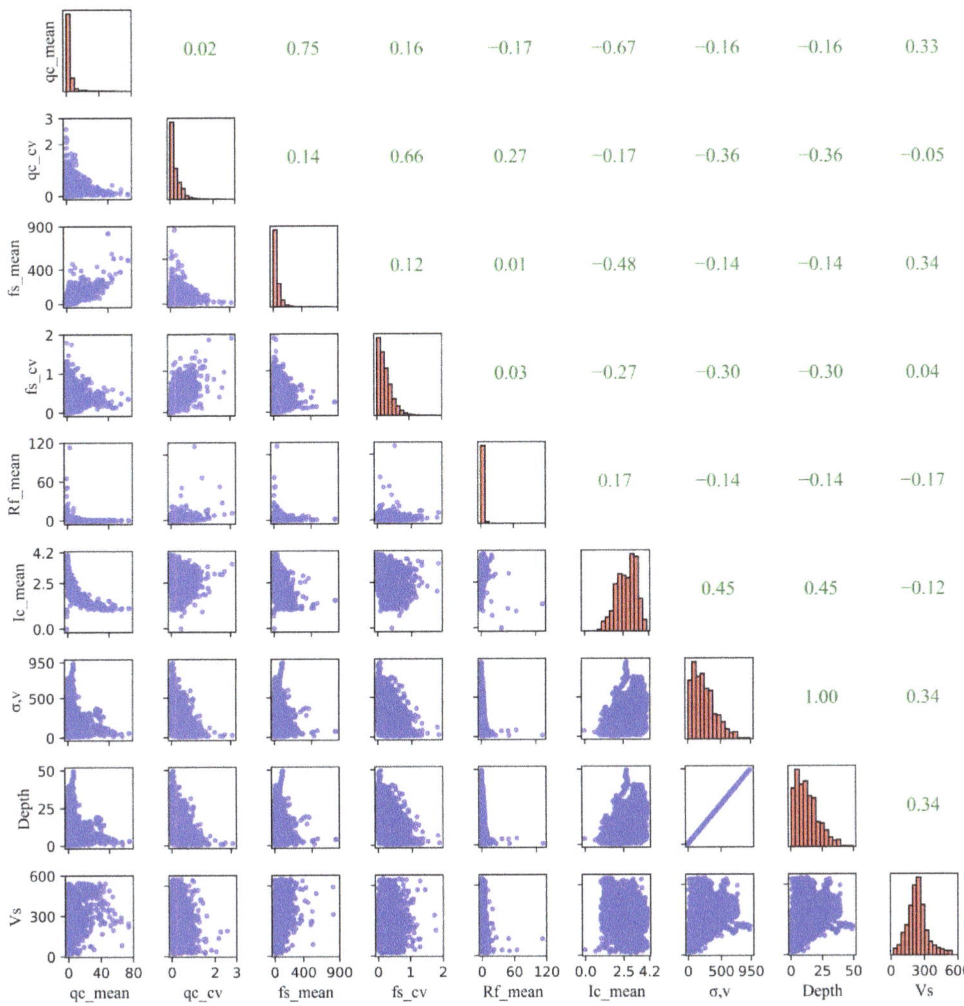

Figure 2. Correlation matrix illustrating the relationships between input features.

The XGBoost model was trained on 80% of the dataset, while the remaining 20% of the dataset was used to evaluate its performance. To ensure the model's robustness and minimize the risk of overfitting, hyperparameter optimization was conducted using the Optuna framework. This optimization process was paired with 10-fold cross-validation, where the training dataset was divided into 10 equal subsets. In each trial, the model was trained on nine folds and validated on the remaining fold. The primary objective of the optimization was to maximize the coefficient of determination (R^2). The optimized model was used to conduct partial dependence analysis to examine the influence of each feature on Vs. Features with negligible importance were excluded, and the model was retrained using the updated feature set. For the Bayesian GLM, only features exhibiting a linear or nearly linear relationship with Vs were selected. The performance of the final XGBoost model was assessed using multiple statistical metrics. Additionally, the predictive performance of both the Bayesian GLM and XGBoost models was validated on a separate validation dataset.

3.2. Hungarian Seismic Cone Penetration Test (SCPT)

The cone penetration test (CPT) data have been in use by Hungarian geotechnical engineers for quite some time. A wealth of CPT data is readily available for research purposes. This study used SCPT data (see Figure 3) from Hungary (Pak site) to validate the predictive performance of Bayesian GLM and XGBoost models. The local soil deposit at the site is predominantly characterized by fluvial sediments from the Danube River. Details about the SCPT procedures and geological characteristics are available elsewhere and readers are encouraged to refer to [3,36–39].

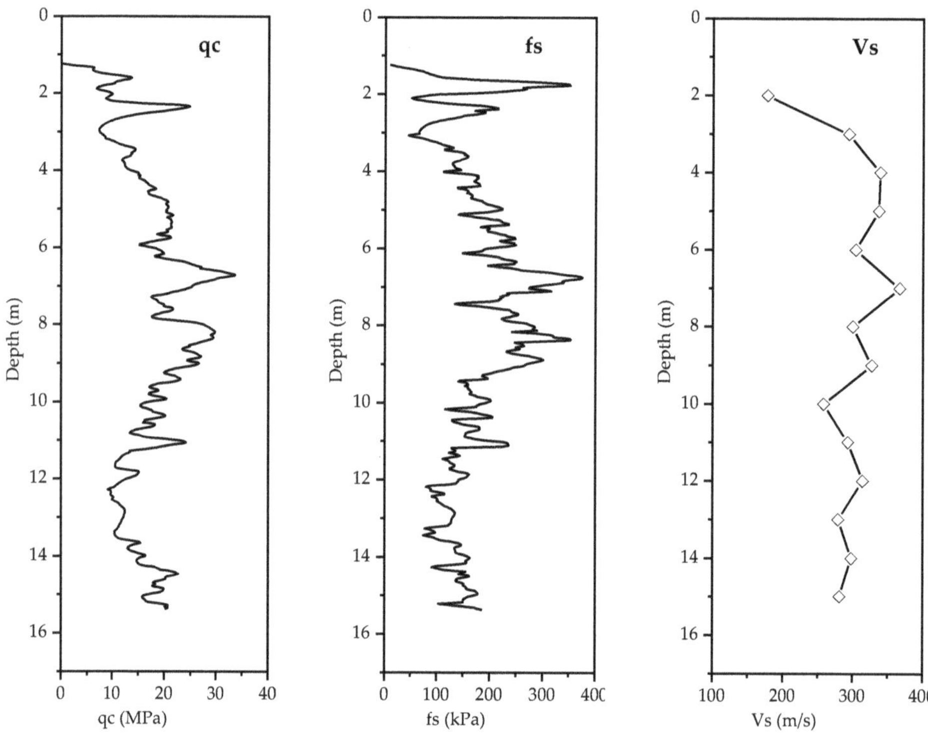

Figure 3. Seismic cone penetration (SCPT) data used for model validation.

3.3. Performance Measurements

The predictive performance of the Bayesian GLM and XGBoost models was evaluated by computing various performance and error metrics (Table 2). For Bayesian GLM, trace plot and R-hat, mean prediction, 95% credible intervals were used to assess the performance. Additionally, the linear correlation coefficients (r) between the measured Vs values and the predicted values for the validation dataset were computed for both models.

Table 2. Performance and error metrics used to evaluate Vs predictive capabilities of XGBoost and Bayesian GLM.

Metrics	Formula	Ideal Value	Equation. No
Correlation coefficient	$r = \frac{\sum_{i=1}^{n}(x_i-\overline{x})(y_i-\overline{y})}{\sqrt{\sum_{i=1}^{n}(x_i-\overline{x})^2}\sqrt{\sum_{i=1}^{n}(y_i-\overline{y})^2}}$	1	(13)
Coefficient of determination	$R^2 = 1 - \sum_{i=1}^{N} \frac{(x_i - y_i)^2}{(x_i - \overline{x})^2}$	1	(14)
Index of Agreement	$IA = 1 - \frac{\sum_{i=1}^{N}(y_i-\hat{y}_i)^2}{\sum_{i=1}^{N}(\lvert y_i-\overline{y}\rvert+\lvert\hat{y}_i-\overline{y}\rvert)^2}$	1	(15)
Kling–Gupta efficiency	$KGE = 1 - \sqrt{(r-1)^2 + (\alpha-1)^2 + (\beta-1)^2}$	1	(16)
Mean squared error	$MSE = \frac{1}{n}\sum_{i=1}^{n}(x_i - y_i)^2$	0	(17)
Root mean squared relative error	$RMSRE = \sqrt{\frac{1}{n}\sum_{i=1}^{n}\left(\frac{x_i - y_i}{x_i}\right)^2}$	0	(18)
Mean absolute error	$MAE = \frac{1}{n}\sum_{i=1}^{n}\lvert x_i - y_i\rvert$	0	(19)
Mean absolute relative error	$MARE = \frac{1}{n}\sum_{i=1}^{n}\left\lvert\frac{x_i - y_i}{x_i}\right\rvert$	0	(20)
Mean square relative error	$MSRE = \frac{1}{n}\sum_{i=1}^{n}\left\lvert\frac{x_i - y_i}{x_i}\right\rvert^2$	0	(21)
Mean bias error	$MBE = \frac{1}{n}\sum_{i=1}^{n}(x_i - y_i)$	0	(22)
Maximum absolute relative error	$MaxARE = Max\left(\left\lvert\frac{x_i - y_i}{x_i}\right\rvert\right)$	0	(23)

Let X represent the measured Vs values and Y represent the predicted Vs values. Then, x_i represents the ith value of variable X, $\overline{x}$ is mean of X, y_i is the ith value of variable Y, $\overline{y}$ is the mean of Y, r represents the linear correlation coefficients, and n is the number of data points.

The predictive performance of XGBoost model was further evaluated using several metrics, including the coefficient of determination (R^2), index of agreement (IA), Kling–Gupta efficiency (KGE), mean squared error (MSE), root mean squared relative error (RMSRE), mean absolute error (MAE), mean absolute relative error (MARE), mean square relative error (MSRE), mean bias error (MBE), and maximum absolute relative error (MaxARE) [40–42]. These metrics provide a comprehensive quantitative evaluation of the models in predicting the Vs. IA, KGE, and R^2 indicate overall agreement and explanatory power, with ideal values close to unity, while error metrics quantify prediction errors (lower values indicate high performance).

4. Discussion

In the following subsections, the Vs prediction capabilities of the proposed models are evaluated. The Bayesian GLM was implemented in R while the XGBoost model was implemented in Python (version 3.11.7).

4.1. Evaluation of the Models

Table 3 presents the optimized critical hyperparameters of the XGBoost model. These hyperparameters were fine-tuned through Optuna with a 10-fold cross-validation technique to enhance the model's predictive performance. The Optuna framework iteratively trained and evaluated models using the mean R^2 score across the cross-validation folds as the objective metric. By focusing on maximizing this metric, Optuna identified the optimal set of hyperparameters (see Table 3) that balanced accuracy and generalization. The results of the hyperparameter importance analysis are illustrated in Figure 4. As can be observed, the relative importance of each parameter is ranked based on its contribution to model performance. Among the hyperparameters, the learning rate (eta) emerged as the most influential factor. This underscores its pivotal role in determining the pace of model updates during training. Following the learning rate, the subsampling ratio and maximum tree depth (max_depth) were found to significantly affect model performance. The max_depth governs the complexity of the decision trees, allowing the model to capture complicated patterns in the dataset. Notably, the regularization term (lambda) exhibited relatively lower importance indicating that the other hyperparameters already provided sufficient control over model complexity.

Table 3. Optimized hyperparameters values of XGBoost model.

Hyperparameters	Search Span	Optimized Values
Number of trees (n_estimators)	50–600	444
Learning rate	0.001–0.5	0.0094
Maximum depth of trees	1–10	10
Subsampling ratio	0.05–1	0.640
L1 Regularization term on weights (reg_alpha)	0.01–1	0.208
L2 Regularization term on weights (reg_lambda)	0.01–1	0.603
Column subsampling ratio per tree (colsample_bytree)	0.5–1	0.707
Minimum loss reduction required to make a further split (gamma)	0–10	8.79
Maximum number of bins for feature quantization (max_bin)	128–512	499
Balancing weight for positive and negative classes (scale_pos_weight)	0.1–10	6.09

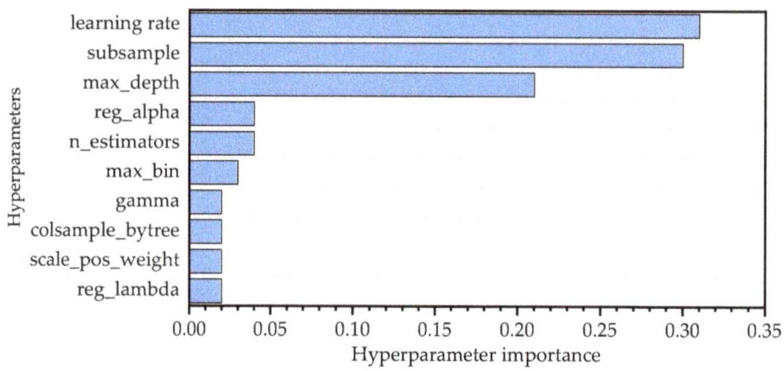

Figure 4. Hyperparameter importance of XGBoost algorithm.

The partial dependence analysis was conducted to explore the relationships between key input features and the predicted Vs. The results (see Figure 5) indicate that Vs exhibits positive correlations with σ,v, qc_mean, fs_mean, depth and qc_cv. The relationship between Ic_mean and Vs, however, is non-linear, indicating a complex influence on Vs predictions. This non-linear interaction can be effectively captured by the XGBoost model but could not be accommodated by the linear framework of the Bayesian GLM. Consequently, Ic_mean was excluded from the Bayesian GLM. In contrast, fs_cv, Rf_mean and Rf_cv show negligible influence on Vs, as observed from their flat PDP curves. These parameters were excluded from both the XGBoost and Bayesian GLMs to optimize performance. As a result, the final optimized XGBoost model was trained using σ,v, qc_mean, fs_mean, depth, qc_cv, and Ic_mean, while Bayesian GLM was fitted using σ,v, qc_mean, fs_mean, depth, and qc_cv.

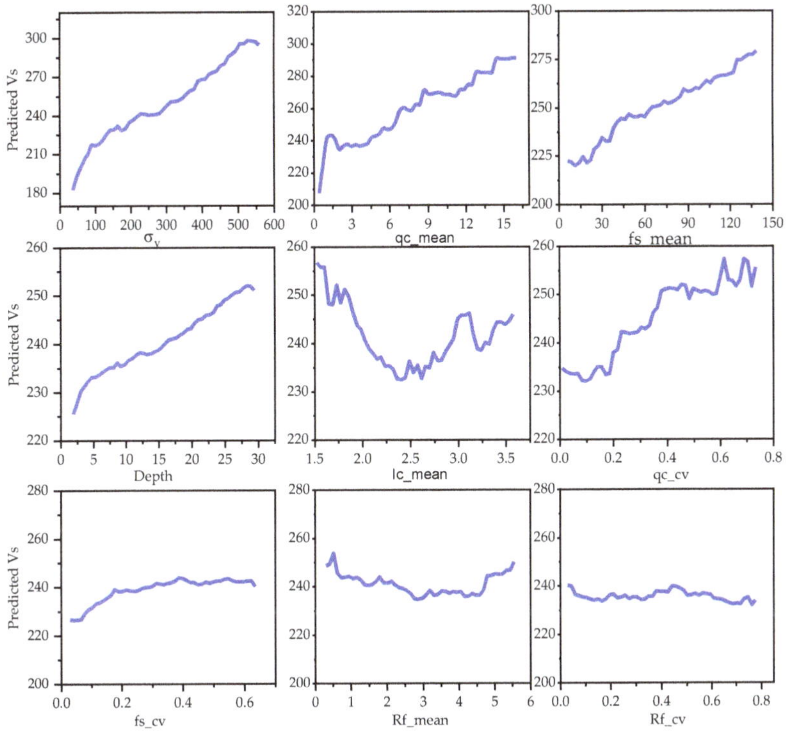

Figure 5. Partial dependence plots.

The predictive performance of the optimized XGBoost model is presented in Figure 6. In Figure 6a, the Vs values predicted by XGBoost model for both the training and testing dataset are plotted against measured Vs values. The diagonal dashed green line (y = x) represents perfect prediction (i.e., the predicted Vs values match the measured Vs values). A good agreement is observed between predicted and measured Vs values. The residual (the difference between predicted and measured Vs) plots for the predictions are illustrated in Figure 6b. It can be observed that random scatters of residuals around zero (horizontal dashed green line) for both training and testing dataset, indicating unbiased predictions. On the right side of the scatter plots, histograms of the residuals are presented. These histograms offer a visual summary of the distribution of residuals. For the training dataset, the residuals appear to be symmetrically distributed around zero, with a relatively narrow spread, indicating strong predictive performance and minimal bias. In contrast, for the testing dataset, the residuals exhibit a slightly wider spread, reflecting the model's reduced performance on unseen data. However, the residuals are still centered around zero, suggesting that the model generalizes reasonably well without significant overfitting. The results demonstrate the robustness of the XGBoost model and emphasize the importance of data preprocessing, feature selection, and hyperparameter optimization in improving model performance.

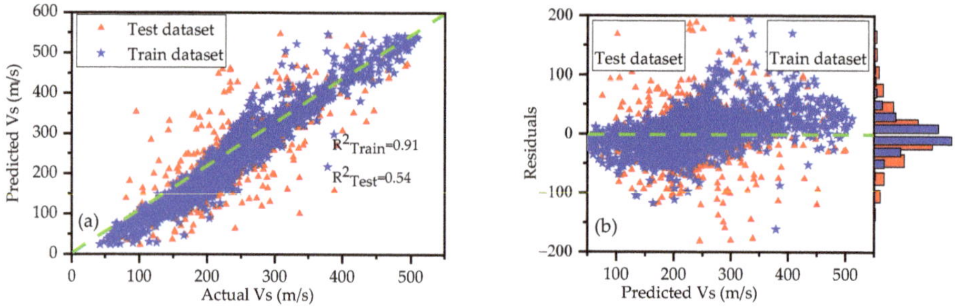

Figure 6. Predictive performance of the XGBoost model: (**a**) scatter plots of predicted Vs versus measured Vs for the training and testing datasets; (**b**) residual scatter and histogram plots for training and testing datasets.

The predictive performance of the model across various performance metrics (R^2, IA, KGE) and error metrics are summarized in Table 4 for both the training and testing dataset. For instance, the model achieved an R^2 of 0.54, an IA of 0.84 and KGE of 0.65 on the test dataset. For the training dataset, the model achieved an R^2 of 0.91, an IA of 0.97 and KGE of 0.82. The difference in performance between the training and testing datasets can be attributed to the complexity and variability inherent in the test data, which were not fully captured during model training. This difference is a common occurrence in ML models, particularly when the training data does not encompass all the potential variations present in the test set. Additionally, while the model was designed to generalize well and hyperparameter optimization techniques were employed to mitigate overfitting, the relatively lower performance on the test dataset reflects the challenges in accurately modeling highly complex or non-linear relationships between input features and the target variable Vs. However, these results are superior to those reported in [20]. The improvements can be attributable to the feature selection and data preprocessing step.

Table 4. Performance of XGBoost model on test and train dataset.

Performance Metrics	Test Dataset	Train Dataset
R^2	0.54	0.91
IA	0.84	0.97
KGE	0.65	0.82
MSE	3792	781
RMSRE	0.39	0.22
MAE	41	19.6
MARE	0.21	0.11
MSRE	0.15	0.05
MBE	1.12	0.13
MaxARE	3.30	3.95

Figure 7 shows SHAP analysis results, where the input features are displayed on the y-axis in descending order based on their contribution to the Vs prediction. The most influential feature appears at the top. The x-axis represents SHAP values, which quantify the contribution of each feature to the model's prediction. A positive SHAP value indicates that the feature increases the predicted Vs, whereas a negative SHAP value indicates a decrease in the predicted Vs. The SHAP values are represented by colored points that indicate the influence of the features on the model (Figure 7a). Each point reflects the impact of specific features on an observation from the entire database. The red color in the figure indicates high feature values while the blue color indicates low feature values. Positive SHAP values (red points) increase the Vs prediction while the negative SHAP value (blue points) decrease Vs prediction. For instance, for cone tip resistance (qc_mean), the red points predominantly on the positive side indicate higher qc_mean value generally increase Vs. On the other hand, the blue points on the negative side indicate lower qc_mean values generally decrease Vs. Interestingly, the instances of red points on the left or blue points on the right indicate a non-linear relationship between the feature and Vs. The global impact of features on Vs is shown by the mean absolute SHAP values of each feature (Figure 7b). Features with higher bar length indicate significant impact on the Vs. It can be observed that total overburden stress σ,v is the most influential feature, followed by cone tip resistance and sleeve frictions. On the other hand, the depth and soil behavior type index have lower impact on Vs. It is important to note that less impactful features were removed from the model, and only these features were incorporated into the final model.

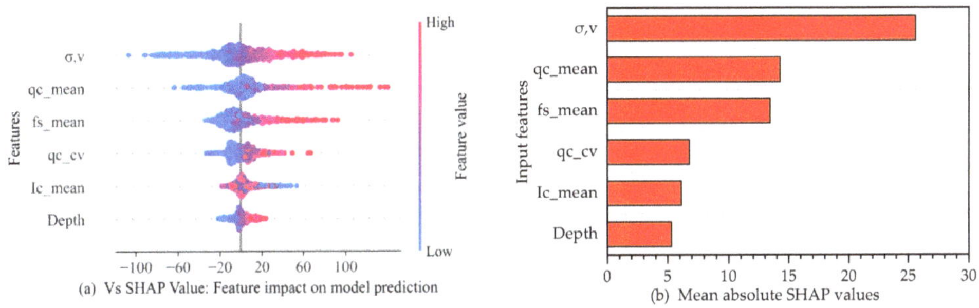

Figure 7. Summary of global feature impacts (**a**) and importance rankings (**b**) for input features.

The Bayesian GLM was fitted on the complete dataset using the final predictors identified through partial dependence analysis. It is important to note that the term "predictors" is more appropriate for the Bayesian GLM, while the term "input features" is suited for the ML models. In this study, both terms refer to the same geotechnical parameters. The model was developed using rstanarm package with four independent Markov chains, each consisting of 10,000 iterations, of which the first 1000 iterations were discarded as burn-in period. Weakly informative prior (normal prior (0, 2.5) for the intercept and coefficients, and an exponential prior ($\lambda = 1$) for the variance of the error term) were used. The performance of the model was evaluated using Rhat convergence diagnostic (evaluation of the agreement between between-chain and within-chain estimates for the model parameters), Monte Carlo Standard Error, MCSE (the ratio of standard deviation of the model parameters to the square root of the effective sample size), and trace plots of the parameters.

Figure 8 illustrates the convergence of Rhat values across the first 6000 iterations (burn-in period excluded) for the model fitted using four independent Markov chains. As can be observed, Rhat values remained consistently close to one (0.996–1.016) across iterations for each model parameter. This indicates the effective convergence of Markov chains. The trace plots (Figure 9) show the sampling behavior of Markov chains for each model parameter. It can be observed that the sampling demonstrates stable posterior distributions and reliable inference for all model parameters, as evidenced by well-mixed chains with no apparent trends. Table 5 summarizes the MCMC diagnostics (mean, standard deviation, MCSE, Rhat, and effective sample size n_eff) for the model parameters. The final Rhat values (computed based on all iteration) are nearly one, further confirming the convergence of the MCMC chains. Additionally, high effective sample sizes and minimal MCSE values validate the robustness of the parameter estimates.

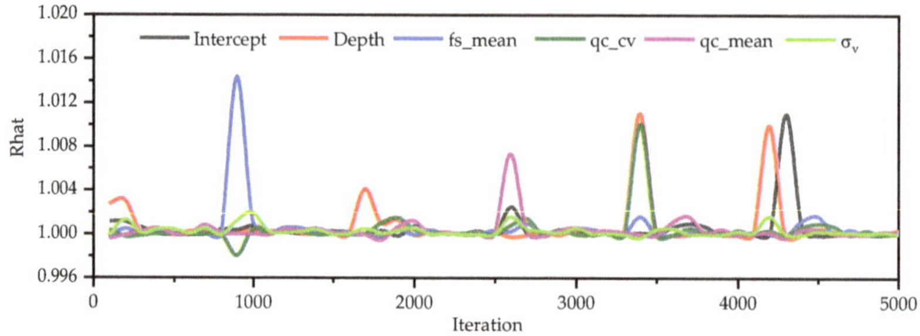

Figure 8. Convergence diagnostic plot showing Rhat values against the number of iterations.

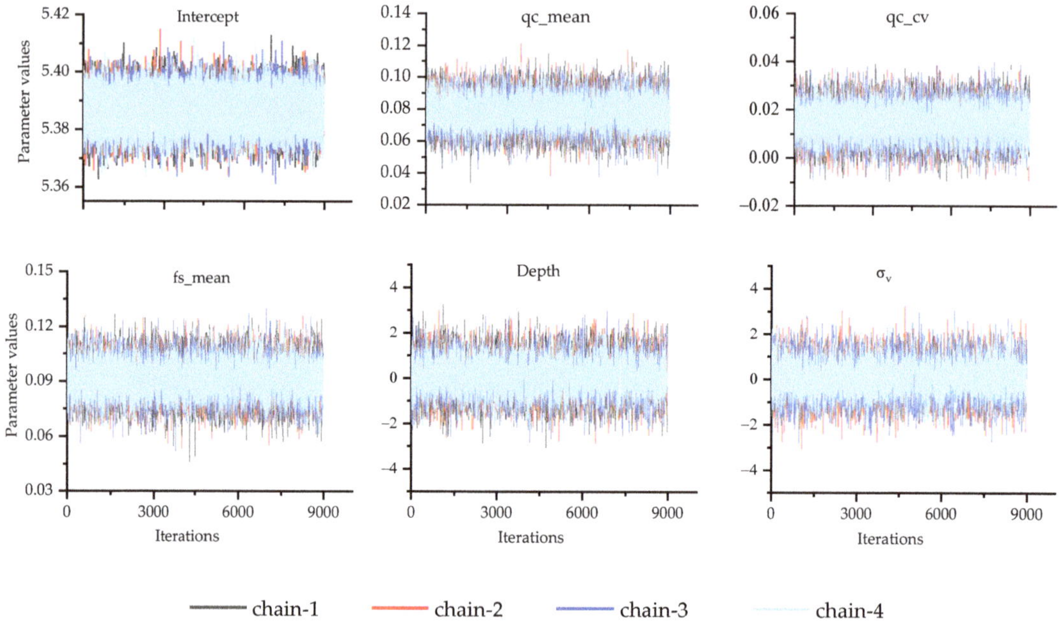

Figure 9. Trace plots depicting the sampling dynamics and convergence behavior of each model parameter in the Bayesian GLM.

Table 5. MCMC diagnostics for the Bayesian GLM.

Parameters	Mean	STD	MCSE (%)	Rhat	n_eff
Intercept	5.387	0.006	0.004	1.000	32,562
qc_mean	0.079	0.010	0.007	0.999	19,341
qc_cv	0.015	0.007	0.004	1.000	28,913
fs_mean	0.092	0.010	0.007	1.000	19,305
σv	0.092	0.786	0.65	0.999	14,770
Depth	0.109	0.787	0.65	1.000	14,771

4.2. Validation of the Models

The Bayesian GLM was evaluated using validation SCPT data from Hungary to assess its predictive performance. Figure 10 presents the comparison between the measured and predicted Vs values across varying depths. Figure 10a displays the mean predictions (blue diamonds) with 95% credible intervals (2.5–97.5%), along with the measured Vs values (red stars). The close alignment of the predicted mean values with the measured values across the entire depths demonstrates the model's accuracy and reliability. The lower plots (Figure 10b,c) provide a detailed analysis of predictions at a depth of 10 m (selected for illustration purpose only). The histogram (Figure 10c) shows the posterior distribution of Vs prediction at this depth. The blue curve represents the fitted probability density curve. It can be observed that the distribution follows lognormal distribution (attributable to data transformation during fitting the model). The vertical red arrow marks the measured value of Vs, while the blue arrows mark the 2.5% (left), mean prediction (middle), and 97.5% (right) credible interval. The 95% credible interval for this depth (142–627 m/s) encompasses the measured value (258 m/s). Additionally, Figure 10b summarizes the predictors used for this prediction. In general, the results demonstrate robustness of Bayesian GLM in capturing uncertainty and its capability to provide reliable predictions for Vs across varying depths. The use of credible intervals and posterior distributions further underscores its significance in quantifying prediction reliability.

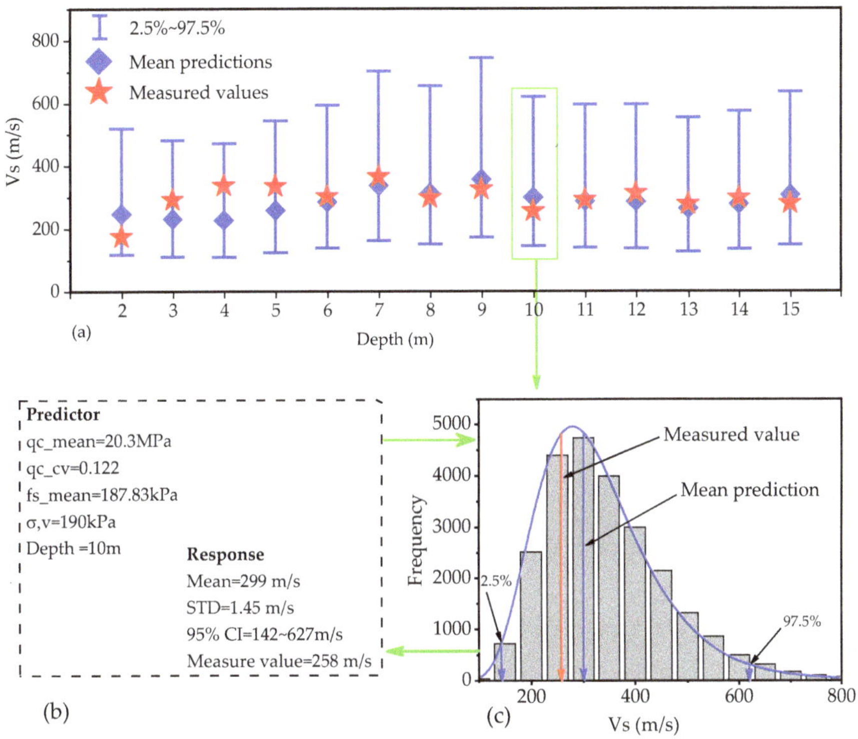

Figure 10. Predictive performance Bayesian GLM, (**a**) comparison of predicted Vs against measured Vs, (**b**) predictor values at 10 m depth and obtained results, and (**c**) histogram of predicted Vs values at 10 m depth.

In Figure 11, the predictive performance of the XGBoost model was compared against measured Vs values and Bayesian GLM predictions for the same validation dataset. The analysis involved comparing the outputs of the XGBoost model and the mean predictions of the Bayesian GLM across the Vs profile depth. For the Bayesian GLM, the 95% credible interval was also included to capture the uncertainty in the Vs predictions. It can be observed that both the XGBoost model's predictions and the Bayesian GLM's mean predictions align well with the measured Vs values across depths. Notably, the XGBoost model's predictions fall within the 95% credible interval across the entire depth, showcasing its excellent predictive performance of the Vs. The uncertainty captured by the Bayesian GLM (the credible intervals) is crucial for site response analysis, where accounting for variability in Vs is essential. These intervals can be used to conduct sensitivity studies by varying the Vs profile within the predicted bounds, enabling the estimation of a range of possible seismic responses for the site. This approach minimizes the risk of underestimating hazards and provides a foundation for prioritizing additional site investigations in areas with significant uncertainty. Furthermore, to quantitatively assess the predictive performance of the two models, linear correlation coefficients (r) were computed. Both models demonstrated good agreement with the measured Vs values, with the XGBoost model achieving a relatively higher r value (0.39).

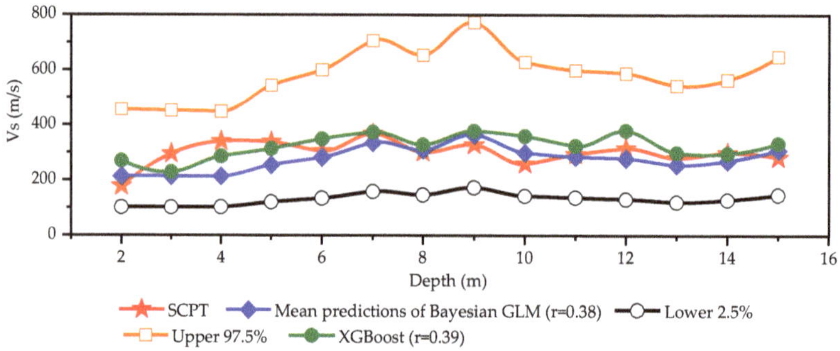

Figure 11. Predictive performance of Bayesian GLM and XGBoost models against validation dataset.

5. Conclusions

This study integrates Bayesian GLM and the XGBoost algorithm, coupled with SHAP and partial dependence analysis, to robustly predict Vs. Six key features were identified as the most influential factors for Vs prediction: cone tip resistance and its coefficient of variation, sleeve friction, soil behavior type index, depth, and overburden stress. These features formed the foundation for the final development of the XGBoost model. For the Bayesian GLM, cone tip resistance and its coefficient of variation, sleeve friction, depth, and overburden stress were identified as key predictor variables to fit the model on the same preprocessed datasets as the XGBoost model.

The XGBoost model demonstrated strong predictive performance, achieving high scores across various performance metrics and maintaining low prediction errors. SHAP analysis enhanced the model's interpretability by quantifying the contribution of each feature to the Vs predictions. Overburden stress was identified as the most influential factor, followed by cone tip resistance and sleeve friction. The validation results on the SCPT dataset revealed a strong agreement between measured and predicted Vs values for both models, with the Bayesian GLM offering credible intervals representing prediction uncertainty.

The integration of Bayesian GLM and explainable ML algorithms signifies a robust framework for predicting Vs while leveraging the strengths of both approaches. Explainable ML model excel at identifying critical input geotechnical parameters and explaining their contributions to the prediction. This capability not only enhances the interpretation of the underlying factors influencing Vs but also delivers high predictive performance. On the other hand, Bayesian GLM offers complementary advantages by explicitly capturing the uncertainties associated with geotechnical variability, measurement errors, and other sources of uncertainty. This approach enhances confidence in Vs predictions and their applicability to seismic response assessments.

To further enhance the applicability of this work, developing a user-friendly computational platform for Vs prediction is recommended. Such a platform could integrate both Bayesian GLM and ML approaches, enabling geotechnical engineers to easily interpret predictions and uncertainties. Additionally, site-specific seismic response analyses using Vs profiles obtained from the two approaches should be conducted to further validate their performance in practical engineering scenarios.

Author Contributions: Conceptualization, A.T.C. and R.R.; methodology, A.T.C.; software, A.T.C.; validation, A.T.C. formal analysis, A.T.C.; writing—original draft preparation, A.T.C.; writing—review and editing, R.R.; visualization, A.T.C.; supervision, R.R. All authors have read and agreed to the published version of the manuscript.

Funding: This research was funded by Széchenyi István University.

Institutional Review Board Statement: Not applicable.

Informed Consent Statement: Not applicable.

Data Availability Statement: The date utilized for this study can be downloaded from the following link: https://www.tugraz.at/en/institutes/ibg/research/computational-geotechnics-group/database/ (accessed on 12 May 2023).

Conflicts of Interest: The authors declare no conflicts of interest.

Abbreviations

CHT	Cross-hole test
CI	Credible interval
colsample _bytree	Column subsampling ratio per tree
CPT	Cone penetration test
cv	Coefficient of variation
DHT	Downhole test
E	Error term
fs	Sleeve friction
fs_cv	Coefficient of variation of sleeve friction
fs_cv	sleeve friction coefficient of variation
fs_mean	Mean of sleeve friction
gamma	Minimum loss reduction
GLM	Generalized linear model
IA	Index of Agreement
Ic	Soil Behavior type index
Ic_mean	Mean of soil behavior type index
IQR	Interquartile range
KGE	Kling–Gupta efficiency
kPa	Kilo pascal
$l(.)$	Loss function
$L(.)$	Likelihood function
$M(X)$	Predictor function
m/s	Meter per second
MAE	Mean absolute error
MARE	Mean absolute relative error
MASRE	Mean square relative error
MASW	Multichannel analysis of surface waves
max_bin	Maximum number of bins
max_depth	Maximum depth of trees
MaxARE	Maximum absolute relative error

MBE	Mean bias error
MCMC	Markov Chain Monte Carlo
MCSE	Monte Carlo Standard Error
ML	Machine learning
MPa	Mega pascal
MSE	Mean squared error
N	Total number of input features
n_eff	Effective sample size
n_estimators	Number of trees
PDPs	Partial dependence plots
qc	Cone tip resistance
qc_cv	Coefficient of variation of cone tip resistance
qc_mean	Mean of cone tip resistance
r	Linear correlation coefficient
R2	coefficient of determination
reg_alpha	L1 Regularization term on weights
reg_lambda	L2 Regularization term on weights
Rf	Friction ratio
Rf_mean	Mean of friction ratio
Rhat	Potential scale reduction factor
RMSRE	Root mean squared relative error
S	Subset of features
scale_pos_ weight	Balancing weight for positive and negative classes
SCPT	Seismic cone penetration test
SCPTu	Cone penetration test with pore pressure measurement
SHAP	Shapley Additive Explanations
STD	Standard deviation
Vs	Shear wave velocity
ω_t	Weight of leaf node
X	Measured shear wave velocity
XGBoost	Extreme Gradient Boosting
Y	Predicted shear wave velocity
σ,v	Total overburden stress
i	Mean of prior distribution
σ_i^2	Variance of prior distribution
β_0	Intercept
β_i	Coefficients of predictor variables
$\mathcal{L}(\phi)$	Objective function
$\Omega(f_t)$	regularization parameter
γ, λ	regularization coefficients
ϕ_0	Base value
ϕ_i	SHAP value

References

1. Bazzurro, P. Ground-Motion Amplification in Nonlinear Soil Sites with Uncertain Properties. *Bull. Seismol. Soc. Am.* **2004**, *94*, 2090–2109. [CrossRef]
2. Rathje, E.M.; Kottke, A.R.; Trent, W.L. Influence of Input Motion and Site Property Variabilities on Seismic Site Response Analysis. *J. Geotech. Geoenviron. Eng.* **2010**, *136*, 607–619. [CrossRef]
3. Chala, A.; Ray, R. Impact of Randomized Soil Properties and Rock Motion Intensities on Ground Motion. *Adv. Civ. Eng.* **2024**, *2024*, 1–12. [CrossRef]
4. Campanella, R.G.; Stewart, W.P. Seismic Cone Analysis Using Digital Signal Processing for Dynamic Site Characterization. *Can. Geotech. J.* **1992**, *29*, 477–486. [CrossRef]
5. Hardee, H.C.; Elbring, G.J.; Paulsson, B.N.P. Downhole Seismic Source. *Geophysics* **1987**, *52*, 729–739. [CrossRef]

6. Robertson, P.K.; Campanella, R.G.; Gillespie, D.; Rice, A. Seismic CPT to Measure in Situ Shear Wave Velocity. *J. Geotech. Eng.* **1986**, *112*, 791–803. [CrossRef]
7. Stokoe, K.H.; Woods, R.D. In Situ Shear Wave Velocity by Cross-Hole Method. *J. Soil Mech. Found. Div.* **1972**, *98*, 443–460. [CrossRef]
8. Park, C.B.; Miller, R.D.; Xia, J. Multichannel Analysis of Surface Waves. *Geophysics* **1999**, *64*, 800–808. [CrossRef]
9. Meisina, C.; Bonì, R.; Bordoni, M.; Lai, C.G.; Bozzoni, F.; Cosentini, R.M.; Castaldini, D.; Fontana, D.; Lugli, S.; Ghinoi, A.; et al. 3D Engineering Geological Modeling to Investigate a Liquefaction Site: An Example in Alluvial Holocene Sediments in the Po Plain, Italy. *Geosciences* **2022**, *12*, 155. [CrossRef]
10. Yang, H.Q.; Chu, J.; Wu, S.; Zhu, X.; Qi, X.; Chiam, K. Advancing Geological Modelling and Geodata Management: A Web-Based System with AI Assessment in Singapore. *Georisk* **2024**. [CrossRef]
11. Robertson, P.K. Interpretation of Cone Penetration Tests—A Unified Approach. *Can. Geotech. J.* **2009**, *46*, 1337–1355. [CrossRef]
12. Mayne, P.W.; Rix, G.J. Correlations Between Shear Wave Velocity and Cone Tip Resistance in Natural Clays. *Soils Found.* **1995**, *35*, 107–110. [CrossRef] [PubMed]
13. Andrus, R.D.; Mohanan, N.P.; Piratheepan, P.; Ellis, B.S.; Holzer, T.L. Predicting shear-wave velocity from cone penetration resistance. In Proceedings of the 4th International Conference on Earthquake Geotechnical Engineering, Thessaloniki, Greece, 24–29 June 2007.
14. Griffiths, S.C.; Cox, B.R.; Rathje, E.M.; Teague, D.P. Surface-Wave Dispersion Approach for Evaluating Statistical Models That Account for Shear-Wave Velocity Uncertainty. *J. Geotech. Geoenviron. Eng.* **2016**, *142*, 04016061. [CrossRef]
15. Matasovic, N.; Hashash, Y. *NCHRP Synthesis 428: Practices and Procedures for Site-Specific Evaluations of Earthquake Ground Motions, a Synthesis of Highway Practice*; National Academies Press: Washington, DC, USA, 2012.
16. Toro, G.R. Probabilistic Models of Site Velocity Profiles for Generic and Site-Specific Ground-Motion Amplification Studies. *Tech. Rep.* **1995**, 779574.
17. Rauter, S.; Tschuchnigg, F. Cpt Data Interpretation Employing Different Machine Learning Techniques. *Geosciences* **2021**, *11*, 265. [CrossRef]
18. Padarian, J.; Minasny, B.; McBratney, A.B. Machine Learning and Soil Sciences: A Review Aided by Machine Learning Tools. *SOIL* **2020**, *6*, 35–52. [CrossRef]
19. Chala, A.T.; Ray, R.P. Machine Learning Techniques for Soil Characterization Using Cone Penetration Test Data. *Appl. Sci.* **2023**, *13*, 8286. [CrossRef]
20. Felić, H.; Marzouk, I.; Tschuchnigg, F.; Peterstorfer, T. Data-Driven Site Characterization—Focus on Small-Strain Stiffness. In Proceedings of the 7th International Conference on Geotechnical and Geophysical Site Characterization—CIMNE, Barcelona, Spain, 18–21 June 2024.
21. Olayiwola, T.; Tariq, Z.; Abdulraheem, A.; Mahmoud, M. Evolving Strategies for Shear Wave Velocity Estimation: Smart and Ensemble Modeling Approach. *Neural Comput. Appl.* **2021**, *33*, 17147–17159. [CrossRef]
22. Taheri, A.; Makarian, E.; Manaman, N.S.; Ju, H.; Kim, T.H.; Geem, Z.W.; Rahimizadeh, K. A Fully-Self-Adaptive Harmony Search GMDH-Type Neural Network Algorithm to Estimate Shear-Wave Velocity in Porous Media. *Appl. Sci.* **2022**, *12*, 6339. [CrossRef]
23. Goodrich, B.; Gabry, J.A.I.; Brilleman, S. *Rstanarm: Bayesian Applied Regression Modeling via Stan*, R Package Version 2.21.4. 2023.
24. Gelman, A.; Carlin, J.B.; Stern, H.S.; Rubin, D.B. *Bayesian Data Analysis*; Chapman and Hall/CRC: Boca Raton, FL, USA, 1995; ISBN 0429258410.
25. Yang, H.Q.; Zhang, L.; Pan, Q.; Phoon, K.K.; Shen, Z. Bayesian Estimation of Spatially Varying Soil Parameters with Spatiotemporal Monitoring Data. *Acta Geotech.* **2021**, *16*, 263–278. [CrossRef]
26. Gong, W.; Tien, Y.M.; Juang, C.H.; Martin, J.R.; Luo, Z. Optimization of Site Investigation Program for Improved Statistical Characterization of Geotechnical Property Based on Random Field Theory. *Bull. Eng. Geol. Environ.* **2017**, *76*, 1021–1035. [CrossRef]
27. Gelman, A.; Rubin, D.B. Inference from Iterative Simulation Using Multiple Sequences. *Stat. Sci.* **1992**, *7*, 457–472. [CrossRef]
28. Wang, X.; Wang, X.S.; Li, N.; Wan, L. Bayesian Inversion of Soil Hydraulic Properties from Simplified Evaporation Experiments: Use of DREAM(ZS) Algorithm. *Water* **2021**, *13*, 2614. [CrossRef]
29. Qin, S.; Song, R.; Li, N. Bayesian Model Updating for Bridge Engineering Applications Based on DREAMKZS Algorithm and Kriging Model. *Structures* **2023**, *58*, 105565. [CrossRef]
30. Liu, G.; Jiang, W. Model Updating of a Prestressed Concrete Rigid Frame Bridge Using Multiple Markov Chain Monte Carlo Method and Dfferential Evolution. *Int. J. Struct. Stab. Dyn.* **2022**, *22*, 2240020. [CrossRef]
31. Chen, T.; Guestrin, C. Xgboost: A Scalable Tree Boosting System. In Proceedings of the 22nd Acm Sigkdd International Conference on Knowledge Discovery and Data Mining, San Francisco, CA, USA, 13–17 August 2016; pp. 785–794.
32. Akiba, T.; Sano, S.; Yanase, T.; Ohta, T.; Koyama, M. Optuna: A Next-Generation Hyperparameter Optimization Framework. In Proceedings of the ACM SIGKDD International Conference on Knowledge Discovery and Data Mining, Anchorage, AK, USA, 4–8 August 2019; ACM: New York, NY, USA, 2019; pp. 2623–2631.

33. Lundberg, S. A Unified Approach to Interpreting Model Predictions. *arXiv* **2017**, arXiv:1705.07874.
34. Friedman, J.H. Greedy Function Approximation: A Gradient Boosting Machine. *Ann. Stat.* **2001**, *29*, 1189–1232. [CrossRef]
35. Oberhollenzer, S.; Premstaller, M.; Marte, R.; Tschuchnigg, F.; Erharter, G.H.; Marcher, T. Cone Penetration Test Dataset Premstaller Geotechnik. *Data Brief* **2021**, *34*, 106618. [CrossRef]
36. Ray, R.P.; Wolf, A.; Kegyes-Brassai, O. Harmonizing Dynamic Property Measurements of Hungarian Soils. In Proceedings of the 6th International Conference on Geotechnical and Geophysical Site Characterization (ISC2020), Budapest, Hungary, 7–11 September 2020.
37. Kegyes-Brassai, O.; Wolf, Á.; Szilvágyi, Z.; Ray, R.P. Effects of Local Ground Conditions on Site Response Analysis Results in Hungary. In Proceedings of the 19th International Conference on Soil Mechanics and Geotechnical Engineering (19th ICSMGE), Seoul, Republic of Korea, 17–21 September 2017; pp. 2003–2006.
38. Szilvágyi, Z.; Panuska, J.; Kegyes-brassai, O.; Wolf, Á.; Tildy, P.; Ray, R.P. Ground Response Analyses in Budapest Based on Site Investigations and Laboratory Measurements. *World Acad. Sci. Eng. Technol. Int. J. Environ. Chem. Ecol. Geol. Geophys. Eng.* **2017**, *11*, 307–317.
39. Wolf, Á.; Ray, R.P. Comparison and Improvement of the Existing Cone Penetration Test Results: Shear Wave Velocity Correlations for Hungarian Soils. *Int. J. Geol. Environ. Eng.* **2017**, *11*, 362–371.
40. Kazemi, F.; Asgarkhani, N.; Jankowski, R. Optimization-Based Stacked Machine-Learning Method for Seismic Probability and Risk Assessment of Reinforced Concrete Shear Walls. *Expert Syst. Appl.* **2024**, *255*, 124897. [CrossRef]
41. Asgarkhani, N.; Kazemi, F.; Jakubczyk-Gałczyńska, A.; Mohebi, B.; Jankowski, R. Seismic Response and Performance Prediction of Steel Buckling-Restrained Braced Frames Using Machine-Learning Methods. *Eng. Appl. Artif. Intell.* **2024**, *128*, 107388. [CrossRef]
42. Wakjira, T.G.; Kutty, A.A.; Alam, M.S. A Novel Framework for Developing Environmentally Sustainable and Cost-Effective Ultra-High-Performance Concrete (UHPC) Using Advanced Machine Learning and Multi-Objective Optimization Techniques. *Constr. Build. Mater.* **2024**, *416*, 135114. [CrossRef]

applied sciences

MDPI

Article

Prediction of Wind Turbine Blade Stiffness Degradation Based on Improved Neural Basis Expansion Analysis

Shuai Yang, Jianxiong Gao *, Yiping Yuan, Jianxing Zhou and Lingchao Meng

School of Mechanical Engineering, Xinjiang University, Urumqi 830046, China; 18693653635@163.com (S.Y.); yipingyuan@163.com (Y.Y.); xju_zhjx@xju.edu.cn (J.Z.)
* Correspondence: jianxionggao888@163.com

Abstract: To reduce the significant time and cost associated with wind turbine blade fatigue testing, the applicability of the deep learning model Neural Basis Expansion Analysis (N-BEATS) for modeling the stiffness degradation of wind turbine blades was investigated. First, on the basis of a traditional blade stiffness degradation model, the stiffness data were expanded to meet the data volume requirements of N-BEATS. Second, the basic block structure of N-BEATS was improved (by treating the sequence-to-sequence prediction problem as a nonlinear multivariate regression problem) to meet the specific prediction requirements of this task, and the Pinball Mean Absolute Percentage Error (Pinball-MAPE) loss function was adopted to further reduce bias during the prediction process. Additionally, two data augmentation methods—time series combination and random noise injection—were applied to mitigate the risk of model overfitting and improve prediction accuracy. Experimental results demonstrated that the model can effectively learn underlying patterns in the stiffness data and successfully predict the remaining stiffness.

Keywords: wind turbine blades; deep learning; data augmentation; stiffness prediction

1. Introduction

Wind energy, as a sustainable power source, has become a key component of energy systems worldwide [1]. To improve the efficiency of wind energy utilization, the size of wind turbine blades has been continuously increasing, and the complexity of their material properties and structural design has also advanced. Enhancing blade performance to withstand extreme loads and extending the fatigue life are key challenges in current wind turbine blade design. Owing to the complexity of blade fatigue performance, the accurate prediction of fatigue life through theoretical calculations alone is difficult, and engineering insights or expertise from other fields frequently lack direct applicability to blade fatigue prediction. Therefore, full-scale fatigue testing is commonly used in engineering to validate whether wind turbine blades can meet the service life requirement of 20 years [2].

To accurately predict the fatigue life, it is essential to establish a reliable model for fatigue damage evolution. In this context, residual strength and stiffness are key parameters for assessing the damage state of composite materials, both of which gradually degrade with the accumulation of fatigue cycles. For engineering components with complex geometry and a large size, residual stiffness can be assessed without the need for destructive testing. Therefore, stiffness is more suitable for evaluating the fatigue damage of the blades [3,4]. If the residual stiffness of the blades can be predicted via early data during fatigue tests, the cost associated with blade evaluation could be significantly reduced.

1.1. Related Work

The fatigue testing process of wind turbine blades generates a large amount of data. These data contain rich patterns and characteristics, requiring effective analytical methods to unlock its potential value. Data-driven approaches can extract features from input data without relying on extensive prior knowledge, thus enabling predictions of Remaining Useful Life (RUL) [5]. Data-driven modeling methods can be categorized into statistical data-driven methods and machine learning methods [6]. Compared to statistical data-driven methods, machine learning methods can effectively extract deep features from monitoring data to achieve life prediction, avoiding the large errors caused by selecting inappropriate models. Furthermore, they do not significantly increase the training difficulty as the data dimension increases, making machine learning a widely explored and applied approach in the life prediction and health management fields in recent years. By training models with historical data, machine learning methods can be used for damage identification, reliability analysis, and prediction [7,8]. Among these, neural networks (NNs) have been extensively explored in the prediction. They possess many appealing characteristics, including the ability to model nonlinear relationships, perform large-scale parallelism, and approximate functions universally. Tao et al. [9] used an unsupervised learning network based on Neural Ordinary Differential Equations (Neural ODE) and Variational Autoencoders (VAE) to predict the fatigue residual stiffness of fiber-reinforced polymer composites, demonstrating that artificial neural networks outperform traditional mechanical and phenomenological models. Jiang [10] proposed a method for predicting residual stiffness on the basis of latent variable interpolation using latent Ordinary Differential Equation (ODE) neural networks and partial data retraining. Yin proposed a BP neural network optimized by the Chimp Optimization Algorithm (ChOA), significantly improving the accuracy of wind turbine tower vibration prediction [11]. Due to these advantages, neural networks have been widely applied in industrial fields, particularly in predictive maintenance strategies for composite structures. In the wind energy sector, the accurate fatigue prediction of fiber-reinforced polymer composites plays a crucial role in optimizing wind turbine blade maintenance schedules and extending service life. Additionally, neural networks have been employed for the real-time structural health monitoring of composite components, enabling condition-based maintenance and reducing operational costs. The ability of neural networks to effectively learn underlying degradation patterns makes them a key tool for enhancing the structural reliability and efficiency of composite-based applications across various industrial sectors.

Over the past few years, deep learning (DL) has proven effective in solving intricate prediction challenges. DL overcomes the inherent limitations of traditional neural networks and can better capture nonlinear features, enhancing RUL prediction accuracy. Modern DL models consist of various combinations of fundamental structures, including Recurrent Neural Network (RNN) and Convolutional Neural Network (CNN). Although RNN can capture dependencies in time series, they often face the problem of vanishing or exploding gradients while dealing with extended sequences, limiting their effectiveness in predicting distant states. To address this issue, the Long Short-Term Memory network (LSTM) was proposed [12]. LSTM has demonstrated superior performance compared to statistical and machine learning approaches, including AutoRegressive Integrated Moving Average (ARIMA), Support Vector Machine (SVM), and conventional neural networks [13]. CNN, which was originally developed for image processing, has recently been applied to time series analysis. The ability of these methods to automatically extract local features has made them highly effective in extracting fatigue damage features, especially when combined with LSTM for long-term dependency modeling. Liu [14] combined CNN and LSTM, using initial fatigue residual stiffness as input parameters, to construct a neural network model that learns characteristics directly from raw stiffness data, enabling subsequent predictions of residual stiffness.

Besides LSTM-based architecture, the subsequent DL models also show potential for time series forecasting. For instance, the Temporal Convolutional Network (TCN), proposed in 2018, can extract information from historical data relevant to the prediction moment and avoid the problems of vanishing and exploding gradients. It is more efficient and more effective at learning long-term dependencies when performing time series predictions [15]. Zha [16] proposed a RUL prediction model based on feature extraction methods and an improved TCN, which achieved a better prediction performance than other mainstream models did. Furthermore, the Encoder–Decoder Attention Mechanism [17] and Transformer [18] provide alternatives to LSTM. The attention mechanism enables the model to concentrate on the most relevant sections of the sequence when making predictions. Owing to their impressive prediction accuracy, transformer models have significantly outperformed expectations across many tasks. One of the key advantages of deep learning in industrial applications is its ability to process vast amounts of heterogeneous data, including real-time sensor measurements, vibration signals, acoustic emissions, and thermal imaging data, while maintaining higher accuracy and adaptability. For instance, in the wind energy sector, deep learning-based predictive maintenance systems leverage multi-source sensor data to detect early-stage damage in wind turbine blades, enabling condition-based maintenance and reducing unplanned downtime [19]. When dealing with small-scale univariate data, these models share common issues: first, the complexity of the model structures often leads to inefficient use of computational resources on small datasets, and limited data size increases the risk of overfitting. Additionally, the computational complexity of models such as the transformer increases quadratically with input size, making them computationally expensive and less advantageous for small datasets.

To address these issues, this study adopts the N-BEATS model. N-BEATS not only demonstrates excellent performance on large-scale time series datasets but also has a simple and efficient architecture that is suited for small-scale univariate data. First, N-BEATS, through forward and backward residual connections, enables cross-learning across multiple time series, which allows it to extract shared features across sequences while capturing the unique patterns of each sequence [20]. This characteristic is particularly crucial in this study's stiffness prediction task under five different loading conditions, helping to better identify the commonalities and specificities of each dataset. Additionally, N-BEATS exhibits high flexibility when dealing with univariate data, providing accurate and efficient prediction results with minimal computational overhead, significantly enhancing the model's applicability in this study.

1.2. Motivation and Contributions

This paper focuses on the prediction of stiffness degradation in wind turbine blade fatigue testing. Many existing studies are based on physical models, but such methods have limitations when dealing with nonlinear features or complex multidimensional data and rarely explore the great potential of DL in this field. This study clearly demonstrates the superior performance of DL in predicting the fatigue life of blades. Advanced nonlinear modeling based on N-BEATS architecture outperforms established models like LSTM and TCN in prediction accuracy. However, the dataset used in this paper is much smaller, which may hinder the application of deep neural networks. Therefore, the fundamental block structure of N-BEATS was improved to enable more flexible handling of datasets of varying scales. This paper clearly demonstrates that the improved N-BEATS model can be effectively applied to such small-scale prediction tasks and exhibits superior performance. Additionally, this study uses two methods—time series combination and random noise injection—to augment the dataset, aiming to mitigate the overfitting problem and further improve the model's generalizability and prediction accuracy. Finally, most existing prediction algorithms focus primarily on improving accuracy while neglecting bias, which plays a critical role in fatigue prediction. This paper adopts a simple and effective mechanism: the Pinball-MAPE loss function is used to control prediction bias, and its effectiveness is demonstrated. Figure 1 shows the proposed framework for blade stiffness degradation prediction.

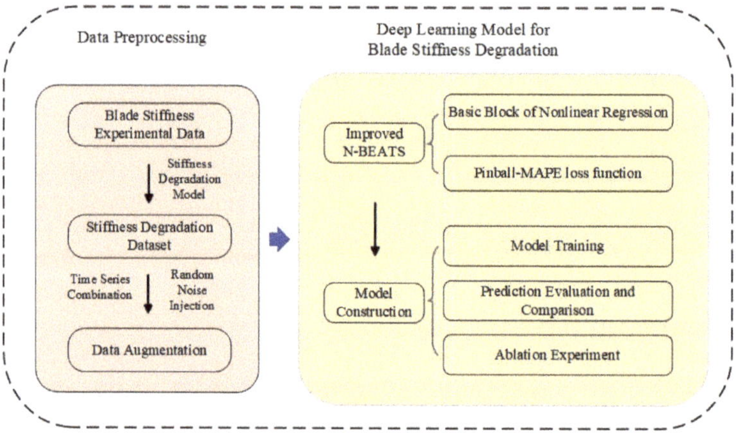

Figure 1. Framework of blade stiffness degradation prediction.

2. Methods

2.1. Improved N-BEATS Architecture

The N-BEATS architecture in this paper differs from the existing architecture. Its fundamental block employs a fully connected nonlinear regressor (Figure 2a), using nonlinear multivariate regression to predict the next step's solution [21]. Moreover, N-BEATS uses the principle of residual stacking to stack multiple layers together (Figure 2b), and its superiority has been fully demonstrated through extensive experimentation. By leveraging this principle, the model can effectively stack layers according to different tasks, thus fully learning the underlying patterns of the time series and showing excellent performance in subsequent predictions. At this stage, the basic module predicts future outputs while backtracking the input sequence to decompose its contribution to future predictions. The sum of the contributions from multiple basic blocks forms the output of the corresponding stack (Figure 2c). Conceptually, each fully connected layer can be regarded as a multivariate linear regression block, followed by a Rectified Linear Unit (ReLU) nonlinear layer. Therefore, improving the structure of the N-BEATS basic block can be viewed as a repeated multivariate regression with nonlinearity.

Mathematically, each block of the improved N-BEATS comprises a sequence of fully connected layers, ending with a forecast/backcast split. This architecture performs residual recursion within each stack and sums the outputs of all stacks to make the final prediction. Assuming that the model has M stacks and R residual blocks, with each residual block containing L hidden layers, the architecture takes $x \in \mathrm{R}^{w}$ as the input. The model predicts a window of length H and backcasts over a window of length ω, using superscripts (r, l) for the residual blocks and layers, and denotes the Fully Connected Layers (FC) with weights $\mathrm{W}^{r,l}$ and $\mathrm{b}^{r,l}$ as follows:

$$\mathrm{FC}_{r,l}\left(\mathrm{h}^{r,l-1}\right) \equiv \mathrm{ReLu}\left(\mathrm{W}^{r,l}\mathrm{h}^{r,l-1} + \mathrm{b}^{r,l}\right) \tag{1}$$

Assume that $\hat{x}^0 \equiv 0, x^0 \equiv x$, where the projection matrix has dimensions $B^r \in R^{\omega \times d_h}, F^r \in R^{H \times d_h}$, The operations for improving the N-BEATS model can be described as follows:

$$\left.\begin{array}{c} x^r = \mathrm{ReLu}\left[x^{r-1} - \hat{x}^{r-1}\right] \\ h^{r,1} = FC_{r,1}(x^r), \ldots, h^{r,L} = FC_{r,L}\left(h^{r,L-1}\right) \\ \hat{x}^r = B^r h^{r,L}, \hat{y}^r = F^r h^{r,L} \end{array}\right\} \quad (2)$$

In the formula, x^r represents the input of the rth block and $\hat{x}^r$ represents the approximate backcast output of the rth block. $\hat{y}^r$ represents the forecast output of the rth block. B^r and F^r represent the backcast matrix and forecast matrix, respectively. $h^{r,L}$ represents the output of the Lth layer in the rth block. d_h represents the dimension of the hidden layer.

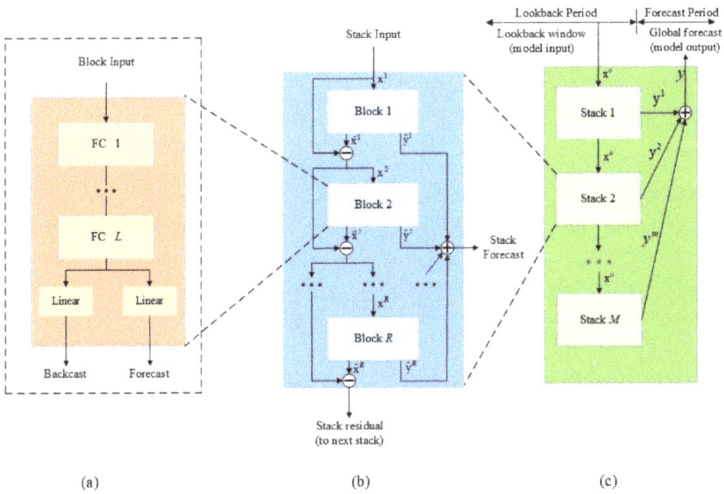

Figure 2. Improved N-BEATS architecture. (**a**) Fully connected nonlinear regressor block. (**b**) Residual stacking with multiple layers. (**c**) Stack aggregation of outputs for the final prediction.

The first residual block serves as the initial input of the model. For the other block, the backward residual branch can be considered as a continuous analysis of the input signal, where the preceding block removes the signal components that it can approximate well $\hat{x}$, thereby making the prediction task for downstream blocks easier. This structure also facilitates smoother gradient backpropagation. More importantly, the partial predictions output by each block are first aggregated at the stack level and then at the entire network level, providing a hierarchical decomposition. The final prediction is the sum of all partial predictions. The model's hierarchical aggregation can be formulated as follows:

$$\left.\begin{array}{c} y_m = \sum\limits_{r=1}^{R} \hat{y}^r \\ y = \sum\limits_{m=1}^{M} y^m \end{array}\right\} \quad (3)$$

In the formula, y_m represents the output of the mth stack, and y represents the total output of the model.

2.2. Pinball-MAPE Loss Function

Most existing prediction algorithms focus primarily on improving prediction accuracy while neglecting bias, which plays a crucial role in fatigue prediction. In this study, the

MAPE [22] is used as the primary accuracy metric. Using it as a loss function ensures maximum consistency between the training process and performance evaluation. However, the bias generated during the prediction process cannot be determined by the MAPE. Smyl [23] suggested that using lower τ can help avoid overprediction by the model during training. To minimize bias as much as possible during prediction, this paper adopts Pinball-MAPE as the loss function:

$$\text{Pinball-MAPE}(y,\hat{y}) = \frac{1}{N}\sum_{i=1}^{N}\begin{cases} 200 \cdot \tau(y_i - \hat{y}_i)/y_i & \text{if} \quad y_i \geq \hat{y}_i \\ 200 \cdot (1-\tau)(\hat{y}_i - y_i)/y_i & \text{otherwise} \end{cases} \tag{4}$$

In the formula, N represents the number of samples, y_i denotes the true value, and $\hat{y}_i$ represents the predicted value. The parameter τ in Pinball-MAPE can be tuned on the validation set to offset bias introduced during training.

2.3. Experimental Setup

2.3.1. Data Preprocessing

The full-scale fatigue test of blades is essentially an accelerated fatigue test based on the principle of damage equivalence. The goal is to replicate the fatigue damage equivalently and complete the fatigue performance assessment within a testing period that is acceptable in engineering practice [3,24]. During the full-scale fatigue test, if the stiffness variation in the blade is less than 10% and no significant damage, cracks, or local instability are observed, the blade is considered to meet design requirements. For simplicity, the blade is treated as a cantilever beam with isotropic material properties under constant load, and its stiffness can be expressed as:

$$E = \frac{F}{L} \tag{5}$$

In this formula, E represents the static stiffness of the blade, F is the concentrated calibration load applied to the blade, and L denotes the measured displacement at the critical section. During testing, the ambient temperature and humidity can influence the variation in blade stiffness. If temperature and humidity are taken into account, it is necessary to train stiffness along with these factors simultaneously. According to Bai's study [25], the changes in processed stiffness can reflect the impacts of temperature and humidity. Thus, stiffness alone is adequate for assessing the fatigue condition of blades. This method simplifies network training and facilitates increasing its depth to enhance performance [14]. In previous work by Kou [3], experimental data at the critical point (32 m) of a 52.5 m glass fiber-reinforced plastic (GFRP) wind turbine blade were obtained. Table 1 lists the specific parameters of the blade. In this paper, this critical point is chosen as an example for stiffness prediction, predicting blade stiffness under different fatigue life ratios. Table 2 presents the measured normalized stiffness under different calibration loads and fatigue life cycle ratios.

Table 1. Basic parameters of the blade.

Name	Value
Power	2.5 MW
Blade length	52.5 m
The length of the blade root	2.4 m
Blade mass	10,968 kg
Maximum chord length	3 m
Airfoil	DU/NACA
Rated speed	23 rpm
Main material	class fiber-reinforced epoxy resin

Table 2. Blade stiffness under different fatigue life ratios.

Loading Level	Percent of Life (n/N)					
	0	0.145	0.250	0.315	0.750	1
20%	1	0.9887	0.9847	0.9821	0.9729	0.9568
40%	1	0.9829	0.9800	0.9782	0.9675	0.9526
60%	1	0.9831	0.9795	0.9779	0.9649	0.9488
80%	1	0.9830	0.9891	0.9767	0.9628	0.9470
100%	1	0.9826	0.9788	0.9771	0.9616	0.9421

To better train the model for residual stiffness degradation prediction, this study adopts Liu's stiffness degradation model [26] to expand the dataset. According to Liu's model, the blade damage can be described as

$$D(n) = \frac{E_0 - E_n}{E_0 - E_f} = 1 - \sin\left[1 - \left(\frac{n}{N}\right)^b\right]^a \frac{\pi}{2} \tag{6}$$

In this formula, E_0 is the initial Young's modulus, E_f is the failure Young's modulus, $E_{(n)}$ represents the Young's modulus after the nth cycle, n is the corresponding number of cycles, and N is the fatigue life of the blade. Parameters a and b are material constants fitted from experimental data. Through Equation (6), the stiffness data in Table 2 can be converted into damage data while simultaneously fitting the parameters a and b. Table 3 lists the fitted parameters a and b, as well as E_f under five loading levels.

Table 3. Fitted parameters of the stiffness degradation equation under five loading levels.

Parameters	Loading Level				
	20%	40%	60%	80%	100%
a	0.42466	0.42265	0.45750	0.57364	0.45208
b	0.11875	0.08192	0.11092	0.21731	0.12594
E_f	0.9568	0.9526	0.9488	0.9470	0.9421

Based on Liu's model, the model for stiffness at various load levels is formulated as follows:

$$E(n) = E_f\left\{1 - \sin\left[1 - \left(\frac{n}{N}\right)^b\right]^a \frac{\pi}{2}\right\} + \sin\left[1 - \left(\frac{n}{N}\right)^b\right]^a \frac{\pi}{2} \tag{7}$$

Figures 3 and 4 illustrate the blade damage curves and stiffness degradation curves, respectively, under different load conditions.

Owing to the limited stiffness data obtained from experiments, DL methods typically need a substantial amount of samples to ensure the generalizability of the model. Therefore, stiffness degradation curve data serve as testing datasets to validate the effectiveness of this prediction approach. As observed in Figure 3, certain data points (the second green data point) exhibit localized deviations from the overall trend. These deviations primarily originate from experimental uncertainties in the original dataset, such as sensor accuracy, material property variations, and environmental influences. These uncertainties propagate through the stiffness degradation model parameter fitting and the damage data conversion process, gradually accumulating during numerical optimization, which ultimately results in localized data point shifts. However, from an overall perspective, the damage evolution and stiffness degradation curves under different loading levels exhibit expected trends, which are largely consistent with the findings of previous studies. Therefore, despite the presence of localized measurement errors in individual data points, they do not affect the overall degradation trend or compromise the conclusions of this study. The purpose of selecting Liu's model is to fit the experimental data, generating additional data points to expand the dataset. On the basis of the experimental results, five original datasets—$E_{20\%}(n)$, $E_{40\%}(n)$, $E_{60\%(n)}$, $E_{80\%}(n)$, and $E_{100\%}(n)$—were constructed.

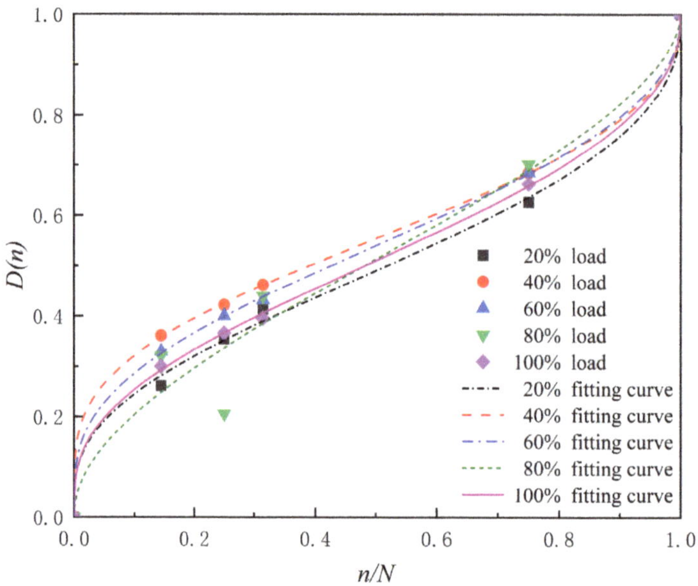

Figure 3. Damage progression curves under different loads.

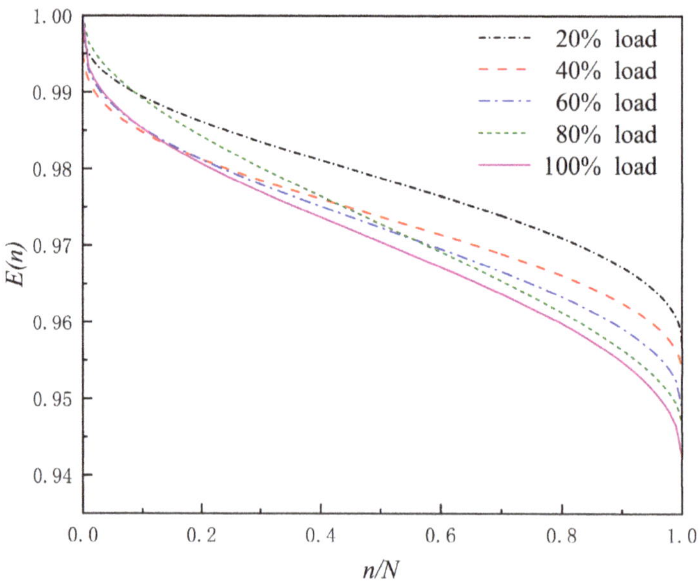

Figure 4. Stiffness degradation curves under different loads.

2.3.2. Data Augmentation

Although DL models rely on large datasets to capture complex patterns, they may tend to memorize the training data when the data diversity is insufficient, leading to overfitting.

The initial data in this study are derived from regularized stiffness degradation curves, which exhibit a relatively smooth degradation trend. Therefore, this paper proposes two data augmentation methods to meet the model's need for a sufficient amount of training data and to enhance the model's generalizability.

Time series combination: In addition to being easy to implement and computationally efficient, this technique generates new samples that maintain some similarity with existing samples while introducing a degree of variability, thus enhancing the overall generalization capability of predictive models. In this study, the stiffness degradation data under five different load conditions were combined into three groups: $E_{20\%}(n)$ and $E_{40\%}(n)$, $E_{40\%}(n)$ and $E_{60\%}(n)$, and $E_{60\%}(n)$ and $E_{80\%}(n)$. A mean operator was applied to these combinations, generating three new datasets, $E_a(n)$, $E_b(n)$, and $E_c(n)$, to help the model learn the mechanism of blade stiffness degradation.

Random noise injection: The use of this technique to create new training samples offers two key advantages. First, in theory, it can produce numerous new samples, thus expanding the training set. Second, when properly adjusted, random noise can introduce variance within the sequence while preserving the fundamental patterns that are crucial for prediction, thereby improving the generalization ability of the predictive model [27]. In this study, the initial stiffness degradation data are obtained through curve fitting, making the data relatively smooth. However, in real-world testing environments for wind turbine blades, stiffness degradation is influenced not only by fatigue accumulation but also by external environmental factors. By introducing random variables, the dynamic behavior and randomness during the blade testing process are described [28]. In this paper, random noise with a fixed signal-to-noise ratio of 65 dB is added to the training dataset.

2.3.3. Dataset Partitioning

Before training the model, the dataset is partitioned as follows: The training set comprises three groups of data with added noise at load levels of 20%, 40%, and 60%, as well as $E_a(n)$, $E_b(n)$, and $E_c(n)$, along with the first 70% of the data under 80% load conditions. The remaining 30% of the data under 80% load conditions are used as a validation set to assess model's predictive performance. Data at the 100% load level are not involved in the initial training and validation of the model; instead, they are evenly split into a training set and a test set. The training set is used for incremental training on the model that has already been trained on the previous four data groups, whereas the test set is used to assess the model's predictive performance on entirely new data. This design allows for assessing the model's generalization on previously unseen 100% load level data, ensuring that the model demonstrates robust predictive ability when faced with new load conditions in practical applications.

2.3.4. Model Training and Prediction

The model is trained using the Adam optimizer. During validation, a segment of data from the end of the training set, matching the length of the backcast window, is combined with the initially defined validation set to form a new validation set. This ensures that the contextual information of the input sequence is preserved, allowing the model to perform a more comprehensive evaluation while considering the temporal nature of the data. The optimal model parameters obtained through the hyperparameter tuning process described in the previous section are shown in Table 4. Hyperparameters were tuned to minimize MAPE on the validation set. The loss function was adjusted separately for different configurations of τ, and ultimately, $\tau = 0.45$ was selected. Finally, the MAPE, RMSE, MAE, and R^2 were selected as the metrics to evaluate the model's predictive performance.

Table 4. Optimal hyperparameters of the improved N-BEATS.

Hyperparameters	Value	Hyperparameters	Value
Backcast Window	16	Layers	5
Prediction Window	5	Width	291
Batch size	152	Epochs	20
Stacks	10	Dropout	0.12
Blocks	4	Learning rate	6.5×10^{-5}

2.3.5. Baseline Models

The baseline models used in the comparative experiments include CNN-LSTM, TCN, Neural Hierarchical Interpolation for Time Series (NHiTS), and N-BEATS (generic architecture). These baseline models follow the same hyperparameter optimization strategy as the improved N-BEATS model. Some of the baseline models utilize the Darts library for their predictions [29].

2.3.6. Effect Evaluation

During the training process, Pinball-MAPE was used as the loss function to control potential biases in the prediction process and the Mean Absolute Error (MAE), Root Mean Squared Error (RMSE), MAPE, and Coefficient of Determination (R^2) were selected as the metrics to evaluate the model's predictive performance. Table 5 lists the calculation functions for the four evaluation metrics.

Table 5. Evaluation functions.

Metric	MAE	RMSE	MAPE	R^2
Function	$\mathrm{MAE} = \frac{1}{N}\sum_{t=1}^{N}\lvert y_t - \widetilde{y}_t\rvert$	$\mathrm{RMSE} = \sqrt{\frac{1}{N}\sum_{t=1}^{N}(y_t - \widetilde{y}_t)^2}$	$\mathrm{MAPE} = \frac{100\%}{N}\sum_{t=1}^{N}\left\lvert\frac{y_t - \widetilde{y}_t}{y_t}\right\rvert$	$R^2 = 1 - \frac{\sum_{t=1}^{N}(y_t-\widetilde{y}_t)^2}{\sum_{t=1}^{N}(y_t-\overline{y})^2}$

In the above formula: N represents the number of samples, y_t denotes the true value, $\overline{y}$ represents the mean of the true values, and $\widetilde{y}_t$ represents the predicted value.

3. Results and Discussion

Figure 5 and Table 6 present the prediction results and metric scores of all the models. The improved N-BEATS model outperforms the other baseline models across all the evaluation metrics. This demonstrates that the improved N-BEATS exhibits outstanding performance in predicting the stiffness degradation sequence of wind turbine blades, particularly excelling in handling small-scale datasets and mitigating the impact of noise. In contrast, although the TCN model excels at long-sequence modeling and capturing dependencies within them, its sensitivity to small-scale datasets and high-noise data often leads to a decline in prediction accuracy. The TCN model still shows limitations in handling short time series and noisy data, making its performance slightly inferior in this experiment. Although both N-BEATS and NHiTS have shown excellent performance in multiple time series forecasting tasks, especially on large-scale datasets, where their layered architectures can effectively capture global features, the complexity of these models becomes a burden when dealing with the small-scale data in this study, resulting in slower training speeds and slightly lower generalization performance. The CNN-LSTM model, while combining the advantages of both to capture local features and model long-term dependencies in sequences, is more limited in extracting useful feature information from small-scale, high-noise datasets. As a result, its performance is inferior to that of the improved N-BEATS model.

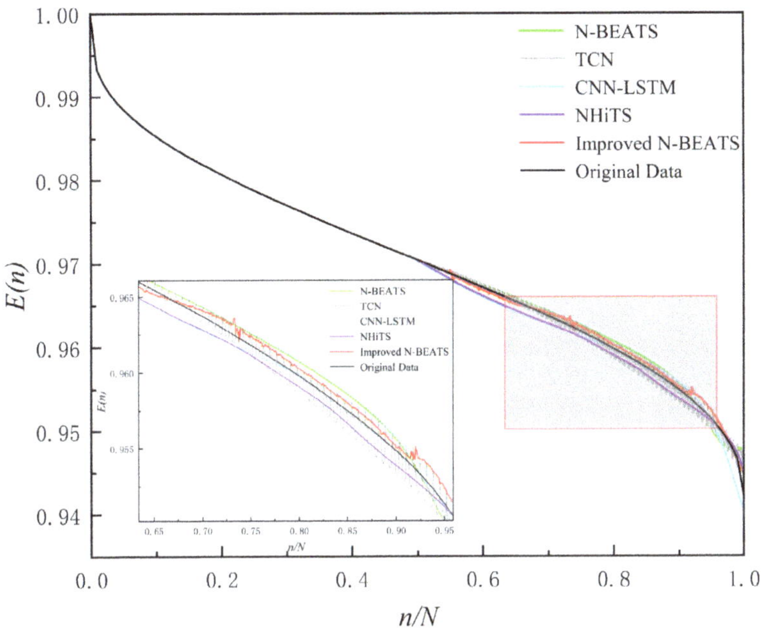

Figure 5. Comparison of prediction performance of different models.

Table 6. Comparison of evaluation metrics for different prediction models.

Model	Evaluation Metrics			
	MAPE(%)	RMSE	MAE	R^2
CNN-LSTM	0.08736	0.00101	0.00083	0.9758
TCN	0.05886	0.00078	0.00056	0.9859
N-BEATS	0.07445	0.00091	0.00071	0.9811
NHiTS	0.04696	0.00058	0.00045	0.9923
Improved N-BEATS	0.04152	0.00049	0.00041	0.9945

Moreover, each model exhibited a certain degree of bias during the forecasting process. This study uses the Mean Percentage Error (MPE) to measure the bias in each model's predictions. Table 7 presents the bias associated with each model's predictions. Among all models, the improved N-BEATS model, which introduces the Pinball-MAPE loss function, achieved the lowest MPE of −0.02%, significantly reducing the risk of over-prediction.

Table 7. Prediction biases of different models.

Model	Improved N-BEATS	NHiTS	N-BEATS	TCN	CNN-LSTM
MPE	−0.02%	0.04%	−0.06%	−0.04%	−0.05%

To evaluate the impact of different modules on the predictive performance of the improved N-BEATS model, this study conducted two ablation experiments. Figure 6a,b, respectively, illustrate the comparison of prediction results under different loss functions and data processing methods in two sets of experiments. The first experiment examined the effect of changing the loss function, demonstrating that the adoption of the Pinball-MAPE loss function reduced the model's MAPE from 0.04435 to 0.04152. The second experiment compared the impact of using original data versus employing a data augmentation strategy. The results indicated that the data augmentation strategy reduced the model's MAPE from 0.07223 to 0.04152. These findings collectively demonstrate that the Pinball-MAPE loss function and the data augmentation strategy significantly enhance the predictive accuracy and generalization capability of the improved N-BEATS model.

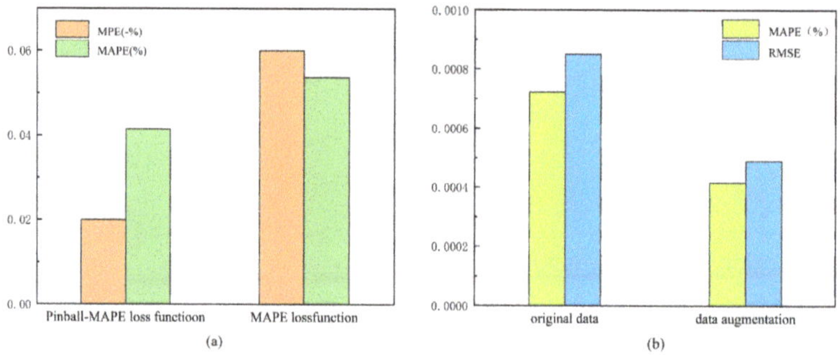

Figure 6. Ablation experiments. (**a**) Comparison of prediction results under Pinball-MAPE and MAPE loss functions. (**b**) Comparison of prediction results using original data and data augmentation.

The improved N-BEATS model adopts advanced architecture, including a nonlinear multivariate regression structure and forward–backward residual connections, significantly enhancing its adaptability to the small-scale dataset used in this study. Even in the presence of noise, the model can still identify fundamental patterns that are crucial for sequence prediction, demonstrating high robustness. This architecture not only captures shared patterns of stiffness sequences under different loading conditions but also learns the unique characteristics of each predictive sequence, thereby improving model adaptability. Additionally, the Pinball-MAPE loss function parameters can be flexibly adjusted to accommodate different prediction tasks, reducing bias and optimizing prediction accuracy. Traditional methods for predicting the stiffness degradation of wind turbine blades are primarily based on physics-driven and empirical models. While these methods provide a certain degree of theoretical interpretability, they exhibit significant limitations in practical applications. For instance, finite element analysis (FEA) offers high-fidelity local stress and stiffness analysis, but its computational cost is extremely high. Due to their strong dependence on predefined material properties and boundary conditions, these methods exhibit limited generalization capabilities under dynamically changing environments. For data-driven physics-based models, extensive experimental calibration is often required, making them highly dependent on the availability and quality of data. Furthermore, these models typically fail to capture complex multi-factor interactions, leading to reduced generalization performance when encountering previously unseen working conditions. In contrast, the improved N-BEATS model can autonomously learn degradation patterns from historical data without requiring predefined complex physical assumptions, thereby enhancing its generalization capability.

4. Conclusions

Accurately predicting the stiffness degradation of wind turbine blades is crucial for ensuring their long-term reliability and reducing high experimental costs. This study proposes an effective method to predict the stiffness degradation of wind turbine blades by improving the N-BEATS model, and its superior performance is validated through multiple experiments.

First, the N-BEATS basic block structure was improved, and the Pinball-MAPE loss function was introduced to make the model more suitable for the stiffness degradation prediction task in this study. In addition, a dataset was constructed based on the stiffness degradation model and used for model validation. Compared to other well-established deep learning models, the improved N-BEATS demonstrated the best prediction accuracy. Meanwhile, partial data re-training was conducted on unseen sequences to verify the model's generalization capability under unknown working conditions.

Second, this study employs two data augmentation methods—time series combination and random noise injection—that successfully expand the dataset and effectively enhance the model's generalization ability. This method effectively addresses the common issue of data in wind turbine blade testing data, allowing the model to adapt to diverse data environments and tackle challenges posed by limited original data.

In engineering applications, the proposed model, combined with historical blade stiffness degradation data and early-stage experimental data, enables the prediction of the subsequent stiffness evolution of wind turbine blades. This allows for the assessment of whether the blades still meet operational requirements. The proposed approach reduces the need for physical testing during blade evaluation, thereby lowering assessment costs and improving engineering efficiency. Furthermore, this method can be extended to the fatigue life prediction of composite materials, providing data-driven decision support for structural health monitoring. It is applicable to wind energy, aerospace, and other composite material structures requiring long-term reliability assessment.

Author Contributions: Conceptualization, S.Y. and J.G.; methodology, S.Y.; software, S.Y.; validation, S.Y., Y.Y., J.Z. and L.M.; formal analysis, Y.Y.; investigation, J.G.; data curation, J.Z.; writing—original draft preparation, S.Y.; writing—review and editing, S.Y., J.G., Y.Y. and L.M.; visualization, S.Y.; supervision, J.G.; project administration, Y.Y. and J.Z.; funding acquisition, J.G. All authors have read and agreed to the published version of the manuscript.

Funding: This work was supported by the National Natural Science Foundation of China (52465019), Natural Science Foundation of Xinjiang Uygur Autonomous Region (2024D01C27), Xinjiang Uygur Autonomous Region Central Guidance Local Science and Technology Development Fund Project (ZYYD2025JD07), and Changji Hui Autonomous Prefecture Science and Technology Plan Project (2023Z02).

Institutional Review Board Statement: Not applicable.

Informed Consent Statement: Not applicable.

Data Availability Statement: Data is contained within the article.

Conflicts of Interest: The authors declare no conflicts of interest.

References

1. Davis, M.; Dejesus, E.N.; Shekaramiz, M.; Zander, G.; Memari, M. Identification and localization of wind turbine blade faults using Deep Learning. *Appl. Sci.* **2024**, *14*, 6319. [CrossRef]
2. Su, H.M.; Kam, T.Y. Reliability analysis of composite wind turbine blades considering material degradation of blades. *Compos. Struct.* **2020**, *234*, 111663. [CrossRef]
3. Kou, H.X. Study on Stiffness Degradation Model of Composite Wind Turbine Blades. Ph.D. Thesis, Lanzhou University of Technology, Lanzhou, China, 2019.
4. Liu, X.T.; Zhang, X.Y.; Wang, S.C. An improved model of fatigue life for stitched carbon fiber composites considering stiffness degradation. *J. Braz. Soc. Mech. Sci. Eng.* **2023**, *45*, 184. [CrossRef]
5. Li, H.L.; Liu, Y.Z.; Zou, Y.S.; Liu, Y.T.; Song, X.X. A method for predicting the remaining life of bearings based on TC-CAE. *J. Vib. Shock* **2022**, *41*, 105–113.

6. Zhao, Z.H.; Li, Q.; Yang, S.P.; Li, L.H. Research on remaining useful life prediction based on BiLSTM and attention mechanism. *J. Vib. Shock* **2022**, *41*, 44–50.
7. Jordan, M.I.; Mitchell, T.M. Machine learning: Trends, perspectives, and prospects. *Science* **2015**, *349*, 255–260. [CrossRef] [PubMed]
8. Shi, Z.Y.; Lu, Z.Z.; Zhang, X.B.; Li, L.Y. A novel adaptive support vector machine method for reliability analysis. *Proc. Inst. Mech. Eng. Part O J. Risk Reliab.* **2021**, *235*, 896–908. [CrossRef]
9. Tao, C.C.; Zhang, C.; Ji, H.L.; Qiu, J.H. Application of neural network to model stiffness degradation for composite laminates under cyclic loadings. *Compos. Sci. Technol.* **2021**, *203*, 108573. [CrossRef]
10. Jiang, D.; Qian, H.; Wang, Y.S.; Zheng, J.C.; Zhang, D.H.; Li, Q.Y. Data driven prediction of fatigue residual stiffness of braided ceramic matrix composites based on Latent-ODE. *Compos. Struct.* **2023**, *323*, 117504. [CrossRef]
11. Yin, X.J.; Mu, Q.Z.; Zhou, L.; Li, B.; Shao, G.C.; Du, Z.L. Vibration modeling and simulation analysis of wind turbine flexible tower based on ChOA algorithm optimization. *J. Appl. Sci. Eng.* **2024**, *28*, 109–118.
12. Hochreiter, S.; Schmidhuber, J. Long short-term memory. *Neural Comput.* **1997**, *9*, 1735–1780. [CrossRef]
13. Yan, K.; Wang, X.D.; Du, Y.; Jin, N.; Huang, H.C.; Zhou, H.X. Multi-Step Short-Term Power Consumption Forecasting with a Hybrid Deep Learning Strategy. *Energies* **2018**, *11*, 3089. [CrossRef]
14. Liu, H.W.; Zhang, Z.C.; Jia, H.B.; Li, Q.Y.; Liu, Y.J.; Leng, J.S. A novel method to predict the stiffness evolution of in-service wind turbine blades based on deep learning models. *Compos. Struct.* **2020**, *252*, 112702. [CrossRef]
15. Wang, H.R.; Li, D.J.; Li, Y.; Zhu, G.F.; Lin, R.X. Method for Remaining Useful Life Prediction of Turbofan Engines Combining Adam Optimization-Based Self-Attention Mechanism with Temporal Convolutional Networks. *Appl. Sci.* **2024**, *14*, 7723. [CrossRef]
16. Zha, W.T.; Ye, Y.H. An aero-engine remaining useful life prediction model based on feature selection and the improved TCN. *Frankl. Open* **2024**, *6*, 100083. [CrossRef]
17. Chorowski, J.; Bahdanau, D.; Cho, K.; Bengio, Y. End-to-end continuous speech recognition using attention-based recurrent nn: First results. *arXiv* **2014**, arXiv:1412.1602.
18. Vaswani, A.; Shazeer, N.; Parmar, N.; Uszkoreit, J.; Jones, L.; Gomez, A.N.; Kaiser, L.; Polosukhin, I. Attention is all you need. In Proceedings of the 31st Conference on Neural Information Processing Systems, Long Beach, CA, USA, 4–9 December 2017.
19. Lu, Q.; Ye, W.; Yin, L. ResDenIncepNet-CBAM with principal component analysis for wind turbine blade cracking fault prediction with only short time scale SCADA data. *Measurement* **2023**, *212*, 112696. [CrossRef]
20. Oreshkin, B.N.; Carpov, D.; Chapados, N.; Bengio, Y. N-BEATS: Neural basis expansion analysis for interpretable time series forecasting. In Proceedings of the 2020 International Conference on Learning Representations, Addis Ababa, Ethiopia, 26 April–1 May 2020.
21. Oreshkin, B.N.; Dudek, G.; Pełka, P.; Turkina, E. N-BEATS neural network for mid-term electricity load forecasting. *Appl. Energy* **2021**, *293*, 116918. [CrossRef]
22. Makridakis, S.; Hibon, M. The M3-Competition: Results, conclusions and implications. *Int. J. Forecast.* **2000**, *16*, 451–476. [CrossRef]
23. Smyl, S. A hybrid method of exponential smoothing and recurrent neural networks for time series forecasting. *Int. J. Forecast.* **2020**, *36*, 75–85. [CrossRef]
24. Li, G.J.; Hu, P.; Xie, C.Y.; Ge, R.W.; Wang, Y.S. Research on life prediction of rubber vibration damping structure based on mechanical-thermal double stress accelerated test. *Equip. Environ. Eng.* **2021**, *18*, 25–30.
25. Bai, X.Z.; An, Z.W.; Hou, Y.F.; Ma, Q. Health assessment and management of wind turbine blade based on the fatigue test data. *Microelectron. Reliab.* **2017**, *75*, 205–214. [CrossRef]
26. Liu, H.W.; Zhang, Z.C.; Jia, H.B.; Liu, Y.J.; Leng, J.S. A modified composite fatigue damage model considering stiffness evolution for wind turbine blades. *Compos. Struct.* **2020**, *233*, 111736. [CrossRef]
27. Semenoglou, A.A.; Spiliotis, E.; Assimakopoulos, V. Data augmentation for univariate time series forecasting with neural networks. *Pattern Recognit.* **2023**, *134*, 109132. [CrossRef]
28. Zhang, F.; Nan, H.; Yao, H.J.; Lyu, C.C. Analysis of dynamic wear reliability of mechanism under stochastic noise. *Acta Armamentarii* **2014**, *35*, 2104–2108.
29. Herzen, J.; Lässig, F.; Piazzetta, S.G.; Neuer, T.; Tafti, L.; Raille, G.; van Pottelbergh, T.; Pasieka, M.; Skrodzki, A.; Huguenin, N.; et al. Darts: User-friendly modern machine learning for time series. *J. Mach. Learn. Res.* **2022**, *23*, 1–6.

applied sciences

MDPI

Communication

Measurement Uncertainty and Compliance Evaluation Applied to Natural Gas Moisture

Rosana Medeiros Moreira [1,2], Cesar Luís Biazon [3] and Elcio Cruz de Oliveira [4,5,*]

1 Product and Process Engineering, National Institute of Technology, Rio de Janeiro 20081-312, Brazil; rosana.moreira@int.gov.br
2 Postgraduate Programme in Metrology, Brazilian Institute of Metrology, Quality, and Technology (INMETRO), Duque de Caxias 25250-020, Brazil
3 Exploration and Production (UN-BC/SE/LFCEN), PETROBRAS S.A., Macaé 27913-350, Brazil; cbiazon@petrobras.com.br
4 Postgraduate Programme in Metrology, Pontifical Catholic University of Rio de Janeiro, Rio de Janeiro 22451-900, Brazil
5 Land Transportation and Storage, Measurement and Product Inventory Management, Logistics, PETROBRAS S.A., Rio de Janeiro 20231-030, Brazil
* Correspondence: elciooliveira@puc-rio.br

Abstract: The reliability of natural gas moisture measurements is crucial in preventing corrosion in pipelines and equipment, ensuring burning efficiency and mitigating operational risks, and is often mandated by standards and regulations. An important quality parameter that aids in conformity assessment, particularly in risk assessment, is measurement uncertainty. The assessment of measurement uncertainty, based on the Law of Uncertainty Propagation, depends on the mathematical model used to calculate this physicochemical property. This study aimed to compare different algorithms for calculating moisture content in natural gas, estimate and validate the measurement uncertainty based on the algorithms implemented in the Portable Moisture Analyzer (PM880) equipment, a portable hygrometer manufactured by Panametrics in Wilmington, NC, USA, and evaluate compliance with Brazilian legislation using guard bands as a decision rule. The moisture results in natural gas varied by a maximum of 1% among the three approaches presented. Furthermore, based on the dew point and pressure results, the expanded uncertainties of moisture were about 20%, which did not compromise the risk assessment for the consumer, as the moisture results were well below the specification value. Consequently, the upper tolerance limit of 58.4 ppmv H_2O was established.

Keywords: consumer's risk; Portable Moisture Analyzer; PM880; ppmv H_2O; dew point; specification limit; acceptance limit

1. Introduction

The water vapor in natural gas, commonly called moisture, must be monitored with high reliability to ensure the gas's safe processing and transportation. Moisture control is critical in processing plants, as it can significantly impact the quality and efficiency of transportation and lead to corrosion and other types of damage to equipment and pipelines.

The water content in natural gas is the primary factor influencing internal corrosion. Even at low levels, variations in pressure and temperature can cause water to condense, resulting in corrosion, hydrate formation (a semi-solid combination of hydrocarbons and water), or ice formation. To mitigate these problems, natural gas companies invest in dehydration units. Based on international standards such as ASTM D5454 [1], ASTM

D1142 [2] and ISO 18453 [3], the design and cost of these installations are influenced by the need for accurate knowledge of the water content at the dew point and the contractually required water content, as natural gas pipelines have specifications that limit the maximum permissible water vapor concentration.

For instance, a standard pipeline specification in the Petrobras offshore production system is a maximum moisture content of 60 ppmv. The water dew point in gas refers to the temperature, at a specified pressure, at which liquid water will begin to condense from the water vapor present. Consequently, water concentration levels can be measured based on natural gas systems' dew point and pressure.

As stated in ISO 6327, "the partial pressure of water vapor in gas samples is the saturated vapor pressure corresponding to the observed dew point, provided that the gas in the hygrometer is at the same pressure as the gas at the time of sampling" [4]. Therefore, to conduct moisture measurements in gas, it is essential to establish a correlation between the water dew point and the water content.

Shkarovskij and Volikov improved gas quality management, proposing a procedural framework for accurately determining the moisture of natural gas using dew point temperature indicators found in fuel certificates [5].

Studies conducted on hygrometers, generally, have shown to be an area of interest by researchers. One classic study evaluated the uncertainty in measuring the dew point of natural gas using the traditional cooling mirror condensation hygrometers [6]; another one compared various hygrometers used for measuring moisture in gases that are utilized with varying frequency in the natural gas industry [7]; the last study centers around hygrometers being used for measuring water vapor (moisture) in natural gas, focusing on their response to co-exposure to ethylene glycol [8].

Gallegos et al. [9] compared eleven hygrometers at natural gas facilities in Spain utilizing various measurement techniques (including chilled-mirror, electrolytic sensors, spectroscopic analyzers, and polymeric and metal oxide humidity sensors).

Chinese researchers have developed an optimized Gaussian process model that effectively evaluates parameters to predict the water dew point in natural gas dehydration units, enhancing the efficiency and accuracy of the dehydration process [10].

A conference paper illustrated that an online dew point analyzer operates within the lowest error range after a six-month recalibration period compared to other calibrated and reliable devices [11].

The National Physical Laboratory (United Kingdom) presented a method to isolate and adjust the influence of background gas composition on trace moisture measurements by employing two distinct frequencies in a trace moisture analyzer that utilizes a ball surface acoustic wave (SAW) sensor [12].

A recent study outlines approaches for increasing the quantity and ranges of reproducible values while enhancing the functionality of the state primary standard for relative humidity of gases, molar (volume) fraction of moisture, and dew/frost point temperature (DPT) as specified in GET 151-2014 [13].

Last year, Chinese researchers published an interesting study regarding an interdigital conductance sensor designed to enhance electric field distribution for the precise measurement of uneven liquid film thickness in inclined gas–liquid two-phase flows. Through experimental validation, the sensor showcased improved effectiveness over conventional sensors [14].

For an accurate assessment of natural gas moisture, it is crucial to employ standardized methodologies, ensure regular calibration of measuring instruments, and consider the uncertainty associated with each process stage.

This study evaluated the measurement uncertainty of natural gas moisture calculated from the dew point temperature and gauge pressure results using the algorithms implemented in the Portable Moisture Analyzer—PM880. An upper acceptance limit has been proposed to ensure that the moisture content in natural gas does not compromise the operational efficiency of gas pipelines. This limit is based on uncertainty data and employs the concept of guard bands to mitigate the risk of false conformity concerning moisture levels in natural gas.

2. Methodology

Several methodologies are used to determine the water vapor content in natural gas from the dew point temperature and pressure measurements. The first two parts of this section describe two methodologies, and the next part describes the methodology used here. The last sections are designated for calculating and using the measurement uncertainty for compliance assessment.

2.1. ASTM D1142 Test Method, Approach 1

The ASTM D1142-95 (2021) [2] test method includes two correlations, the first of which (herein referred to as ASTM D1142 test method, approach 1) is a variation of Equation (1) that expresses water content in terms of the weight of saturated water vapor, Equation (2):

$$y_w = \frac{P_w^{sat}}{P} \tag{1}$$

where y_w is the mole fraction of water in the vapor phase, P is the total pressure, and P_w^{sat} is pure water's saturation (vapor) pressure at the dew point temperature T.

$$W = \left(w \times 10^6\right) \frac{P_b \times T}{P \times T_b} \tag{2}$$

where W is the water content (pounds water/million standard cubic feet natural gas) at reference conditions, T_b (base gas measurement temperature, $t_b + 400$), P_b (base gas measurement pressure, psia), w is the weight of saturated water vapor (lb/ft^3), P is the pressure at which the dew point was determined (psia), and T is the observed Rankine dew point temperature ($t + 460$). The reciprocal of w, or the specific volume of saturated water vapor (ft^3/lb), is listed as a function of temperature in Table 1 of ASTM D1142-95 (2021) for temperatures ranging from 0 °F to 100 °F. Although not explicit in temperature due to the temperature dependence of w, given the water content, the corresponding dew point temperature can be solved iteratively.

2.2. ASTM D1142 Test Method, Approach 2

Bukacek proposed a relatively simple modification of Raoult's law approach, where the water content of the sweet gas is calculated using the ideal expression of Equation (1) supplemented by a deviation factor [15], Equation (3):

$$W = (760.4)\frac{P_w^{sat}}{P} + 0.016016 \times B \tag{3}$$

where W is the water content (g/Nm^3), P_w^{sat} is the saturation vapor pressure of pure water (MPa), P is the total system pressure (MPa), and B is given by Equation (4):

$$logB = \frac{-1713.66}{T} + 6.69449 \tag{4}$$

where T is the dew point temperature (K) and the saturation vapor pressure can be calculated by Equation (5) [16]:

$$ln\left(\frac{P_w^{sat}}{P_c}\right) = \frac{T_c}{T}\left(-7.85823\tau + 1.83991\tau^{1.5} - 11.7811\tau^3 + 22.6705\tau^{3.5} - 1539393\tau^4 + 1.77516\tau^{7.5}\right) \quad (5)$$

where T is the temperature (K), T_c is the critical water temperature (647.14 K), P_c is the critical water pressure (22.064 MPa), and $\tau = 1/T - T_c$. Thus, a simplified version of Equation (3) is Equation (6):

$$W = \frac{A}{P} + B \quad (6)$$

Referred to here as Section 2.2, ASTM D1142 Method 2 is included in ASTM Test Method 1142-95 (2021). The coefficients A and B (referenced to T_b = 520 R and $P_b = 14.7$ psia) are listed as a function of temperature in Table 2 for dew point temperatures ranging from −40 °F to 440 °F. Although not explicitly stated in temperature due to the temperature dependence of A and B, the corresponding dewpoint temperature can be solved iteratively given the water content. Although conveniently simplistic, neither ASTM method considers the actual gas composition. Furthermore, the range of data available for the specific volume of saturated water vapor (ASTM D1142 Method 1) or for the coefficients A and B (ASTM D1142 Method 2) is somewhat limited.

2.3. *Method Used by the Portable Moisture Analyzer—PM880*

The Panametrics PM880, as shown in Figure 1, is a portable hygrometer manufactured by Panametrics in Wilmington, NC, USA. It employs an aluminum oxide sensor to assess the moisture content in liquids and gases. This measurement is based on the temperature at which condensation occurs under specific barometric pressure conditions.

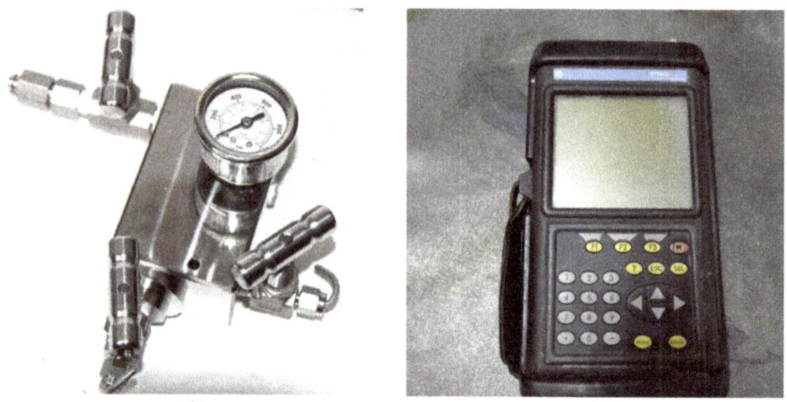

(i) (ii)

Figure 1. (**i**) Sampling system, and (**ii**) PM880 meter.

The vapor pressure of water and ice, Pv, depends on the dew point in °C, DP, and can be calculated from Equations (7) and (8):

$$Pv_{water} = 611.21^{\left(\frac{17.502 \times DP}{240.97 + DP}\right)} \quad DP > 0 \quad (7)$$

$$Pv_{ice} = 611.15^{\left(\frac{22.542 \times DP}{273.48 + DP}\right)} \quad DP \leq 0 \quad (8)$$

The 1958 IGT8 (Institute of Gas Technology research bulletin #8) calculation based on the dew point value is expressed by Equation (9):

$$IGT8 = 10^{\left(\frac{-1711}{DP+273.15}+6.687\right)} \tag{9}$$

The non-ideality of natural gas, B, depends on the dew point in °C, DP, and can be calculated by Equation (10):

$$B = \frac{IGT8 \times P_v}{Pv_{water}} \tag{10}$$

The total pressure converted to Pascals, $Pressure_{PA}$, is calculated by Equation (11):

$$Pressure_{PA} = \frac{Pressure_{psig} + 14.697}{14.697} \times 101325 \tag{11}$$

The ratio of water vapor pressure to total pressure, X_{water}, is calculated by Equation (12):

$$X_{water} = \frac{P_v}{Pressure_{PA}} \tag{12}$$

Finally, W is the water content (pounds water/million standard cubic feet natural gas) at reference conditions T_b (base gas measurement temperature, $t_b + 400$), P_b (base gas measurement pressure, psia), Equation (13), considering the conversion of lbs/MMSCF to a fixed value, $\frac{10^6 \times 30.48^3}{\left(22414 \times \frac{519.67}{491.67}\right)} \times \frac{18.0152}{453.59} = 47473.05721$.

$$W = X_{gua} \times \left[\frac{10^6 \times 30.48^3}{\left(22414 \times \frac{519.67}{491.67}\right)} \times \frac{18.0152}{453.59}\right] + B \tag{13}$$

For some industrial applications, the expression of moisture in parts per million by volume of water, $ppm_v\ H_2O$, can be quite useful, as showcased in Equation (14):

$$ppm_v\ H_2O = \frac{Pv_{ice} \times 10^6}{Pressure_{PA}} \tag{14}$$

To highlight the innovations of this study, we emphasize the comparative advantages of the proposed method over existing approaches. The temperature range of the ASTM D1142 test method, approach 1, is limited from 0 °F to 100 °F, which is completely different from the Brazilian real samples of natural gas. The ASTM D1142 test method, approach 2, simplifies the final algorithm to calculate the water vapor content, in addition to not being appropriate where the dew point is measured in conditions close to the critical temperature of gaseous fuels. These limitations are not present in the method used by the Portable Moisture Analyzer—PM880.

2.4. Measurement Uncertainty

The procedure proposed by JCGM 100:2008 [17] for estimating the measurement uncertainty of a given measurand consists of the following steps: definition of the measurand and its mathematical model, identification of the most relevant sources of uncertainty; quantification of these sources of uncertainty; reduction of uncertainties to standard deviation; combination of standard uncertainties and their respective sensitivity coefficients; declaration of the expanded uncertainty, coverage factor, and degrees of freedom for an appropriate confidence level, 95.45%, used in this work. The combined standard uncertainties can be calculated by Equations (15)–(22):

$$u_c^2(Pv_{water}) = \left(\frac{12888759281487 \exp\left(\frac{8751 \times DP}{500 \times \left(DP + \frac{24097}{100} \right)} \right)}{500 \times (100 \times DP + 24097)^2} \times u(DP) \right)^2 \quad DP > 0 \tag{15}$$

$$u_c^2(Pv_{ice}) = \left(\frac{941902265421 \exp\left(\frac{11271 \times DP}{500 \times \left(DP + \frac{6837}{25} \right)} \right)}{400 \times (25 \times DP + 6837)^2} \times u(DP) \right)^2 \quad DP \leq 0 \tag{16}$$

$$u_c^2(IGT8) = \left(\frac{1711 \ln(10) \times 10^{\frac{6687}{1000} - \frac{1711}{DP + \frac{5463}{20}}}}{\left(DP + \frac{5463}{20} \right)^2} \times u(DP) \right)^2 \tag{17}$$

$$u_c^2(B) = \left(\frac{Pv}{Pv_{water}} \times u(IGT8) \right)^2 + \left(\frac{IGT8}{Pv_{water}} \times u(Pv) \right)^2 + \left(-\frac{IGT8 \times Pv}{Pv_{water}^2} \times u(Pv_{water}) \right)^2 \tag{18}$$

$$u_c^2(Pressure_{PA}) = (33775000/4899 \times u(Pressure_{psig}))^2 \tag{19}$$

$$u_c^2(X_{water}) = \left(\frac{1}{Pressure_{PA}} \times u(Pv) \right)^2 + \left(-\frac{Pv}{Pressure_{PA}^2} \times u(Pressure_{PA}) \right)^2 \tag{20}$$

$$u_c^2(W) = \left(\frac{10^6 \times 30.48^3}{\left(22414 \times \frac{519.67}{491.67} \right)} \times u(X_{water}) \right)^2 + (u(B))^2 \tag{21}$$

$$u_c^2(ppm_v\ H_2O) = \left(\frac{10^6}{Pressure_{PA}} \times u(Pv_{ice}) \right)^2 + \left(-\frac{Pv_{ice} \times 10^6}{Pressure_{PA}^2} \times u(Pressure_{PA}) \right)^2 \tag{22}$$

2.5. Measurement Uncertainty in Compliance Evaluation

Based on the information about the measurement uncertainty, it is possible to verify whether the results obtained are within the specified value. Consequently, it is essential to have a decision rule that considers the risks associated with an incorrect decision, which is determined by guard bands, g, that define acceptance and rejection regions (Figure 2).

Based on JCGM 106:2012, the guard band is established to ensure that the probability of a false acceptance or rejection occurring is less than or equal to a previously defined confidence level α when a measurement occurs in the acceptance region [18].

Each measurement result's standard uncertainty, u, is added to the established guard band value. According to the Eurachem guide [19], a value of $1.64 \times u$ is determined at a significance level of 5%.

In this study, we considered the risk of the consumer, which is defined as the probability of accepting a batch that ought to be rejected.

The histograms illustrated the most likely value, associated uncertainties, guard bands, and upper acceptance limits. To assess the risk to consumers, Monte Carlo simulations were generated from 100,000 pseudorandom values, ensuring a stability of 0.001%, for the moisture content in natural gas based on the mean experimental value and its expanded uncertainty.

Here, it was calculated using the MS-Excel function "=NORM.INV(RAND();y_i;u_{y_i})", where y_i is the mean of the measured value and u_{y_i} is its expanded uncertainty for the ith parameter, considering a Gaussian or normal distribution.

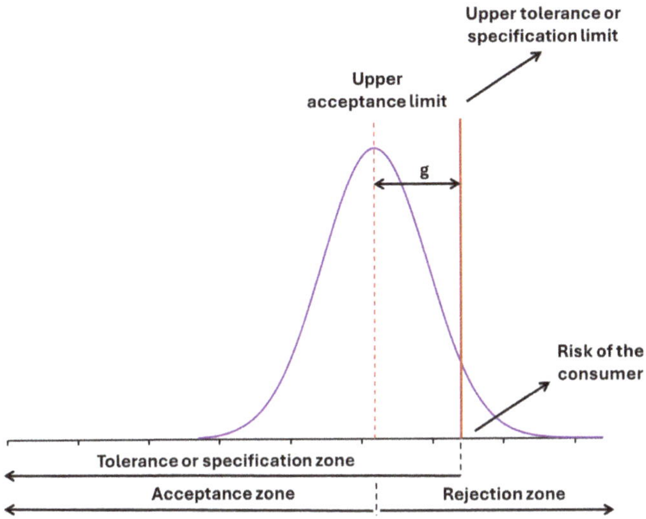

Figure 2. The guard band tool is used to reduce consumer risk, considering an upper tolerance or specification limit.

In recent decades, several studies have involved calculating uncertainty in the physicochemical properties of natural gas. Using measurement uncertainty, Oliveira demonstrated the extent to which the variation in molar fractions obtained from these two methods metrologically influences the higher calorific value of natural gas. Additionally, it affects the compressibility factor of natural gas [20]. Brazilian researchers compared five approaches to assess the uncertainty of natural gas composition by gas chromatography [21].

However, the use of measurement uncertainty in assessing the conformity of these parameters in natural gas is still very incipient. This is apart from a recent study where acceptance thresholds were established to ensure no risk of incorrect compliance in the carbon isotopic analysis of natural gas based on this uncertainty data [22]. On the other hand, this approach can be highlighted in other areas of knowledge, such as environmental pollution [23], drug and medicine analysis [24], and microbiology [25].

3. Results and Discussion

This section is divided into three parts. The introductory part compares the results from the Sections 2.1–2.3. approaches. The second part addresses the calculated uncertainties, while the third discusses using these uncertainties in conformance assessment. Section 3.2 compares the results from the Portable Moisture Analyzer (PM880) with the calculations described in Section 2.3.

3.1. Comparing Different Approaches

Although the objective of this study is not to compare different approaches for calculating moisture in natural gas, we compared three methods described in Sections 2.1–2.3. for determining the water vapor content based on dew point temperature measurements to assess any significant disparities.

We considered the water dew point to be 37 °F at a pressure of 15.0 psia as input data. After converting the temperature, 2.7778 °C, and the pressure, 0.304055 psig, the water vapor content could be calculated in pounds of H_2O/million standard cubic feet of natural gas, at a base of 60 °F and 14.7-psia pressure, Table 1.

Table 1. Comparing the approaches from Sections 2.1–2.3.

ASTM D1142 Test Method, Approach 1	ASTM D1142 Test Method, Approach 2	Portable Moisture Analyzer—PM880
342.8 lbs H_2O/MMSCF NG	346.1 lbs H_2O/MMSCF NG	345.6 lbs H_2O/MMSCF NG

In addition to the results of the approach in question, PM880, among others, the maximum difference found between them is 1%. This is a very interesting scientific result when considering the possible variabilities of each approach.

3.2. Uncertainty Evaluation

Between 13 September 2024 and 20 September 2024, 10 dew point (°C) and pressure (bar) results of natural gas from the Portable Moisture Analyzer—PM880 were collected to validate the algorithms of the developed MS-Excel spreadsheet. After the pressure data were converted to psig, the MS-Excel spreadsheet developed, which is attached as Supplementary Material in this study, calculated the moisture values and their respective expanded uncertainties, as shown in Tables 2 and 3. In the spreadsheet, the input quantities in yellow, cells H2 and H3, are the dew point in Celsius degree and the pressure in psig, respectively, of the natural gas.

Table 2. Moisture results and their respective expanded uncertainties in lbs H_2O/MMSCF NG.

Dew Point (°C)	Pressure (psig)	Moisture Calculated by PM880 *	Moisture Calculated by the Spreadsheet *	Moisture Expanded Uncertainty Calculated by the Spreadsheet *
−27.4	1165.91	0.70	0.70	0.14
−38.0	455.37	0.41	0.42	0.08
−47.5	436.37	0.17	0.16	0.04
−32.1	555.30	0.64	0.65	0.12
−23.1	1129.80	1.01	1.02	0.19
−25.3	1121.384	0.84	0.85	0.16
−24.5	1113.99	0.91	0.91	0.17
−30.9	888.31	0.57	0.58	0.11
−41.0	443.62	0.31	0.31	0.06
−39.8	848.28	0.26	0.26	0.06

* lbs H_2O/MMSCF NG = pounds of H_2O/million standard cubic feet of natural gas.

Table 3. Moisture results and their respective expanded uncertainties in ppmv H_2O.

Dew Point (°C)	Pressure (psig)	Moisture Calculated by PM880 *	Moisture Calculated by the Spreadsheet *	Moisture Expanded Uncertainty Calculated by the Spreadsheet *
−27.4	1165.91	6.08	6.1	1.3
−38.0	455.37	2.40	5.0	1.1
−47.5	436.37	1.64	1.7	0.40
−32.1	555.30	7.47	7.8	1.7
−23.1	1129.80	9.42	9.7	1.9
−25.3	1121.384	7.63	7.8	1.6
−24.5	1113.99	8.32	8.5	1.7
−30.9	888.31	5.40	5.6	1.2
−41.0	443.62	3.46	3.6	0.90
−39.8	848.28	2.14	2.2	0.50

* ppmv H_2O.

To calculate the expanded uncertainty of moisture, in the conservative approach, the expanded uncertainties of 2.0 °C, cell I2, and 1% of the full scale (300 to 3000 psig), cell I3, were considered as input data for dew point temperature and pressure, respectively, considering a coverage factor, k = 2. These conservative values of uncertainty sources already encompass other contributing factors, such as fluctuations in ambient temperature and moisture and long-term sensor drift.

The moisture results calculated by the developed spreadsheet are compatible with those generated by the portable moisture analyzer PM880 regarding the unit in lbs H_2O/MMSCF NG; however, it presents some discrepancies regarding the ppmv H_2O unit. The expanded uncertainty results, k = 2, with a confidence level of 95.45%, are in the range of 20%.

Therefore, a significant advantage is that the manuscript explicitly presents the algorithms, whereas commercial software operates as a "black box" without demonstrating the steps, calculating measurement uncertainty, or assessing compliance.

3.3. Use of the Information on the Measurement Uncertainty in Compliance Evaluation

In the MS-Excel spreadsheet, only the yellow cell H6 corresponding to the upper specification limit of 60 ppmv H_2O, which the Brazilian oil and gas industry is currently considering (Figure 3), needs to be filled in.

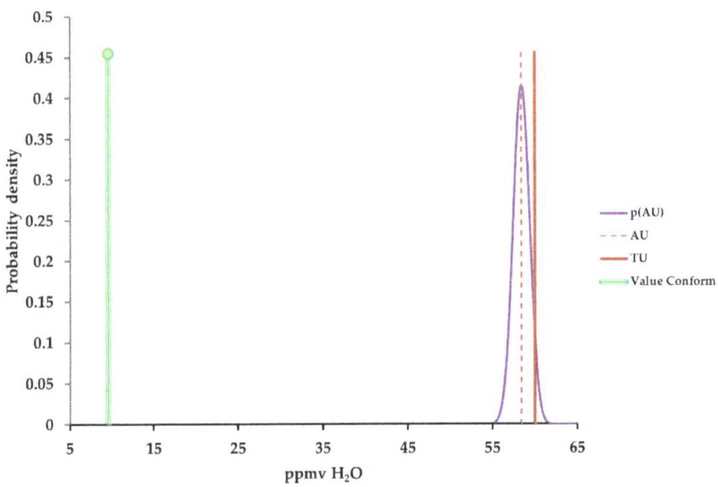

Figure 3. Histogram for 5% significance level, moisture in natural gas.

The guard band was calculated considering a significance level of 0.05 and an upper specification limit, USL. USL calculated the upper tolerance limit for the moisture in natural gas minus $1.64 \times u$.

Figure 3 provided the histogram for the measurement value, the specification, and the guard band limit, considering p(AU)—probability density at the upper acceptance limit, AU—upper acceptance limit, and TU—upper tolerance limit.

The natural gas' measured moisture value, 9.7 ± 1.9 ppmv H_2O, was significantly below the upper tolerance limit, suggesting compliance with the specification. Consequently, this indicates an estimated consumer risk of 0.0% associated with this measurement, as the upper tolerance limit is 58.4 ppmv H_2O.

4. Conclusions

This study developed and validated an MS-Excel spreadsheet for assessing the uncertainty of natural gas moisture based on the Portable Moisture Analyzer—PM880 algorithms.

In this preliminary study, the sources of uncertainty considered were the dew point temperature and the natural gas pressure.

Even with a measurement uncertainty value of around 20%, there is no risk of false compliance assessment, as the moisture levels in natural gas are significantly lower than the upper specification limit.

Regarding future research, it is important to evaluate the impact of sampling variability on measurement uncertainty, as this factor can influence risk assessment. Furthermore, the calculations for measuring the dew point and pressure could be elaborated upon by incorporating more realistic estimates of their uncertainties. The sensitivity analysis on the uncertainty sources (dew point temperature and pressure) could also be included in cases where measurement uncertainty can compromise the compliance evaluation. Another area for future work is comparing the results between different methods.

Supplementary Materials: The following supporting information can be downloaded at: https://www.mdpi.com/article/10.3390/app15052482/s1, Table S1: Uncertainty and Compliance.

Author Contributions: Conceptualization, R.M.M., C.L.B. and E.C.d.O.; methodology, R.M.M., C.L.B. and E.C.d.O.; software, R.M.M., C.L.B. and E.C.d.O.; validation, R.M.M., C.L.B. and E.C.d.O.; formal analysis, R.M.M., C.L.B. and E.C.d.O.; investigation, R.M.M., C.L.B. and E.C.d.O.; resources, R.M.M., C.L.B. and E.C.d.O.; data curation, R.M.M., C.L.B. and E.C.d.O.; writing—original draft preparation, R.M.M., C.L.B. and E.C.d.O.; writing—review and editing, R.M.M., C.L.B. and E.C.d.O.; visualization, R.M.M., C.L.B. and E.C.d.O.; supervision, R.M.M., C.L.B. and E.C.d.O.; project administration, R.M.M., C.L.B. and E.C.d.O.; funding acquisition, R.M.M., C.L.B. and E.C.d.O. All authors have read and agreed to the published version of the manuscript.

Funding: The authors thank for the financial support provided by the scholarship from the Brazilian agency CNPq (305479/2021-0). This study was financed in part by the Coordenação de Aperfeiçoamento de Pessoal de Nível Superior—Brasil (CAPES)—Finance Code 001.

Institutional Review Board Statement: Not applicable.

Informed Consent Statement: Not applicable.

Data Availability Statement: The original contributions presented in this study are included in the article. Further inquiries can be directed to the corresponding author.

Conflicts of Interest: Authors Elcio Cruz de Oliveira and Cesar Luís Biazon were employed by the company PETROBRAS S.A. The remaining author declares that the research was conducted in the absence of any commercial or financial relationships that could be construed as a potential conflict of interest.

References

1. *ASTM D5454-11(2020)*; Standard Test Method for Water Vapor Content of Gaseous Fuels Using Electronic Moisture Analyzers. ASTM International: West Conshohocken, PA, USA, 2020.
2. *ASTM D1142-95(2021)*; Standard Test Method for Water Vapor Content of Gaseous Fuels by Measurement of Dew-Point Temperature. ASTM International: West Conshohocken, PA, USA, 2021.
3. *ISO 18453*; Natural Gas—Correlation Between Water Content and Water Dew Point, 1st ed. ISO: Geneva, Switzerland, 2004.
4. *ISO 6327*; Gas Analysis—Determination of the Water Dew Point of Natural Gas—Cooled Surface Condensation Hygrometers, 1st ed. ISO: Geneva, Switzerland, 1981.
5. Shkarovskij, A.; Volikov, A. Influence of Gas Quality on Efficiency of its Use. *World Appl. Sci. J.* **2013**, *23*, 45–49. [CrossRef]
6. Zou, W. Evaluation of Uncertainty in Detecting Dew Point of Natural Gas by Cooling Mirror Condensation Hygrometer. *Jiliang Xuebao Acta Metrol. Sin.* **2019**, *40*, 150–153.
7. Løkken, T.V. Comparison of hygrometers for monitoring of water vapour in natural gas. *J. Nat. Gas Sci. Eng.* **2012**, *6*, 24–36. [CrossRef]
8. Løkken, T.V. Water Vapour Measurements in Natural Gas in the Presence of Ethylene Glycol. *J. Nat. Gas Sci. Eng.* **2013**, *12*, 13–21. [CrossRef]

9. Gallegos, J.G.; Benyon, R.; Avila, S.; Benito, A.; Gavioso, R.M.; Mitter, H.; Bell, S.; Stevens, M.; Böse, N.; Ebert, V.; et al. An investigation of the comparative performance of diverse humidity sensing techniques in natural gas. *J. Nat. Gas Sci. Eng.* **2015**, *23*, 407–416. [CrossRef]
10. Ren, H.; Yin, A.; Dai, Z.; Liu, X.; Tan, Z.; Zhang, B. Parameter Screening and Optimized Gaussian Process for Water Dew Point Prediction of Natural Gas Dehydration Unit. *Process Saf. Environ. Prot.* **2023**, *170*, 259–266. [CrossRef]
11. Bahmannia, G. The Best Practices in Moisture Control of Sour Natural Gas as Export Gas: Case Study on Line Water Content Analyzer Assessment in a Gas Plant. In *SPE Asia Pacific Oil and Gas Conference and Exhibition*; SPE: Richardson, TX, USA, 2011; p. SPE-147855.
12. Takeda, N.; Carroll, P.; Tsukahara, Y.; Beardmore, S.; Bell, S.; Yamanaka, K.; Akao, S. Trace Moisture Measurement in Natural Gas Mixtures with a Single Calibration for Nitrogen Background Gas. *Meas. Sci. Technol.* **2020**, *31*, 104007. [CrossRef]
13. Vinge, M.A.; Morozov, S.A. State Primary Standard of Relative Humidity of Gases, Molar (Volume) Fraction of Moisture, Dew/Frost Point Temperature, Hydrocarbon Condensation Temperature Units Get 151-2020. *Meas. Tech.* **2023**, *65*, 703–711. [CrossRef]
14. Ren, W.; Jin, N.; Wang, T. An Interdigital Conductance Sensor for Measuring Liquid Film Thickness in Inclined Gas–Liquid Two-Phase Flow. *IEEE Trans. Instrum. Meas.* **2024**, *73*, 9505809. [CrossRef]
15. Bukacek, R.F. *Equilibrium Moisture Content of Natural Gases*; Research Bulletin Institute of Gas Technology: Des Plaines, IL, USA, 1955; Volume 8, ISSN 0097-2886.
16. Saul, A.; Wagner, W. International equations for the saturation properties of ordinary water substance. *J. Phys. Chem. Ref. Data* **1987**, *16*, 893–901. [CrossRef]
17. Joint Committee for Guides in Metrology. *Evaluation of Measurement Data—Guide to the Expression of Uncertainty in Measurement*; International Bureau of Weights and Measures (BIPM): Sèvres, France, 2008; BIPM, IEC, IFCC, ILAC, ISO, IUPAC, IUPAP, and OIML, JCGM 100:2008, GUM 1995 with minor corrections.
18. BIPM; IEC; IFCC; ILAC; ISO; IUPAC; IUPAP; OIML. Evaluation of Measurement Data—The Role of Measurement Uncertainty in Conformity Assessment, JCGM 106:2012. BIPM. 2012. Available online: https://www.bipm.org/documents/20126/2071204/JCGM_106_2012_E.pdf/fe9537d2-e7d7-e146-5abb-2649c3450b25?version=1.7&t=1659083025736&download=true (accessed on 15 January 2025).
19. Williams, A.; Magnusson, B. EURACHEM/CITAC Guide: Use of Uncertainty Information in Compliance Assessment. EURACHEM, 2021. Available online: https://www.eurachem.org/index.php/publications/guides (accessed on 15 January 2025).
20. De Oliveira, E.C. Simplified calibration methodology of chromatographs used in custody transfer measurements of natural gas. *Metrol. Meas. Syst.* **2012**, *19*, 405–416. [CrossRef]
21. De Oliveira, E.C.; De Aguiar, P.F. Comparison of different approaches to evaluate the uncertainty of gas chromatography for natural gas. *Quim. Nova* **2009**, *32*, 1655–1660.
22. Leal, F.G.; de Andrade Ferreira, A.; Silva, G.M.; Freire, T.A.; Costa, M.R.; de Morais, E.T.; Guzzo, J.V.P.; de Oliveira, E.C. Measurement Uncertainty and Risk of False Compliance Assessment Applied to Carbon Isotopic Analyses in Natural Gas Exploratory Evaluation. *Molecules* **2024**, *29*, 3065. [CrossRef] [PubMed]
23. Pennecchi, F.R.; Kuselman, I.; da Silva, R.J.N.B.; Hibbert, D.B. Risk of a false decision on conformity of an environmental compartment due to measurement uncertainty of concentrations of two or more pollutants. *Chemosphere* **2018**, *202*, 165–176. [CrossRef] [PubMed]
24. Separovic, L.; Simabukuro, R.S.; Couto, A.R.; Bertanha, M.L.G.; Dias, F.R.S.; Sano, A.Y.; Caffaro, A.M.; Lourenço, F.R. Measurement Uncertainty and Conformity Assessment Applied to Drug and Medicine Analyses—A Review. *Crit. Rev. Anal. Chem.* **2023**, *53*, 123–138. [CrossRef] [PubMed]
25. Dias, F.R.S.; Lourenço, F.R. Measurement uncertainty evaluation and risk of false conformity assessment for microbial enumeration tests. *J. Microbiol. Methods* **2021**, *189*, 106312. [CrossRef] [PubMed]

applied sciences

MDPI

Article

A Reliability-Oriented Framework for the Preservation of Historical Railway Assets Under Regulatory and Material Uncertainty

Thomas Wailes [1], Muhammad Khan [2,*] and Feiyang He [2,*]

[1] Network Rail, London SE1 8SW, UK
[2] Centre for Life-Cycle Engineering and Management, Cranfield University, Bedford MK43 0AL, UK
* Correspondence: muhammad.a.khan@cranfield.ac.uk (M.K.); feiyang.he@cranfield.ac.uk (F.H.)

Abstract: Preserving historical railway assets presents a complex systems challenge, in which uncertainties in material performance, structural degradation, and regulatory requirements directly impact long-term reliability and operational continuity. Traditional maintenance practices often limit the use of modern materials, introducing inefficiencies, increased lifecycle costs, and higher failure risk due to material ageing and environmental exposure. This study proposes a reliability-informed preservation framework that supports the integration of contemporary materials into historical railway infrastructure while accounting for legal, material, and procedural uncertainties. The framework is validated through two industrial case studies, each reflecting different regulatory and operational constraints. The first case demonstrates the successful substitution of timber with certified PVC cladding on a non-listed signal box, achieving improved durability, reduced maintenance intervals, and enhanced system reliability. The second case explores an unsuccessful attempt to replace decayed timber gables with aluminium, in which late-stage planning misalignment, underestimated risks, and uncertainty in approval outcomes led to a significant cost increase and reduced reliability regarding delivery. By systematically applying and evaluating the framework under real-world conditions, this research contributes to engineering asset management by introducing a structured method for mitigating regulatory and material uncertainties.

Keywords: historical asset preservation; railway infrastructure maintenance; contemporary materials; listed building consent; planning approval

1. Introduction

The railway system of Great Britain has a long history, starting in the 18th century [1,2]. After over 300 years of development, its infrastructure asset portfolio and range have become substantial. The Historical Railways Estate (HRE) has a collection of over 3100 structures and assets that were once part of Britain's rail network. More than 1450 items have survived from all periods of the railway's history and are listed as being of architectural or historic interest—including over 380 station buildings, 639 viaducts and bridges, 65 tunnels, 67 signal boxes, four hotels, 40 memorials, gates, and clocks, and over 250 other buildings or structures [3,4], which contribute to social and economic development through transport and tourism [5,6].

Given the substantial size and age of the asset portfolio, managing and maintaining these historical assets pose significant challenges. The historical nature of many of these structures, combined with the lack of comprehensive records, operational restrictions, and preservation requirements for repairs and renewals, presents a complex scenario.

Academia and industry [7–12] have made some progress in maintaining historical assets. For instance, Borges (2020) introduced an indicator-based approach to assess railway heritage, highlighting systemic preservation challenges [13]. Regarding laws and regulations [14], the national planning policy framework, Section 12, defined over 1400 railway assets as listed buildings and structures. Nearly all modifications to them require some form of listed building and planning consent, as detailed in the Planning (Listed Buildings and Conservation Areas) Act 1990 [15]. In addition, the railway is the only industry with specific legislation protecting its heritage. The government has introduced the Railway Heritage Act [16] to enable historically significant artefacts and records operational within today's railway industry to be 'designated'—protected and saved for the nation so that future generations can understand the railway's history in Britain. The industry also has various maintenance standards, such as NR/GN/CIV/100/05 [3]. They stipulate that short-term solutions or repairs and cosmetic changes likely to cause long-term damage, such as strengthening, signage, propping, and servicing installations, should be avoided according to the law and standards.

Nevertheless, real maintenance activities commonly contradict the law and standards [17,18]. For example, additional installations of modern monitoring systems such as CCTV can cause physical damage to the fabric of a building as well as produce a visual eyesore. NR/L3/CIV/023 Work Instruction [19] is another standard for the assessment of footbridges. As a result of the assessment standard, it denotes the need for immediate action as soon as a defect occurs, regardless of whether it is listed or not. However, the study conducted by Network Rail [20] found that all of the reviewed Level 2 assessments of underbridges demonstrated that the capacity calculated using Level 1 techniques from the standard can be significantly improved using a different standards analysis. Following current assessment standards, the analysis involves complex equations that are not always applied consistently and tend to deliver a low assessed capacity, leading to unnecessary, unsightly, and costly recommendations for strengthening works, contradicting planning laws and the NR/GN/CIV/100/05 standard. It is important to address these contradictions as early as possible. When high-quality heritage and design advice is received at the beginning of each project or repair, unexpected costs can be minimised, with a lower risk of lost time in consultations with local planning authorities or statutory national heritage bodies.

Some research studies have considered ageing materials in heritage structures [21–23]. Iron and steel were frequently used for stations and buildings, incorporating both types of materials with the addition of timber. However, unlike masonry, iron, steel structures, and wooden elements are more susceptible to corrosion and stress, which affects their longevity. The Railway Heritage Trust (RHT) [24] highlights that the materials' extensive maintenance requirements and their associated structural and aesthetic deterioration pose significant issues for railway preservation today. However, only limited studies have been undertaken to address the replacement of unique or minor elements of original material construction using modern emerging materials such as polymers [25–28], fibre-reinforced polymer (FRP), and glass reinforced plastic (GRP) [29–33].

Baxter [24] focused on the history and maintenance requirements of the humble canopy daggerboard. The study investigated the use of FRP for station canopy fascia daggerboards. Timber daggerboard is a design element that serves the vital, practical function of carrying run-off rainwater away from the canopy structure and improving ventilation inside the canopy. Its ornamental potential was quickly recognised, and it can now be considered as one of the archetypal features of a traditional British railway station in the popular imagination. As a result, there are more than one thousand stations on the British rail network with canopy fasciae and daggerboards. The boards exposed to weathering result in the appearance of degradation over time. It can also lead to timber rotting and material failure, resulting in individual boards falling away, posing a health and safety risk to both trains and the public. Maintenance work on timber daggerboards requires tracking 'possessions', impacting the infrastructure availability and asset reliability, incurring considerable costs, and increasing the health and safety risks to which staff are exposed, both from working at heights and from electrocution from overhead contact systems. However, such work is frequently deferred or cancelled due to cost and time implications, resulting in continued deterioration.

Another important consideration is the authenticity or originality of the construction material of the historic asset. Timber daggerboards have an estimated lifespan of thirty years. According to the standard, repair and maintenance are required every ten to twelve years, but it is unlikely that any nineteenth-century timber can survive in a valance canopy.

This highlights another challenge when managing historical assets. Due to the age, longevity, and period in which many historical assets were designed, constructed, or manufactured, full and comprehensive asset data are not always available. Renewals or previous repairs of minor elements are frequently not recorded or available. This leads to difficulties in predicting how structures and assets with many existing major structural defects will behave, even under ideal conditions, in which monitoring and damage prediction remain inherently challenging [34–36]. Network Rail [20] stated that modelling historic assets is exceptionally complex and prone to inaccuracy, as a combination of corrosion, weathering, and previous poor-quality materials or repairs compromises materials. Mcmahon, Zhang, and Dwight [37] identified the main categories of missing data types in rail systems, applicable to the historical assets found at Network Rail (NR), such as from the commissioning of a system and from maintaining and operating an asset. The research proposed, analysed, and discussed AI-led advanced models and algorithms applied to recover the missed data. This is a good approach and an in-depth study in filling performance and data gaps for historical assets, although it has not been tested for specific asset types, materials, or ages. Similarly, Adams et al. [38] understood the impact of weather events on asset performance and the failure of the historic sea wall at Dawlish. They produced a multi-hazard risk model with cascading FPW to remove gaps in the historical data.

The existing maintenance and asset management literature primarily focuses on isolated asset types or failure mechanisms, often neglecting the broader system-level considerations associated with historical infrastructure. Recommendations are typically presented as linear proposals for material strengthening or replacement, with limited attention to the regulatory uncertainties and planning processes specific to heritage assets. Moreover, the potential reliability benefits and lifecycle cost reductions associated with the integration of contemporary materials and design techniques remain underexplored.

This study addresses this gap by evaluating the relationship between material innovation, lifecycle cost performance, and the uncertainty involved in securing planning approvals for changes to historically significant railway assets. A reliability-oriented theoretical framework is developed to support the strategic preservation of historical railway infrastructure, accounting for material, regulatory, and procedural uncertainties in a structured and scalable manner.

2. Methodology

The methodology was staged into three parts to understand further the challenges faced when preserving heritage assets, which were highlighted in the literature review, and to develop a framework for modern preservation in the future.

As illustrated in Figure 1, a targeted survey was initially sent to various subject matter experts and practitioners in the railway industry. The feedback received helped develop the theoretical framework. It also led to a more targeted set of personal interviews. Then, the theoretical framework was developed and applied to two industrial case studies via construction projects delivered by Network Rail UK. The effectiveness of the framework was investigated via personal interviews.

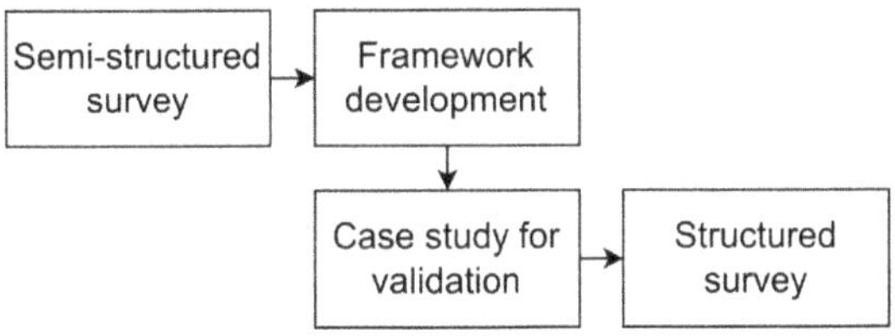

Figure 1. Methodology flow chart.

2.1. Semi-Structured Survey

Given the impracticality of interviewing all employees at NR, a semi-structured targeted survey was deemed the most effective approach for ensuring a representative sample while minimising bias. This method was selected to identify existing gaps in real industrial applications and support the development of a framework to address contemporary challenges in historical asset preservation. The survey was distributed through Microsoft Forms (Microsoft 365 Version 2405), and the responses were analysed using Excel (Microsoft 365 Version 2405).

Ten participants were chosen from key departments relevant to this study, including asset management, the technical authority, town planning for listed consent, and maintenance and project delivery teams responsible for design and implementation. This multidisciplinary approach ensured that the perspectives captured reflected historical asset preservation's practical, technical, and regulatory aspects.

Although this selection does not encompass all roles within NR, it was expected to enhance the sample's representativeness and the findings' reliability. Given the diverse nature of NR's asset portfolio and the regulatory frameworks governing historical structures, comparisons with local planning authorities (LPAs) were also considered valuable for identifying procedural inconsistencies and best practices in historical asset management.

2.2. Framework Development

Based on the survey results, a theoretical framework has been developed to establish a strategic approach for repairing and maintaining historical assets. The framework facilitates continuous improvement, ensuring that buildings and structures retain their functional and historical significance without becoming static heritage artefacts.

Given the operational demands of historical railway assets, adaptation is essential for long-term viability. This framework supports sustainable preservation by balancing historical integrity with contemporary performance enhancements. It provides a structured methodology for implementing changes that uphold the asset's architectural and historical value while optimising performance, standardising data collection, and reducing uncertainties in the listed consent process.

2.3. Case Study Validation

The proposed theoretical framework was applied to real-world asset management projects within NR. Two case studies implemented the framework to assess cost/benefit outcomes and identify inconsistencies in the listed consent process across different local authorities.

3. Initial Survey Results and Discussion

A semi-structured questionnaire was developed based on gaps identified in the literature. To process the collected data, responses were categorised into simplified groups, making it easier to interpret the experiences and challenges faced by participants in the field. The results were quantified in percentage terms to identify trends and patterns.

3.1. Common Causes of Asset Failure and Implications for Material Substitution

Firstly, respondents were asked about the leading causes of asset failure affecting listed and historical structures within NR. The responses were analysed and categorised into overarching failure mechanisms, as visualised in Figure 2.

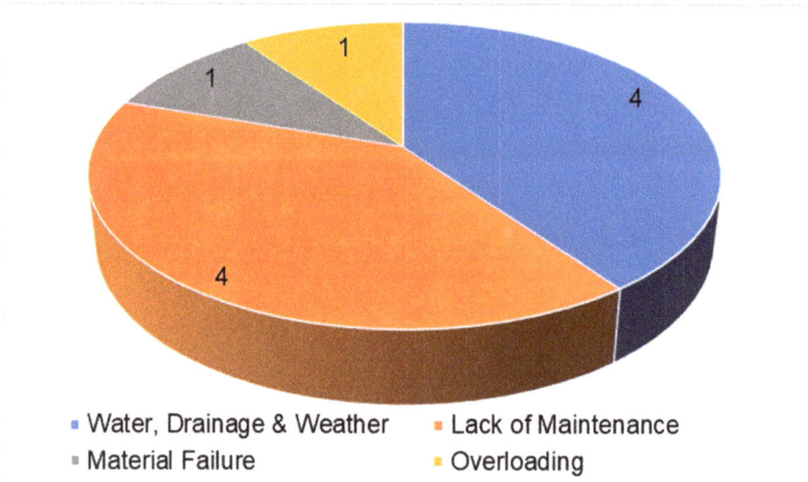

Figure 2. Participants' response to the survey to establish the leading type of asset failure of historical assets.

The results indicate that the most significant contributors to asset deterioration are insufficient and irregular maintenance throughout the asset's lifecycle and damage caused by water ingress and weathering, both of which frequently lead to structural failure.

These findings align with the literature [39], highlighting historical assets' vulnerability to degradation processes such as corrosion and erosion due to the materials used in their original construction. Additionally, inadequate maintenance of drainage systems and protective coatings was identified as a key factor, supporting previous studies, such as one that suggested replacing station canopy daggerboards of timber with FRP to mitigate weather-related deterioration [24].

Comparing these responses with participants' views on material changes, there is a clear correlation between the elements most affected by weathering, degradation, and failure and their architectural and aesthetic significance. This shows the importance of preserving these components to ensure both structural integrity and compliance with legal protection requirements.

The RHT commissioned a report on using fibre-reinforced plastic (FRP) to replace station fascia daggerboards, which has significant advantages over traditional wooden elements prone to material failure. Building on this, the survey included a question seeking suggestions on which elements of a listed asset would benefit from replacement with contemporary materials.

3.2. Identification of Asset Components Suitable for Material Substitution

The responses in Figure 3 identify mainly two elements, glazing systems and timber elements, that benefit the most. They have several key features that experience rapid or frequent degradation, primarily due to the inherent properties of their original materials. This deterioration affects asset performance and leads to increased maintenance frequency and higher long-term costs. This higher maintenance requirement can be mitigated if material and manufacturing processes are changed.

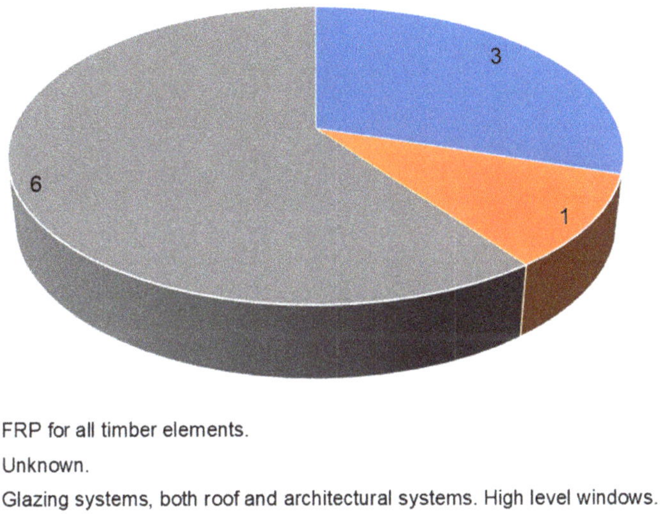

Figure 3. Participant response to the survey to help establish which elements of an asset could benefit most from a contemporary change in material or manufacturing process.

3.3. Industry Awareness and Knowledge Gaps

Following this question, the participants were asked whether they perceived a lack of awareness within the industry regarding contemporary techniques and emerging technologies. The responses included six "yes" responses, three "no" responses, and one response indicating "unknown". Combining the responses from Figure 3 indicates that most of the participants agreed on which elements of historical assets would benefit from a material change to improve their preservation. However, there was limited awareness of the specific contemporary materials and techniques available or previously implemented in similar contexts. This is a critical gap for which further research and industry focus could lead to measurable improvements in cost efficiency, maintenance time, and long-term asset sustainability.

A broader set of questions was included to explore potential correlations between industry competencies, the lack of comprehensive knowledge regarding contemporary materials and manufacturing techniques, and the absence of accessible records on previous successful interventions for historical assets. The absence of such information may negatively impact the planning approval and consent process.

Additional questions were posed to understand why planning applications proposing changes to historical assets are frequently rejected. Participants were asked whether they perceived a connection between proposed modifications and planning rejections. The survey responses are illustrated in Figure 4. Many of the respondents (four out of ten) frequently perceive a relationship between proposed material or technological changes and the rejection of planning applications. This suggests that stakeholders experience significant challenges in gaining approval for modifications to historic assets.

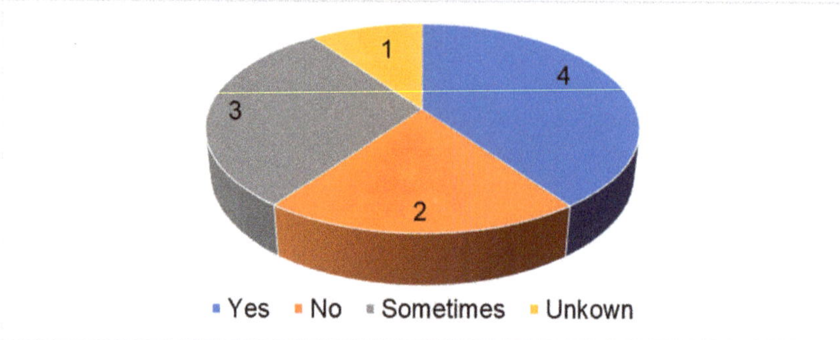

Figure 4. Participants' response to the survey to establish relationships between proposed changes and the success of planning applications.

Additionally, three respondents selected "Sometimes", indicating that while this issue is not always a barrier, it occurs often enough to be a concern. The two "No" responses suggest that, in some cases, proposed changes do not necessarily lead to rejection, possibly depending on the nature of the modification or the local authority involved. These responses reinforce the challenges of integrating contemporary materials into historical asset preservation. The results suggest that planning authorities may have varying acceptance levels for technological and material updates.

The pie chart in Figure 5 presents responses to the following question: 'Do you feel that due to a gap in the Network Rail processes or design documentation, there is a need to duplicate activity when making a planning application to the local authority when working on a listed asset?'

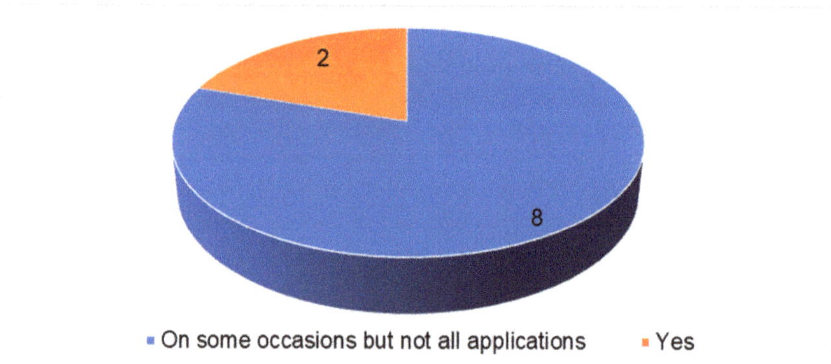

Figure 5. Participant response to the survey to establish any gaps in NR processes or documentation and whether activity is duplicated as a result.

The results indicate that the duplication of efforts in regards to planning applications for listed assets is a recurring issue, though not universally experienced. Eight respondents stated that duplication occurs in some cases, suggesting that while the planning and approval process does not always require redundant work, inconsistencies in Network Rail's internal processes or documentation gaps can lead to unnecessary repetition.

A smaller portion of respondents indicated that duplication is a frequent or systematic issue, implying that inefficiencies exist in aligning Network Rail's documentation with local authority requirements. This may be due to differences in design submission formats, planning application procedures, or coordination between departments. The findings suggest that a more streamlined and standardised approach to planning applications is needed to minimise redundant work. Improving the coordination between the rail asset management body and local authorities could help ensure the design documentation meets planning expectations. Further investigation into specific cases in which duplication is required may help identify common factors contributing to inefficiencies in the process.

3.4. Coordination Barriers and Documentation Misalignment in the Planning Process

The misalignment is further reflected in the specific challenges identified by respondents in Table 1. Several key factors contribute to the duplication of work when obtaining design approval from the railway infrastructure authority and planning consent from local authorities. These issues primarily stem from inconsistencies in documentation requirements, differing expectations between regulatory bodies, and a lack of coordination during the approval process.

Table 1. Participants' responses to the survey to provide reasons for why duplication of work has been performed by industry practitioners, offering initial insights into what needs to be addressed.

What Do You Feel Is the Main Reasoning Behind the Duplication of Work Required to Gain Both Design Approval with Network Rail and Obtain Local Authority Planning Consent?
Response 1.—Planning authorities do not always like the documents in NR format
Response 2.—Designer does not provide the level of information that a non-designer needs, detailed drawings are not always required in some cases, so we have to do more submissions, albeit simplified
Response 3.—Rather than simply stating what they will/will not accept, we end up going through multiple versions of a design until we reach one that they are happy with. Sometimes it takes as many as 5/6 iterations of modifications to make them happy. This being in spite of them saying they would have originally accepted it.
Response 4.—A lack of understanding of heritage issues by the supply chain.
Response 5.—Approval by NR does not necessarily accord with the approval by the LPA.
Response 6.—It feels like there are a number of layers to obtain consent. In some cases, we needs to obtain approval/comments from design review panel and heritage trust. Comments can be subjective and even conflicting
Response 7.—Depends very much on the engagement between asset team and town planning team early on in the remitting/scoping process.

One recurring theme is the misalignment between the design documentation used within the railway sector and the expectations of LPAs. Several respondents noted that planning authorities often do not accept documents in the standard format used by the infrastructure management organisation, leading to additional submissions or modifications. Similarly, the level of detail required in planning applications varies, with some submissions requiring highly detailed drawings while others may need more generalised design information. This discrepancy leads to inefficiencies and additional work to tailor submissions based on differing requirements.

Another challenge identified is the iterative nature of the planning process, and approvals are often contingent on multiple revisions. One respondent noted that planning authorities sometimes fail to provide clear guidelines up front, resulting in multiple design iterations—sometimes up to five or six revisions—before reaching an acceptable solution. This lack of clear direction increases the time and effort required for approvals, adding unnecessary complexity to the process.

A lack of understanding of heritage-specific issues within the supply chain was also cited as contributing to additional work. If those involved in the design and approval process lack sufficient knowledge of historical asset preservation, applications may fail to address key heritage considerations, leading to further reworks. This is compounded by the fact that approvals from the railway authority do not always align with those of the LPAs, creating additional layers of consent that must be navigated.

Some respondents also pointed to the complexity of the multi-layered approval process, in which additional stakeholders, such as design review panels and heritage trusts, must provide comments or grant their approval before planning consent is granted. Sometimes, these comments are subjective or conflicting, further complicating the approval process and requiring additional iterations to satisfy all involved parties.

Finally, the level of early engagement between asset management teams and town planning teams plays a crucial role in minimising duplication. One response emphasised that the extent of early collaboration in the remitting and scoping phase can significantly impact the efficiency of the approval process. Insufficient engagement at the outset may result in misalignment between design proposals and planning requirements, leading to delays and additional revisions.

Overall, the responses suggest that improving coordination between regulatory bodies, standardising documentation requirements, and ensuring early stakeholder engagement could help reduce duplication and inefficiencies in the planning approval process for historical railway assets.

4. Development of a Preservation Framework for Historical Railway Assets

A theoretical framework has been developed, as shown in Figure 6, to establish a structured strategy for repairing and maintaining historical assets. The framework consists of five iterative stages—Empathise, Define, Ideate, Prototype, and Test—which together form a continuous improvement cycle. It is designed to facilitate managed change, balancing preserving historically significant features with the practical requirements of continued use.

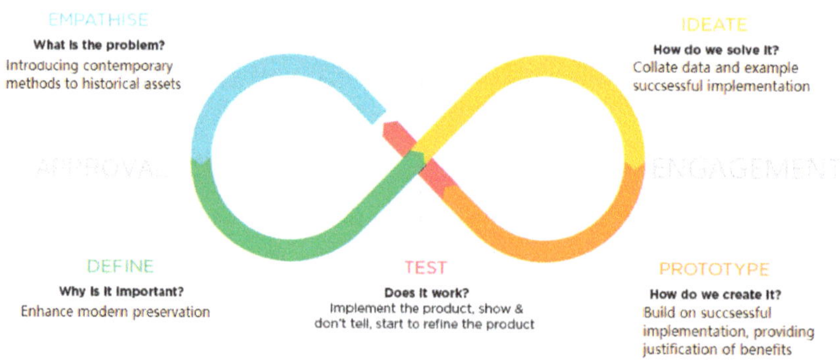

Figure 6. A framework for enhancing preservation through innovation.

Given that structures within the operational railway network must adapt to remain viable, the framework aims to ensure that any proposed modifications retain key historical characteristics while simultaneously enhancing asset performance and data collection processes. This approach reduces uncertainties in the listed consent approval process by improving the accuracy of recorded information.

4.1. Empathise

The preservation of historical assets often lacks a standardised process or clear guidance, leading to subjectivity and conflicting opinions regarding proposed changes. This results in multiple layers of approvals, adding complexity to the consent process. The absence of documented evidence supporting the benefits of contemporary methods can further restrict their adoption in maintaining and conserving historical assets.

4.2. Define

When interventions are required for a historical asset, it is essential to review prior applications of contemporary methods in comparable scenarios and explore existing materials and techniques used in other industries that could enhance modern preservation practices and extend the asset's lifecycle performance [40].

Engagement with independent third-party experts at an early stage of the design and planning process can help establish a strong foundation for project proposals. By incorporating lifecycle benefits into the initial submission to LPAs, it becomes possible to address uncertainties and the lack of supporting evidence, ultimately improving approval rates for contemporary material applications.

4.3. Ideate

It is recommended that the railway infrastructure authority expand its existing property and land database to document historical assets systematically. This database should incorporate records of successfully completed preservation projects, ensuring that relevant data are linked to similar assets. By coding these records effectively, the system could provide a comprehensive reference for projects that have successfully integrated material changes, facilitating informed decision making in future planning and conservation efforts.

4.4. Prototype

As design approvals required under railway infrastructure standards do not always align with the approval documents mandated by LPAs, discrepancies often arise due to differences in documentation formats or a lack of mutual understanding. To address this, the design submission process should be customised for each historical asset, integrating the proposed material changes within a format that aligns with LPAs' expectations. This adaptation could reduce design and documentation redundancies while increasing approval rates for contemporary interventions.

To further reduce planning revisions, the framework proposes a repeatable strategy that includes early engagement with planning officers during the scoping phase and the use of previously approved cases as reference. By anticipating approval expectations and tailoring submissions accordingly, it is possible to streamline the approval process and minimise the likelihood of multiple design iterations.

4.5. Test

As projects incorporating contemporary materials are successfully delivered, these records should be systematically documented within the historical asset database. Accumulating proven case studies will help eliminate subjective and conflicting opinions regarding material changes by providing a repository of accepted products with quantifiable benefits in modern preservation.

The framework is designed to function as a continuous improvement cycle, in which each completed project contributes to refining material performance strategies for historical assets. By maintaining a loop of iterative enhancement, this approach supports the progressive integration of contemporary preservation methods, ensuring both regulatory acceptance and improved long-term asset resilience. It supports enterprise-level decisions by improving the reliability of material planning and reducing uncertainty in the maintenance of historically significant assets.

5. Case Studies

Two case studies were examined to evaluate the proposed framework's applicability.

5.1. Case Study 1: Willesden Carr South Signal Box Renovation

A signal box awaiting refurbishment was identified, with an existing repair design dating back to 2014. The structure exhibited extensive rot in its primary timber structural members and side panelling, as documented in Figures 7 and 8. The observed deterioration is consistent with the research findings on common asset failure mechanisms, particularly the impact of insufficient maintenance and material degradation over time.

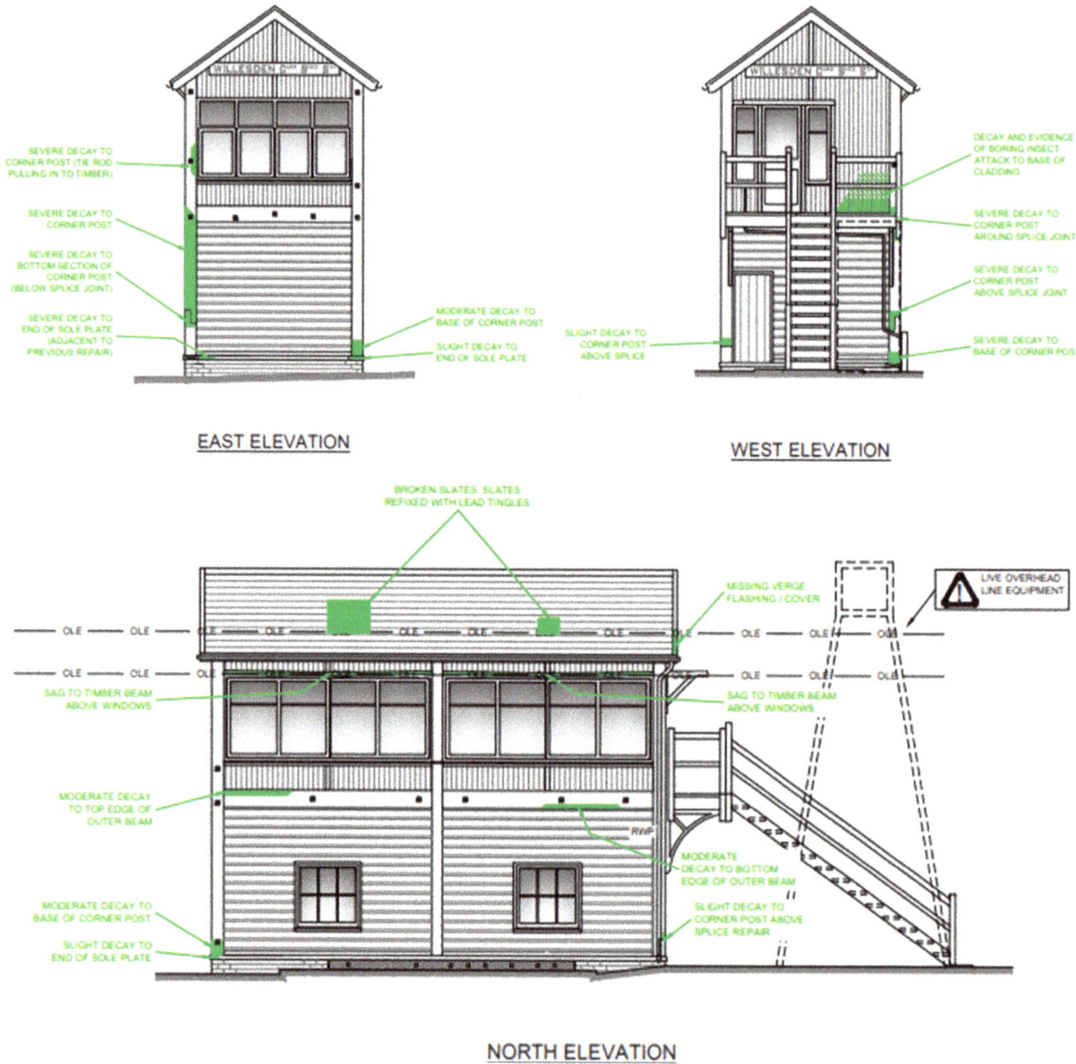

Figure 7. *Cont.*

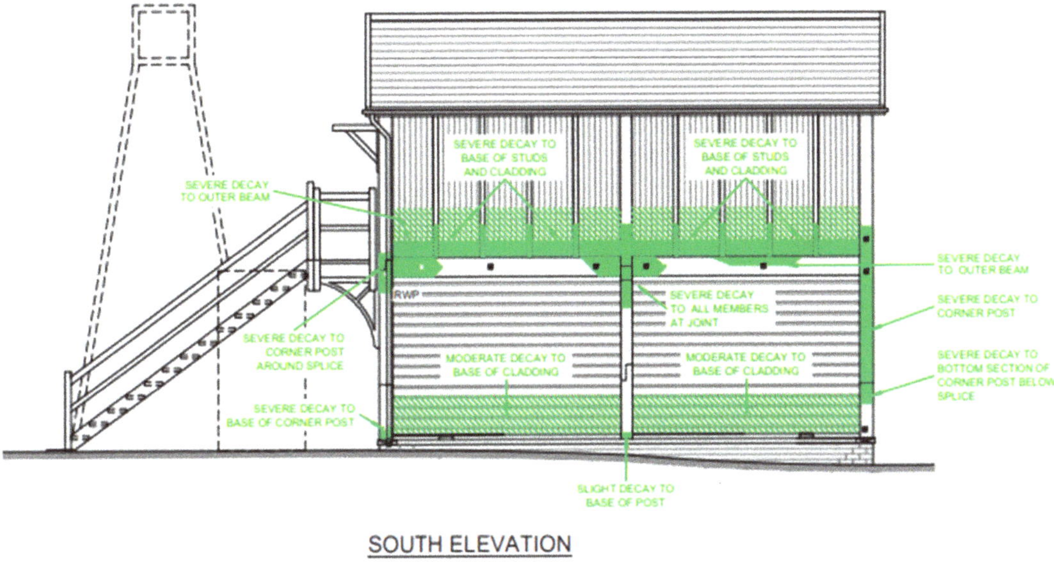

Figure 7. Drawings of details and extent of failed material elements of the signal box.

Figure 8. *Cont.*

Figure 8. Photograph of the inherited state of the signal box asset with prominent material decay and failure and the temporary support scaffolding.

A low-load assessment was conducted and confirmed that short-term, inadequately designed temporary works consisting of scaffolding had been previously implemented to maintain the asset's operational status. This approach, frequently employed as a reactive solution, often results in further degradation and unintended damage to historical railway assets.

The existing repair design requires the following:

- The existing paint finish should be stripped throughout. Renovate or renew all defective timbers, including (but not limited to) the replacement of the areas indicated in the drawings. All renewals to the main frame members should be hardwood with a 'like for like' section size. Replace main timber members and infill framing. Prepare and paint all existing and new timber.
- Remove existing cladding. Renovate the timber structure (see item 1.3). Reclad on a like-for-like basis. North, east, and west elevation: Prepare all cladding for redecoration. Renovate/renew any defective timbers like-for-like (including the replacement of an area of cladding affected by boring insects at an elevation of W above the landing). Paint all existing and new timber.

This case study provided an opportunity to apply the proposed framework to an industrial project in a live railway asset environment. The wider construction industry offers a broad range of contemporary building materials, many of which have the potential to enhance asset management within the rail sector.

By applying the framework, a material change could be implemented without full local LPA approval. A review of the product acceptance standard confirmed that building materials were outside the scope of this document. Following this review, a reassessment of the 2014 design was initiated, incorporating wood-grain-effect PVC cladding as an alternative to traditional timber. As a commercial off-the-shelf product, the PVC cladding met all necessary British Standards certification and, through an inquiry with the original equipment manufacturer (OEM), could be customised to match the original RAL colour scheme of the signal box. The use of substitute materials such as PVC in heritage contexts is increasingly supported in international conservation practice [41–43]. This aligns not only with the findings of the Railway Heritage Trust [24] but also with international preservation guidance, such as that issued by the U.S. National Park Service (2023) [44], which supports the use of substitute materials in historic buildings when appropriately matched. The revised designs for replacing the previous wooden cladding with PVC cladding like-for-like and implementing the new material on site are illustrated in Figure 9.

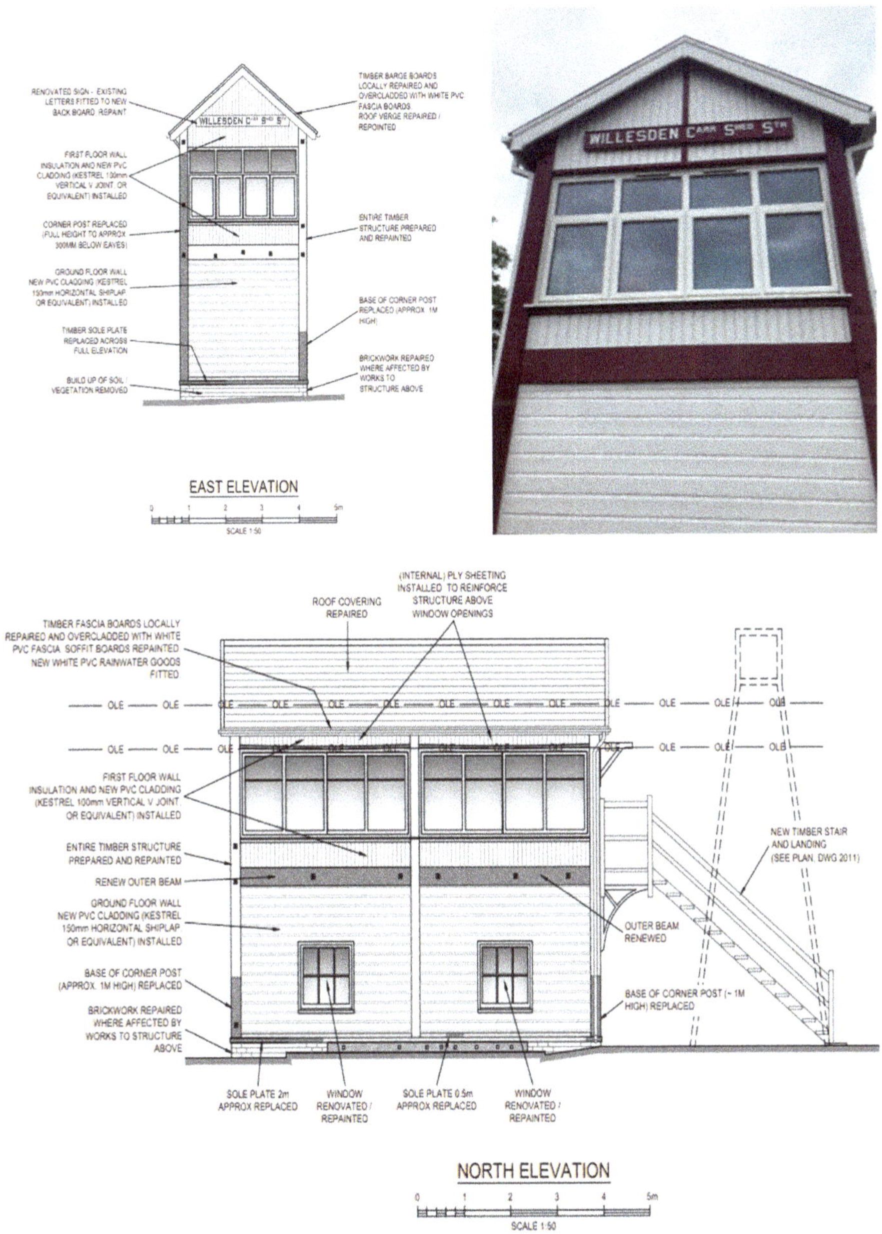

Figure 9. *Cont.*

Figure 9. Drawings of approved repairs for the signal box and on-site application of the new cladding system.

In contrast, PVC cladding offered significant advantages in both material consistency and installation efficiency, as illustrated in Table 2. Unlike timber, it did not require seasonal restrictions for painting, as painting timber can only be undertaken in specific weather conditions. This requirement extended the overall installation timeline, adding to project costs. The adoption of PVC cladding eliminated these constraints, streamlining the installation process and contributing to long-term cost savings in both maintenance and material performance.

The material change also contributes to a substantial reduction in ongoing maintenance costs for the asset. According to the OEM's paint system requirements, shiplap-treated timber cladding requires the stripping and reapplication of protective coatings every 15 years. However, due to budgetary constraints, there is no guarantee that these maintenance cycles are carried out within the specified timeframe. As a result, had timber been used, the signal box could have deteriorated once again, leading to the same state of disrepair.

In contrast, the PVC cladding provides a 60-year maintenance-free guarantee, eliminating the need for four costly and labour-intensive maintenance cycles over the same period. This ensures the long-term preservation and continued operation of the asset while significantly reducing lifecycle costs.

The final outcome was aesthetically consistent with the historical asset, successfully preserving its significant features, as demonstrated in Figure 9. The project also facilitated the removal of temporary structural supports, effectively extending the operational lifespan of the asset by 60 years. Following the completion of the works, the RHT was contacted, and the project has been accepted as a candidate for the National Heritage Awards.

In alignment with the proposed framework, the completed project documentation, product data sheets, and design details can now be incorporated into NR's property and land database. This will serve as a reference catalogue of successfully executed material changes, enabling data integration for similar assets. By establishing a record of tangible benefits, this resource will support future proposals for material substitutions, particularly for listed assets, reducing subjective objections to contemporary materials and reinforcing the rationale for selecting alternatives over traditional materials.

Table 2. Cost comparison of the 2014 and 2024 designs.

Cost to Implement 2014 Repair Design		Cost to Implement Design Incorporating Contemporary Material in 2024	
Shiplap treated Redwood 150 mm × 15 mm—134 @ GBP 16	GBP 2144	300 mm × 15 mm Textured Shiplap Cladding White × 67 @ GBP 39	GBP 2613
Dulux wood primer undercoat GBP 22.89 for 13.5 m^2 coverage—100 m^2 in total	GBP 183.12		N/A
Dulux Trade Weathershield Quick-Drying Exterior Satin Tinted Colours GBP 36.84 for 13.5 m^2 coverage 100 m^2 in total	GBP 294.72		N/A
Fixings and Sundries	GBP 350	Fixings and Sundries	GBP 124
SX Contractors Silicone × 25 Tubes LMN—Neutral—White @ GBP 2.75	GBP 68.75	SX Contractors Silicone × 25 Tubes LMN—Neutral—White @ GBP 2.75	GBP 68.75
Installation costs—3 skilled men × 24 weeks	GBP 112,000	Installation costs—2 skilled men × 16 weeks	GBP 74,000
Total =	GBP 115,041		GBP 76,806

5.2. Case Study 2: Stoke-on-Trent Station Gable End Replacement

Another case was initiated in March 2020 in response to the deterioration of two gable ends at Stoke-on-Trent Station as shown in Figure 10, which posed a significant safety risk to passengers and train operations. Due to the structural instability, the issue was escalated to the project team, who conducted an initial cost forecast. The estimated budget was set at three million GBP, primarily based on a comparable project previously undertaken at Preston Station. However, this estimate was largely analogous, relying on similarities in scope rather than a detailed site-specific assessment.

Figure 10. Condition of Stoke-on-Trent station gable ends before commencement of the replacement.

One of the significant challenges faced in this case was the discharge of Listed Building Consent (LBC). During the design development phase, it was initially assumed that an aluminium frame would be an acceptable solution, as it had been approved in a similar project. However, discussions with the planning officer confirmed that only a like-for-like timber replacement would be permitted, and the proposal to use aluminium would be rejected. Duguid et al. documented similar regulatory challenges in the Ordsall Chord project [45].

The asset engineer explored various softwood alternatives, but Accoya® was ultimately selected to maximise design life, as shown in Figure 11. Accoya® is an engineered timber, for which pine trees are harvested at maturity (approximately 30 years old) and subjected to an acetylation process using acetic acid. This modification enhances the dimensional stability and durability of the wood [46].

Figure 11. The sections of the Accoya® frame and the new wood frame under construction.

The procurement of Accoya® contributed approximately one million GBP to the total project cost. Had alternative materials been accepted by the planning officer, significant cost savings could have been achieved. Additionally, the requirement for a like-for-like timber replacement resulted in a two-year delay as the design underwent an extended approval process.

Beyond the material selection challenge, the project also encountered difficulties in terms of the construction methodology. The original plan proposed lifting the wooden frame in one piece through a piecemeal approach. However, this method proved to be infeasible due to site constraints, including overhead line equipment (OLE) and limited headroom, which prevented the supplier from executing this approach as a cost-saving measure. These constraints necessitated an alternative construction strategy, further complicating project execution.

The challenges associated with the material change, design delays and modifications, and the constructability of a single wooden span over an operational railway have resulted in significant cost implications, as detailed in Table 3. Beyond the initial financial impact, the long-term maintenance obligations of the wooden frame must also be considered.

The timber structure requires repainting every 15 years, posing the same logistical and operational challenges as the original installation, particularly due to the complexities of working over a live railway track. Had aluminium been approved, these maintenance interventions could have been avoided for 40 years, reducing disruptions, costs, and long-term risks while ensuring greater asset longevity.

Table 3. Project cost profile throughout the project from its conception to its current milestone.

September 2023			
Authority Granted	GBP 3.175m	Approved Final Cost	GBP 6.755
December 2023			
Authority Granted	GBP 4.540m	Approved Final Cost	GBP 7.556m
June 2024			
Authority Granted	GBP 7.371m	Approved Final Cost	GBP 7.371m

In summary, the case study review and interviews with the project team led us to conclude that alternative procurement and town planning strategies could have mitigated many of the challenges encountered. It was further estimated that an approved final cost (AFC) of approximately six million GBP could have been achievable with a different approach, highlighting the potential for cost efficiencies through improved material selection, planning engagement, and procurement processes.

5.3. Reflections on Material Selection and Planning Challenges in Asset Conservation

5.3.1. Regulatory Challenges in Historical Asset Preservation

Renewal and repair work on historical assets require separate legal consent, particularly when modifications reach a certain degree of change. However, the term "specified degree" remains subjective, with no clear regulatory framework defining what alterations are permissible on a case-by-case basis. Approval from local planning authorities (LPAs) is essential for any modification. Still, the planning system is structured to present multiple barriers to change, affecting project costs, deliverability, sustainability, and innovation.

A key issue in the preservation of historical assets is the prevailing focus on ensuring that alterations remain "in keeping" with the existing architectural character of an area. While this principle protects historical integrity, it has negatively impacted material innovation and design advancements for over 300 years. The challenges identified in this study align with survey participants' experiences managing Network Rail's historical assets, in which restrictions on material choices often hinder the adoption of modern, more durable solutions.

5.3.2. Potential of Contemporary Materials in Historical Preservation

Despite these challenges, the research findings demonstrate promising initial results in adopting contemporary materials to enhance historical asset performance. The case studies indicate that it is possible to preserve aesthetic and significant historical features while simultaneously improving total lifecycle costs and aligning with both legal requirements and governmental sustainability policies.

5.3.3. Improving Technical Communication for Material Substitutions

One of the key challenges in modern asset preservation is the complexity of communicating engineering benefits to decision makers. The case study findings indicate that those responsible for granting approvals are often not engineers, leading to potential misinterpretations of technical data. Overly detailed design drawings and empirical performance data may fail to convey the practical benefits of material substitutions in a way that aligns with preservation objectives and continued asset functionality.

The study suggests that documentation similar to that in FRP daggerboard research [24] could effectively demonstrate the viability of contemporary materials in asset renewal. Reports endorsed by respected professional organisations can provide the necessary justification for material substitutions, increasing the likelihood of obtaining LPA consent.

5.3.4. Case Study 1: A Positive Outcome for Material Innovation

The validation of the framework through Case Study 1 yielded a positive outcome. The signal box in this case had the advantage of not being a listed asset, although its historical and operational significance suggests a high probability of future listing. This status enabled a simplified approval process, allowing the introduction of contemporary materials without the additional challenges of obtaining LPA consent.

The use of PVC cladding, a widely recognised commercial off-the-shelf product in the construction industry, provided a practical alternative to traditional timber. This material meets all relevant product performance, assurance, and accreditation requirements, ensuring compliance with building regulations. Integrating approved materials from other industries into railway asset renewal, in line with Network Rail's design standards, resulted in improved asset performance and extended preservation through modern techniques.

By applying the framework, the successful implementation of PVC cladding provides a data-driven case for similar applications in the future. The recorded benefits can now serve as a reference for other listed assets requiring renewal, improving the likelihood of LPA approval by reducing uncertainties and offering evidence-based justifications for material changes.

5.3.5. Case Study 2: A Missed Opportunity for Modernisation

The framework application in Case Study 2 yielded less positive results, with the project facing significant setbacks due to incorrect assumptions and overconfidence in prior planning decisions. The unverified expectation that an aluminium alternative would be permitted—based on its approval for a similar project at Preston Station—was a critical misjudgement. This oversight stemmed from a failure to properly define the benefits of the proposed material change and an inability to present clear, substantiated data from the Preston project to the planning officers. As a result, the project faced underestimated costs and an overly optimistic delivery forecast.

Additionally, delayed engagement with town planners led to major modifications to the original proposal, as the LPA only approved a like-for-like material substitution. The selection of engineered modified timber aligned with the preservation requirements but significantly increased both implementation costs and future maintenance demands. These maintenance obligations introduce additional financial and operational burdens, particularly due to the challenges of working over an active railway network. The consequence of these constraints is that the gables will likely deteriorate more rapidly than anticipated, ultimately reducing the asset's lifecycle efficiency.

While some minor positives can be drawn from introducing modified timber and a high standard of craftsmanship, the overall financial and logistical impact has been detrimental. The high costs and project delays have strained the budget, diverting resources from other necessary maintenance projects. This, in turn, increases the risk of accelerated deterioration in other assets, highlighting a missed opportunity for modernisation.

5.3.6. Key Comparative Findings from the Case Studies

A notable trend identified through comparing case studies is the financial and operational advantages of eliminating the need for applied coatings. When a contemporary material can match the appearance of the original construction while delivering superior performance, the total lifecycle costs are significantly reduced.

Many protective coatings are labour-intensive, with a shorter lifespan than the materials they protect. Their effectiveness is entirely dependent on their frequent reapplication and strict maintenance schedules. Failure to undertake these maintenance tasks results in not only aesthetic deterioration but also material degradation, leading to more frequent and costly repairs.

This research demonstrates that significant long-term cost savings can be achieved by replacing traditional materials requiring additional coatings with modern alternatives that inherently meet colour and aesthetic requirements. Similar benefits have been documented in the structural engineering literature, in which CFRP and GFRP have been shown to significantly enhance the flexural strength and stiffness of aged timber elements, indicating the broader applicability of fibre-reinforced materials in asset renewal contexts [47,48].

6. Conclusions

The presented work demonstrates that applying modern preservation techniques within a structured, reliability-oriented framework can significantly enhance the longevity, performance, and cost efficiency of historical railway assets. The proposed framework provides a practical approach to managing material substitution under regulatory and system constraints, enabling the introduction of contemporary methods while maintaining historical and functional integrity.

The research also highlights critical communication barriers in the planning approval process. Many decision makers involved in authorising changes to heritage infrastructure lack engineering backgrounds, which introduces subjective uncertainties into approval outcomes. To address this, the study advocates for a data-driven, system-level strategy, including the development of a national asset renewal database to offer quantifiable evidence supporting the reliability and regulatory acceptance of contemporary materials.

Through the analysis of two industrial case studies, the study reveals both the potential benefits and limitations of integrating modern materials into railway infrastructure. Case Study 1 demonstrates how the successful application of PVC cladding, supported by reliable performance data, can reduce lifecycle uncertainty and improve long-term asset reliability. In contrast, Case Study 2 highlights how delayed planning engagement and unvalidated assumptions regarding material approvals can lead to cost overruns, extended delays, and reduced delivery reliability.

Future research will aim to expand the empirical scope of this study by examining a broader range of historical asset types and contemporary material applications. Further case-based validations of successful modern material integration across various heritage contexts will also help strengthen the practical application of the framework.

Author Contributions: Conceptualization, T.W.; methodology, T.W.; software, T.W.; validation, T.W.; formal analysis, T.W.; investigation, T.W.; resources, T.W.; data curation, T.W.; writing—original draft preparation, T.W. and F.H.; writing—review and editing, F.H. and M.K.; visualization, T.W. and F.H.; supervision, M.K.; project administration, M.K.; funding acquisition, M.K. All authors have read and agreed to the published version of the manuscript.

Funding: This research received no external funding.

Institutional Review Board Statement: Not applicable.

Informed Consent Statement: Not applicable.

Data Availability Statement: The original contributions presented in this study are included in the article. Further inquiries can be directed to the corresponding author.

Conflicts of Interest: Author Thomas Wailes was employed by the company Network Rail. The remaining authors declare that the research was conducted in the absence of any commercial or financial relationships that could be construed as a potential conflict of interest.

References

1. Casson, M. *The World's First Railway System*; OUP: Oxford, UK, 2009.
2. Gunn, S. *The History of Transport Systems in the UK*; Government Office for Science Foresight: London, UK, 2018.
3. Network Rail. *Heritage: Care and Development*; Design Manual NR/GN/CIV/100/05; Network Rail: London, UK, 2020.
4. Coulls, A.; Divall, C.; Lee, R. *Railways as World Heritage Sites*; Report; ICOMOS: Paris, France, 1999.
5. Peira, G.; Lo Giudice, A.; Miraglia, S. Railway and Tourism: A Systematic Literature Review. *Tour. Hosp.* **2022**, *3*, 69–79. [CrossRef]
6. Ristić, D.; Vukoičić, D.; Ivanović, M.; Nikolić, M.; Milentijević, N.; Mihajlović, L.; Petrović, D. Transformation of Abandoned Railways into Tourist Itineraries/Routes: Model of Revitalization of Marginal Rural Areas. *Land* **2024**, *13*, 321. [CrossRef]
7. Tawfik, T.; M Khodeir, L.; Fathy, F. Identifying Retrofit Technology to Improve Building Energy Performance: A Review. *Eng. Res. J.* **2023**, *178*, 174–200. [CrossRef]
8. Rama, D.; Andrews, J.D. Railway Infrastructure Asset Management: The Whole-System Life Cost Analysis. *IET Intell. Transp. Syst.* **2016**, *10*, 58–64. [CrossRef]
9. Jiménez Rios, A.; Petrou, M.L.; Ramirez, R.; Plevris, V.; Nogal, M. Industry 5.0, towards an Enhanced Built Cultural Heritage Conservation Practice. *J. Build. Eng.* **2024**, *96*, 110542. [CrossRef]
10. Liang, W.; Ahmad, Y.; Mohidin, H.H.B. The Development of the Concept of Architectural Heritage Conservation and Its Inspiration. *Built Herit.* **2023**, *7*, 21. [CrossRef]
11. Merciu, F.C.; Păunescu, C.; Dorobanţu, M.; Merciu, G.L. Assessing the Value of Railway Heritage for Sustainable Development: The Case Study of the Oraviţa–Anina Railway, Romania. *Sustainability* **2022**, *14*, 13262. [CrossRef]
12. Tanguay, G.A.; Berthold, E.; Rajaonson, J. *A Comprehensive Strategy to Identify Indicators of Sustainable Heritage Conservation*; Les Cah. Du CRTP (Work Paper); CRTP: Gleize, France, 2014.
13. Borges, B.A.B. A New Approach to the Concepts of Conservation to Identify and Evaluate Railway Heritage through Indicators. *HoST—J. Hist. Sci. Technol.* **2020**, *14*, 183–195. [CrossRef]
14. *Historic England Conservation Principles, Policies and Guidance*; Historic England: London, UK, 2008.
15. Planning (Listed Buildings and Conservation Areas) Act 1990. *Encyclopedic Dictionary of Landscape and Urban Planning*; Springer: Berlin/Heidelberg, Germany, 2010.
16. Railway Heritage Act 1996. Available online: https://www.legislation.gov.uk/ukpga/1996/42/contents (accessed on 22 April 2025).
17. Wells, J.C.; Lixinski, L. Heritage Values and Legal Rules: Identification and Treatment of the Historic Environment via an Adaptive Regulatory Framework (Part 1). *J. Cult. Herit. Manag. Sustain. Dev.* **2016**, *6*, 345–364. [CrossRef]
18. Wells, J.C.; Lixinski, L. Heritage Values and Legal Rules: Identification and Treatment of the Historic Environment via an Adaptive Regulatory Framework (Part 2). *J. Cult. Herit. Manag. Sustain. Dev.* **2017**, *7*, 345–363. [CrossRef]
19. Network Rail. NR/L3/CIV/023 Assessment of Footbridges. Available online: https://www.bridgeforum.org/guidance/nr-l3-civ-023-assessment-of-footbridges/ (accessed on 22 April 2025).
20. Network Rail. *Determination of Load Carrying Capacity of Half-Through Girders*; Recommendation Report; Network Rail: London, UK, 2016.
21. Molina, M.T.; Cano, E.; Ramírez-Barat, B. Protective Coatings for Metallic Heritage Conservation: A Review. *J. Cult. Herit.* **2023**, *62*, 99–113. [CrossRef]
22. Muhit, I.B.; Masia, M.J.; Stewart, M.G. Failure Analysis and Structural Reliability of Unreinforced Masonry Veneer Walls: Influence of Wall Tie Corrosion. *Eng. Fail. Anal.* **2023**, *151*, 107354. [CrossRef]
23. Blavier, C.L.S.; Huerto-Cardenas, H.E.; Aste, N.; Del Pero, C.; Leonforte, F.; Della Torre, S. Adaptive Measures for Preserving Heritage Buildings in the Face of Climate Change: A Review. *Build. Environ.* **2023**, *245*, 110832. [CrossRef]
24. Baxter, A. *Examination of the Use of FRP for Station Canopy Fascia Daggerboards*; Railway Heritage Trust: London, UK, 2018.
25. Zai, B.A.; Khan, M.A.; Khan, S.Z.; Asif, M.; Khan, K.A.; Saquib, A.N.; Mansoor, A.; Shahzad, M.; Mujtaba, A. Prediction of Crack Depth and Fatigue Life of an Acrylonitrile Butadiene Styrene Cantilever Beam Using Dynamic Response. *J. Test. Eval.* **2020**, *48*, 20180674. [CrossRef]
26. He, F.; Khan, M. Effects of Printing Parameters on the Fatigue Behaviour of 3d-Printed Abs under Dynamic Thermo-Mechanical Loads. *Polymers* **2021**, *13*, 2362. [CrossRef]
27. He, F.; Khan, M.; Aldosari, S. Interdependencies between Dynamic Response and Crack Growth in a 3D-Printed Acrylonitrile Butadiene Styrene (ABS) Cantilever Beam under Thermo-Mechanical Loads. *Polymers* **2022**, *14*, 982. [CrossRef]
28. Doni, M.; Fierascu, I.; Fierascu, R.C. Recent Developments in Materials Science for the Conservation and Restoration of Historic Artifacts. *Appl. Sci.* **2024**, *14*, 11363. [CrossRef]
29. Albuja-Sánchez, J.; Damián-Chalán, A.; Escobar, D. Experimental Studies and Application of Fiber-Reinforced Polymers (FRPs) in Civil Infrastructure Systems: A State-of-the-Art Review. *Polymers* **2024**, *16*, 250. [CrossRef]
30. Naser, M.Z.; Hawileh, R.A.; Abdalla, J.A. Fiber-Reinforced Polymer Composites in Strengthening Reinforced Concrete Structures: A Critical Review. *Eng. Struct.* **2019**, *198*, 109542. [CrossRef]

31. Khalel, H.H.Z.; Khan, M. Modelling Fibre-Reinforced Concrete for Predicting Optimal Mechanical Properties. *Materials* **2023**, *16*, 3700. [CrossRef]
32. Qureshi, J. Fibre-Reinforced Polymer (FRP) in Civil Engineering. In *Next Generation Fiber-Reinforced Composites–New Insights*; IntechOpen: Rijeka, Croatia, 2023.
33. Uddin, N. *Developments in Fiber-Reinforced Polymer (FRP) Composites for Civil Engineering*; Elsevier: Amsterdam, The Netherlands, 2013.
34. Zedan Khalel, H.H.; Khan, M.; Starr, A. Dynamic Response-Based Crack Resistance Analysis of Fibre Reinforced Concrete Specimens under Different Temperatures and Crack Depths. *J. Build. Eng.* **2023**, *66*, 105865. [CrossRef]
35. Kamei, K.; Khan, M.A. Current Challenges in Modelling Vibrational Fatigue and Fracture of Structures: A Review. *J. Braz. Soc. Mech. Sci. Eng.* **2021**, *43*, 77. [CrossRef]
36. Zai, B.A.; Khan, M.A.; Khan, K.A.; Mansoor, A. A Novel Approach for Damage Quantification Using the Dynamic Response of a Metallic Beam under Thermo-Mechanical Loads. *J. Sound Vib.* **2020**, *469*, 115134. [CrossRef]
37. McMahon, P.; Zhang, T.; Dwight, R.A. Approaches to Dealing with Missing Data in Railway Asset Management. *IEEE Access* **2020**, *8*, 48177–48194. [CrossRef]
38. Adams, K.; Heidarzadeh, M. A Multi-Hazard Risk Model with Cascading Failure Pathways for the Dawlish (UK) Railway Using Historical and Contemporary Data. *Int. J. Disaster Risk Reduct.* **2021**, *56*, 102082. [CrossRef]
39. Sivaraja, S.S.; Thandavamoorthy, T.S.; Vijayakumar, S.; Moses Aranganathan, S.; Dasarathy, A.K. Preservation of Historical Monumental Structures Using Fibre Reinforced Polymer (FRP)—Case Studies. *Procedia Eng.* **2013**, *54*, 472–479. [CrossRef]
40. Martínez-Corral, A.; Cárcel-Carrasco, J.; Carnero, M.C.; Aparicio-Fernández, C. Analysis for the Heritage Consideration of Historic Spanish Railway Stations (1848–1929). *Buildings* **2022**, *12*, 206. [CrossRef]
41. Substitute Materials in Historic Restoration: When to Consider Alternatives. Available online: https://www.hoffarch.com/wp-content/uploads/Substitute-Materials-in-Historic-Restoration.pdf (accessed on 16 April 2025).
42. Konstantinov, A.; Mukhin, A. Architectural Possibilities of Using PVC Window Units in Historical Buildings. *MATEC Web Conf.* **2018**, *193*, 04018. [CrossRef]
43. Jahreis, M.; Rautenstrauch, K. Rehabilitation, Upgrading and Repair of Historic Timber Structures with Polymer Concrete and FRP-Reinforcement. In *Materials and Joints in Timber Structures*; RILEM Bookseries; Springer: Dordrecht, The Netherlands, 2014; pp. 485–492. [CrossRef]
44. Sandor, J.; Trayte, D.; Uebel, A.E. *The Use of Substitute Materials on Historic Buildings Exteriors*; Preservation Briefs 16; National Park Service: London, UK, 1988.
45. Duguid, B.; Bone, B.; Birdi, J. The Ordsall Chord, UK: Conservation Architecture and Engineering—Part 1: Project Overview. *Proc. Inst. Civ. Eng.–Eng. Hist. Herit.* **2019**, *173*, 43–59. [CrossRef]
46. What Is Accoya? Available online: https://www.accoya.com/uk/what-is-accoya/ (accessed on 26 March 2025).
47. Saad, K.; Lengyel, A. Strengthening Timber Structural Members with CFRP and GFRP: A State-of-the-Art Review. *Polymers* **2022**, *14*, 2381. [CrossRef]
48. Brol, J.; Wdowiak-Postulak, A. Old Timber Reinforcement with FRPs. *Materials* **2019**, *12*, 4197. [CrossRef] [PubMed]

applied sciences

MDPI

Article

An Efficient Uncertainty Quantification Approach for Robust Design of Tuned Mass Dampers in Linear Structural Dynamics

Thomas Most [1,*], Volkmar Zabel [2], Rohan Raj Das [1] and Abridhi Khadka [1]

[1] Institute of Structural Mechanics, Bauhaus-Universität Weimar, 99423 Weimar, Germany; rohan.raj.das@uni-weimar.de (R.R.D.); abridhi.khadka@uni-weimar.de (A.K.)

[2] Structural Analysis and Structural Dynamics, University Rostock, 18059 Rostock, Germany; volkmar.zabel@uni-rostock.de

* Correspondence: thomas.most@uni-weimar.de

Abstract

The application of tuned mass dampers (TMDs) to high-rise buildings or slender bridges can significantly decrease the dynamical vibrations due to external excitation, such as wind or earthquake loads. However, the individual properties of a TMD such as mass, stiffness and damping have to be designed carefully with respect to the dynamical properties of the investigated structure. In real-world structures, the influence of uncertain system properties might be critical for the performance of a TMD and thus the whole structure. Therefore, the design under uncertainty of such systems is an important issue, which is addressed in the current paper. For our investigations, we consider linear single-degree-of-freedom (SDOF) systems, where analytical formulas for the deterministic design already exist, and linear multi-degree-of-freedom (MDOF) systems, where a time integration and numerical optimization algorithms are usually applied to obtain the optimal TMD parameters. If the numerical optimization should be coupled with a sampling-based uncertainty quantification method, such as Monte Carlo sampling, the design procedure would require the evaluation of a coupled double-loop approach, which is very demanding from the computation point of view. Therefore, we focus the following paper on an efficient analytical uncertainty quantification approach, which estimates the mean and scatter from a Taylor series expansion. Additionally, we introduce an efficient mode decomposition approach for MDOF systems with multiple TMDs, which estimates the maximum displacements using a modal analysis instead of a demanding time integration. Different optimal design problems are formulated as single- or multi-objective optimization tasks, where the statistical properties of the maximum displacements are considered as safety margins in the optimization goal functions. The application of numerical optimization algorithms is straightforward and not limited to specific algorithms. As numerical examples, we investigate an SDOF system with single TMD and a multi-story frame with multiple TMDs. The presented procedure might be interesting for the design process of structures, where the dynamical vibrations reach a critical threshold.

Keywords: dynamical systems; structural dynamics; tuned mass damper; uncertainty quantification; robust design

1. Introduction

The application of tuned mass dampers (TMDs) for the reduction in structural vibrations is a well-known procedure that started with the early investigations by Den Hartog [1].

Typically, high-rise buildings under earthquake and wind excitations require the TMD technology to reach new heights or to enable the construction under special loading conditions. An overview of existing famous buildings with TMD applications is given in [2]. Early studies for structures subjected to wind loads can be found in [3,4], as well as for earthquake excitations in [5,6].

Analytical solutions for optimal TMD parameters have been derived by Den Hartog [1] by simplifying the original structure to a linear single-degree-of freedom (SDOF) system and consider the TMD as an additional DOF. The optimality criteria were formulated in this approach in order to minimize the maximum displacements by considering the dynamic amplification function of the 2-DOF system under harmonic excitation. Additional analytical design criteria for harmonic excitations are summarized in [7]. Further investigations on analytical solutions can be found in [8], for harmonic and white noise excitation, and in [7,9–11], where different design criteria in the time and frequency domain have been investigated.

The extension for linear and non-linear multi-degree-of-freedom (MDOF) systems requires in the general case numerical procedure such as time integration to calculate the performance criteria, which could be maximum displacements, accelerations, inter-story drift and others. A good summary of the published design criteria and analysis methods is given in [12]. Numerical investigations on this topic can be found, e.g., for MDOF systems with a single TMD [13] and SDOF systems with multiple TMDs [14,15]. The optimal parameters of MDOF systems with multiple TMDs could not be solved analytically in the general case. Therefore, numerical optimization methods such as Particle Swarm optimization [16] and genetic algorithms [17,18] and machine learning methods [19,20] were often used for this task. However, the application of numerical optimization together with time integration methods could be limited due to the significant numerical effort.

Additional to the optimal tuning of the TMDs, the accurate knowledge of the main system properties as well as the TMD parameters are essential for an optimal performance of the TMDs. Therefore, the influence of uncertainties on the TMD performance and reliability is a critical issue. The application of Reliability-based Design and robust design optimization methods summarized in [21] have been published for different types of random excitation [22,23] and random system parameters [24]. Recent applications perform typically numerical time integration methods with random sampling [25,26], which increases the numerical effort even more. Since a coupled robust design optimization would require the evaluation of the statistical performance measure for every design parameter combination, the time integration might be the critical bottle-neck in the numerical analysis. Therefore, approximation methods have been investigated recently for the optimal design of TMDs considering uncertainties [27,28].

In our study, we will focus on efficient methods for the optimal design of TMD parameters for linear SDOF and MDOF systems by considering parameter uncertainty. In the first step, we will introduce a perturbation approach for the uncertainty propagation, where the system responses are linearized by a Taylor series expansion with respect to the random system parameters. This approach is derived and investigated for an SDOF system first, whereby the optimal parameters are obtained by numerical optimization. As excitation, we consider harmonic excitation, which could vary within a defined frequency range, and the maximum value of the corresponding dynamic amplification function is considered as the design criterion. Here, the amplification function of the displacements is chosen as an example similar to early studies by Den Hartog. However, other performance criteria could be considered in the approach in the same manner. We will show that this linearization works well for individual discrete values of the amplification functions of the main system and the TMD relative displacements. Within the optimization objective we consider the safety margin from the estimated mean and standard deviation within a variance-based robustness evaluation approach. The linearization approach may consider continuous scalar random numbers with an arbitrary distribution type, as long as the covariance matrix is known. However, in the numerical examples we will focus on independent and normally distributed random numbers. A Latin Hypercube sampling approach [29] is utilized as a benchmark method to investigate the accuracy of the presented analytical approach. Additional information on the uncertainty source could be obtained by variance-based sensitivity measures, which could be estimated for the linearization approach as a direct post-processing result.

In the second part of the paper, we will focus on linear MDOF systems with multiple TMDs. The damping of the main system is considered as modal damping, which allows for a decoupled modal analysis of the individual vibration modes. No further restrictions are made with respect to the system as long as the mass, stiffness and damping matrix can be defined. The application of the TMDs with arbitrary varying parameters will not fulfill the initial assumption of modal damping, and the more general case of viscous damping has to be considered. In the general case, the displacement solution can be obtained by a numerically demanding time integration procedure only, which makes the presented robust design optimization very impractical. For this reason, we will introduce a decoupled stationary solution of the displacements of the individual DOFs of the main system and the relative displacements of the TMDs. With the help of this very efficient approach, the maximum displacements could be obtained for a given range of harmonic excitations similarly to the SDOF system. The presented approach is of course an approximation, but in the numerical example, we will show that for the critical excitation frequencies, this approach agrees very well with the time integration results. Finally, the presented analytical uncertainty propagation approach based on a linearization of the maximum displacements with respect to the random system parameters is applied and an optimal design under the consideration of a defined safety margin could be obtained in a straightforward manner. As the numerical example, a three-story frame with two TMDs is investigated.

The novelty of the paper is the combination of the efficient uncertainty propagation method and its extension for linear MDOF systems with multiple TMDs. The definition of the optimization goals is straightforward and could consider single- and multi-objective optimization problems with and without constraints. The presented procedure is independent with respect to the choice of the optimization algorithms. Therefore, we focus the paper on the dynamical analysis methods and the uncertainty analysis. The mechanical model and the dynamical analysis including the uncertainty quantification approach were implemented in MATLAB R2024b [30] and are freely available as mentioned in the Data Availability Statement of this paper. As optimization algorithms the simplex Nelder–Mead method [31] was chosen for single-objective optimization and the Non-dominated Sort Genetic Algorithm (NSGA II) according to [32] for multi-objective optimization. For both algorithms the implementation in the Ansys optiSLang 2025R1 optimization software package was used [33].

2. Materials and Methods

2.1. Single-Degree-of-Freedom System

2.1.1. Mechanical Model

In the following section, we assume an SDOF system, which could be a simplified mechanical beam model of a pedestrian bridge as shown in Figure 1.

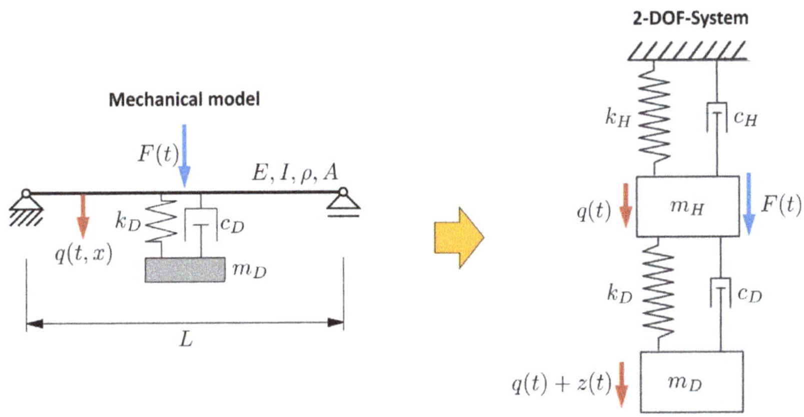

Figure 1. Mechanical beam model with attached damper and simplification as dynamical 2-degree-of-freedom system.

If only a single natural mode of vibration is considered in the analysis, the main system can be modeled as a single-degree-of-freedom (SDOF) system with the following equation of motion [34]:

$$m_H \cdot \ddot{q}(t) + c_H \cdot \dot{q}(t) + k_H \cdot q(t) = F(t), \tag{1}$$

where m_H, c_H and k_H are the mass, the viscous damping and the stiffness of the SDOF system, respectively. $q(t)$ is the corresponding displacement of the simplified degree of freedom. Assuming a harmonic excitation,

$$F(t) = \hat{F} \cdot \sin(\Omega \cdot t), \tag{2}$$

where $\hat{F}$ is the force amplitude and Ω is the circular frequency of the excitation. The particular (stationary) displacement solution $q_p(t)$ for the equation of motion reads [7]

$$q_p(t) = \hat{q}_p \cdot \sin(\Omega \cdot t - \varphi_1), \tag{3}$$

with

$$\begin{aligned} \hat{q}_p &= \frac{\hat{F}}{k_H} V_1, & V_1 &= \frac{1}{\sqrt{(1-\eta^2)^2 + (2\zeta_H \eta)^2}}, \\ \eta &= \frac{\Omega}{\omega_H}, & \tan \varphi_1 &= \frac{2\zeta_H \eta}{1-\eta^2}, \end{aligned} \tag{4}$$

where V_1 is the dynamic amplification function, η is the frequency ratio of the excitation and φ_1 is the phase shift between the excitation and system response. Furthermore, the natural circular frequency of the undamped system and the damping ratio are given as

$$\omega_H = \sqrt{\frac{k_H}{m_H}}, \quad \zeta_H = \frac{c_H}{2 m_H \omega_H}. \tag{5}$$

A performance measure within the design process could be the maximum value of the dynamic amplification function V_1, as it amplifies the ratio of the force amplitude and system stiffness in the resonance case. Figure 2 shows an example of the amplification function assuming a damping ratio of $\zeta_H = 0.5\%$.

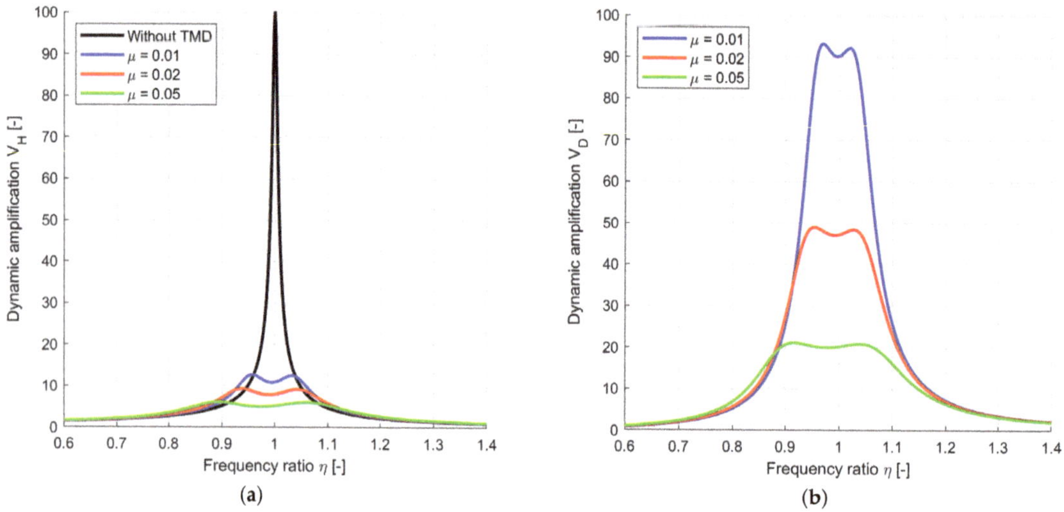

Figure 2. Dynamic amplification functions of an SDOF system with a damping ratio $\zeta_H = 0.5\%$ without TMD and with optimal TMD parameters according to Den Hartog for different mass ratios: (**a**) main system amplification $V_H(\eta)$); (**b**) amplification of relative displacements $V_D(\eta)$.

If a tuned mass damper (TMD) is taken into account as the second degree of freedom, as shown in Figure 1, the following equations of motion are obtained:

$$\begin{aligned} m_H \cdot \ddot{q}(t) + c_H \cdot \dot{q}(t) + k_H \cdot q(t) - c_D \cdot \dot{z}(t) - k_D \cdot z(t) &= F(t), \\ m_D(\ddot{q}(t) + \ddot{z}(t)) + c_D \cdot \dot{z}(t) + k_D \cdot z(t) &= 0, \end{aligned} \tag{6}$$

where m_D, c_D and k_D are the mass, the viscous damping and the stiffness of the additional TMD, and $z(t)$ describes the relative displacement between the main system and the tuned mass damper. Assuming a harmonic excitation $F(t)$ on the main system, the stationary displacement solutions $q_p(t)$ for the main system and $z_p(t)$ for the secondary system read as follows [7]:

$$\begin{aligned} q_p(t) &= \hat{q}_p \cdot \sin(\Omega \cdot t - \varphi_H), \qquad \hat{q}_p = \frac{\hat{F}}{k_H} V_H, \\ z_p(t) &= \hat{z}_p \cdot \sin(\Omega \cdot t - \varphi_D), \qquad \hat{z}_p = \frac{\hat{F}}{k_H} V_D, \end{aligned} \tag{7}$$

where V_H and V_D are the dynamic amplification functions of the main and secondary system and φ_H and φ_D are the phase shifts with respect to the excitation. In [7] the following equations are given:

$$\begin{aligned} V_H &= \sqrt{\frac{b_1^2 + b_2^2}{b_3^2 + b_4^2}}, \qquad \tan\varphi_H = \frac{b_1 b_4 - b_2 b_3}{b_1 b_3 + b_2 b_4}, \\ V_D &= \sqrt{\frac{\eta^4}{b_3^2 + b_4^2}}, \qquad \tan\varphi_D = \frac{b_4}{b_3}, \end{aligned} \tag{8}$$

with the coefficients

$$\begin{aligned}
b_1 &= \kappa^2 - \eta^2, \\
b_2 &= 2\eta\kappa\zeta_D, \\
b_3 &= \eta^4 - \eta^2\left(1 + \kappa^2 + \mu\kappa^2 + 4\kappa\zeta_H\zeta_D\right) + \kappa^2, \\
b_4 &= \eta\left[2\zeta_H(\kappa^2 - \eta^2) + 2\kappa\zeta_D(1 - \eta^2 - \mu\eta^2)\right], \\
\mu &= \frac{m_D}{m_H}, \quad \kappa = \frac{\omega_D}{\omega_H}, \quad \eta = \frac{\Omega}{\omega_H}.
\end{aligned} \tag{9}$$

Additionally, ω_D and ζ_D are introduced as follows:

$$\omega_D = \sqrt{\frac{k_D}{m_D}}, \quad \zeta_D = \frac{c_D}{2m_D\omega_D}. \tag{10}$$

Decisive parameters in the tuning of the TMD are the mass ratio μ, the frequency ratio κ and the damping rate ζ_D of the attached vibration damper.

2.1.2. Optimal Tuning of the TMD Parameters

The optimal frequency ratio κ_{opt} and damping rate $\zeta_{D,opt}$ of a vibration damper with deterministic properties are usually formulated as a function of the mass ratio μ [7]. The optimal parameters for minimum displacements are given according to Den Hartog as

$$\kappa_{opt} = \frac{1}{1+\mu}, \quad \zeta_{D,opt} = \sqrt{\frac{3\mu}{8(1+\mu)^3}}. \tag{11}$$

In Figure 2, the amplification functions from Equation (8) for the displacements of the main system and for the relative displacements are shown for different mass ratios. The figure indicates that the maximum values of both amplification functions decrease with an increasing mass ratio.

Alternatively to the Den Hartog formulas, the optimal values can be obtained by mathematical optimization. If the maximum value of the dynamic amplification function of the main system $V_H(\eta)$ should be minimized for a given maximum mass ratio μ_{limit}, the optimization task can be defined as a single-objective optimization problem as follows:

$$\min_{\mu,\kappa,\zeta_D}\left(\max_{\eta}(V_H(\eta,\mu,\kappa,\zeta_D))\right), \quad \text{subjected to} \quad \mu \leq \mu_{limit}. \tag{12}$$

Since an increase in the mass ratio μ always decreases the amplification function values, the mass ratio can be considered constant as the limit value $\mu = \mu_{limit}$ and the frequency ratio κ and the damping rate ζ_D can be considered as the remaining optimization variables.

As optimization algorithms, gradient-based approaches such as Quasi-Newton methods [35] as well as gradient-free methods can be applied without special adjustments. In this study an extended Nelder–Mead method [31] from the Ansys optiSLang 2025R1 optimization software package [33] is used, which is very efficient for a small number of optimization variables.

The advantage of an optimization approach compared to the analytical formulas according to Den Hartog and others [7] is that additional constraints such as the maximum relative displacement between the main system and the TMD can be considered in a straightforward manner. In the first example in Section 3.1, the amplification function obtained with the optimum parameters according to Den Hartog is compared to the results of a single-objective optimization. For this purpose, different formulations of the objective function are investigated.

However, if we want to minimize the mass ratio of the vibration damper and the maximum values of the amplification function V_H simultaneously, the optimization task can be solved either by gradually adjusting the limit for the mass ratio within a single-objective problem or as a multi-objective problem using Pareto optimization. The objective functions of the Pareto optimization can be formulated as follows:

$$\min_{\mu,\kappa,\zeta_D}\left(\mu,\ \max_{\eta}(V_H(\eta,\mu,\kappa,\zeta_D))\right), \tag{13}$$

where μ, κ and ζ_D are the optimization variables. In case of conflicting objectives, no unique optimal solution exists and the optimal designs build a so-called Pareto frontier as shown in Figure 3. In our study, we use the Non-dominated Sort Genetic Algorithm (NSGA II) according to [32] as the Pareto optimization algorithm. This algorithm is available in the Ansys optiSLang 2025R1 optimization software package [33] and is used in this study without modification. The NSGA II algorithm uses a sorting of the Pareto dominance as the performance criterion as shown additionally in Figure 3. In Section 3.1 the results of the Pareto optimization are investigated in detail for an SDOF system.

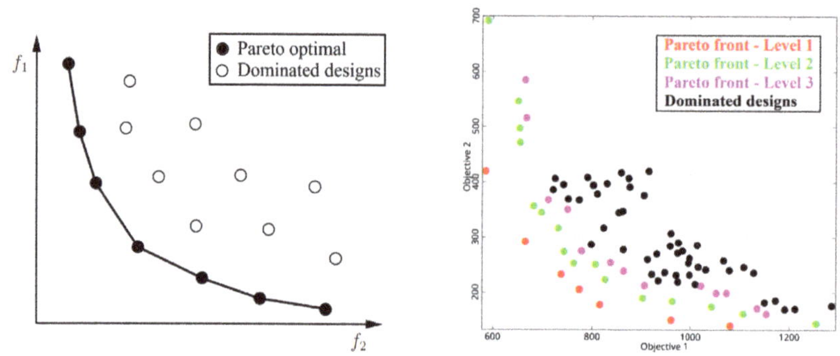

(a) (b)

Figure 3. Pareto optimization: (**a**) Pareto optimal designs on the Pareto frontier for two conflicting objectives; (**b**) basic principle of the non-dominated sort approach used in the NSGA-II algorithm.

2.1.3. Uncertainty Propagation and Quantification

When the propagation of uncertainty should be analyzed, it is advisable to assume the basic variables of the system as scattering variables since their variation can usually be determined directly. This means that the mass and stiffness coefficients as well as the damping ratios of the main system and the TMD are considered as random numbers, which could be assembled in a random vector

$$\mathbf{X} = [m_H, k_H, \zeta_H, m_D, k_D, \zeta_D]. \tag{14}$$

Each random number X_i can be defined as a scalar random variable by a distribution function and statistical moments, such as the mean value and standard deviation. For normally distributed variables, a linear correlation between the random input parameters can be represented with the Gaussian copula in closed form

$$f_{\mathbf{X}}(\mathbf{x}) = \frac{1}{\sqrt{(2\pi)^p|\mathbf{C_{XX}}|}}\exp\left[-\frac{1}{2}(\mathbf{x}-\bar{\mathbf{X}})^T\mathbf{C}_{\mathbf{XX}}^{-1}(\mathbf{x}-\bar{\mathbf{X}})\right], \tag{15}$$

where $f_{\mathbf{X}}(\mathbf{x})$ is the joint probability density function of random vector $\mathbf{X}$ and $\bar{\mathbf{X}}$ is the corresponding mean vector and $\mathbf{C_{XX}}$ the covariance matrix. p is the number of scalar random variables assembled in $\mathbf{X}$. The Nataf model [36] can be used to extend the Gaussian correlation model to non-Gaussian distribution types. Further details on correlation models can be found in [37]. In our study we consider independent normally distributed random numbers.

The resulting values of the amplification functions according to Equation (8) for discrete values of the excitation frequency ratio η_i are defined in the following as scalar random numbers:

$$\begin{aligned} V_{Hi}(\mathbf{X}) &= V_H(\eta_i, \mathbf{X}), \\ V_{Di}(\mathbf{X}) &= V_D(\eta_i, \mathbf{X}). \end{aligned} \tag{16}$$

The statistical properties of the amplification function values can be investigated by sampling methods such as the Monte Carlo Simulation [38], where a certain number of random samples is generated according to the defined distribution of the input parameters. In our study, an improved Latin Hypercube sampling (LHS) according to [29] is applied, where the marginal distributions are represented with respect to the probability contribution, and spurious correlations between the inputs are minimized accordingly as shown in Figure 4. In our study the LHS implementation of the MATLAB R2024b software package [30] is used to generate discrete samples of the random input vector $\mathbf{X}$ from a given mean vector $\bar{\mathbf{X}}$ and a covariance matrix $\mathbf{C_{XX}}$.

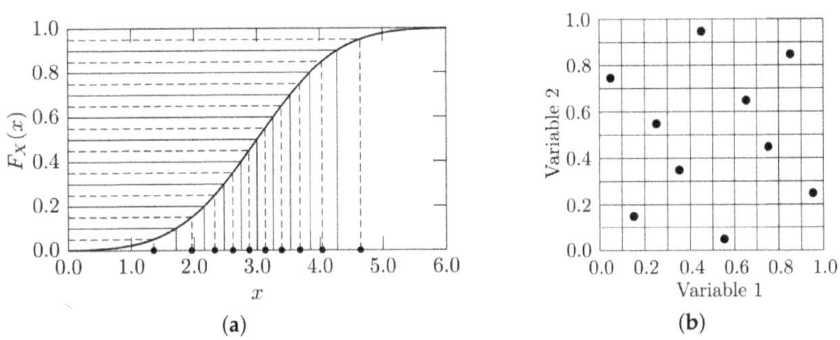

Figure 4. Improved Latin Hypercube sampling: (**a**) representation of the input marginals with equal probability classes; (**b**) minimized spurious correlations.

The statistical properties of the amplification function values could be analyzed by histograms and the estimates of statistical moments. Additional to the properties of a single output, the dependence with respect to the random input parameters might be interesting. A quite common approach for such a sensitivity analysis is the global variance-based method. In this method, the contribution of the variation in the input parameters with respect to the variation in a certain model output is analyzed. For further details the interested reader is referred to [39].

Since the scatter of the random response values, e.g., the maximum amplification function value, for a given nominal design of the input parameters should be considered in an outer optimization loop, the sampling-based estimation of the response scatter might be numerically demanding. Therefore, a more efficient analytical approach is used in this study, where the individual amplification function values are linearized with respect to the random input parameters using a Taylor series expansion at the nominal values. The corresponding mean values and the variance of the linearized response values can be estimated in closed form. By assuming the optimal deterministic parameter values as mean values

$$\bar{\mathbf{X}} = \left[\bar{m}_H, \bar{k}_H, \bar{\zeta}_H, \bar{m}_D, \bar{k}_D, \bar{\zeta}_D\right], \tag{17}$$

the random amplification function values in Equation (16) can be linearized as follows:

$$V_{Hi}(\mathbf{X}) \approx V_{Hi}^{lin}(\mathbf{X}) = V_{Hi}(\bar{\mathbf{X}}) + \left(\frac{\partial V_{Hi}(\mathbf{X})}{\partial \mathbf{X}}\right)^T\Bigg|_{\bar{\mathbf{X}}}(\mathbf{X} - \bar{\mathbf{X}}). \tag{18}$$

Based on this linearization, the mean value and the variance of each amplification function value can be obtained from the standard deviation σ_{X_k} and the correlation coefficients ρ_{kl} of the p random inputs X_k as follows:

$$\begin{aligned} \bar{V}_{Hi}^{lin} &= V_{Hi}(\bar{\mathbf{X}}), \\ \sigma^2_{V_{Hi}^{lin}} &= \sum_k^p \sum_l^p \frac{\partial V_{Hi}(\mathbf{X})}{\partial X_k}\frac{\partial V_{Hi}(\mathbf{X})}{\partial X_l}\sigma_{X_k}\sigma_{X_l}\rho_{kl}. \end{aligned} \tag{19}$$

The required derivatives in Equation (18) are obtained in this study by the central difference method. In case of uncorrelated inputs, Equation (19) simplifies as follows:

$$\bar{V}_{Hi}^{lin} = V_{Hi}(\bar{\mathbf{X}}), \quad \sigma^2_{V_{Hi}^{lin}} = \sum_k^p \left(\frac{\partial V_{Hi}(\mathbf{X})}{\partial X_k}\right)^2 \sigma^2_{X_k}. \tag{20}$$

2.1.4. Optimization Under Uncertainty

By considering uncertain input parameters in the optimization task, we have to distinguish between purely random inputs and design variables, which could be random as well [21]. Let us consider the design variables for the SDOF system similar to the deterministic optimization with the nominal values of μ, κ and ζ_D

$$\mathbf{d} = [\mu_d, \kappa_d, \zeta_{D,d}]. \tag{21}$$

The corresponding random numbers for the TMD coefficients are

$$\mathbf{X}_D = [m_D, k_D, \zeta_D], \tag{22}$$

where the mean values are adapted by the design variables

$$\bar{m}_D = \mu_d \cdot \bar{m}_H, \quad \bar{k}_D = \mu_d \cdot \kappa_d^2 \cdot \bar{k}_H, \quad \bar{\zeta}_D = \zeta_{D,d}. \tag{23}$$

The standard deviation of each parameter could be assumed either as a constant value or could be obtained from the current mean value and a given Coefficient of Variation (CoV)

$$\sigma_{m_D} = \bar{m}_D \cdot CoV(m_D), \quad \sigma_{k_D} = \bar{k}_D \cdot CoV(k_D), \quad \sigma_{\zeta_D} = \bar{\zeta}_D \cdot CoV(\zeta_D). \tag{24}$$

The statistical properties of the pure random parameters of the main system

$$\mathbf{X}_H = [m_H, k_H, \zeta_H] \tag{25}$$

remain constant during the optimization.

In Reliability-based Design optimization, the objective function is usually formulated in terms of the deterministic design variables $\mathbf{d}$. Statistical constraints are introduced to consider certain quality requirements [21]

$$\min_{\mathbf{d}} f(\mathbf{d}), \quad \text{subjected to} \quad P_{F_i} = P[g_i(\mathbf{X}_D, \mathbf{X}_H) \leq 0] \leq P_{F_i}^{target}, \tag{26}$$

where $g_i(\mathbf{X}_D, \mathbf{X}_H)$ are limit state functions depending on the joint set of $\mathbf{X}_D$ and $\mathbf{X}_H$. P_{F_i} is the corresponding failure probability, which must not exceed a given target value. In Figure 5 a single random response Y is shown with an indicated limit Y_{limit}.

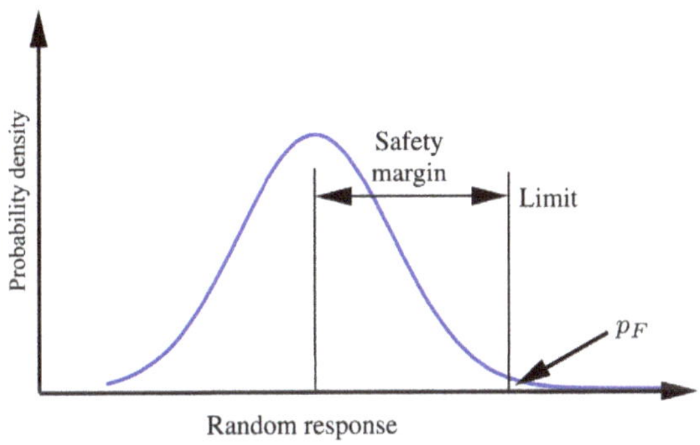

Figure 5. A random response with indicated limit and corresponding exceedance probability and the safety margin between the mean value and the limit.

The limit state function can be formulated in this case as follows:

$$g_Y(\mathbf{X}) = Y_{limit} - Y(\mathbf{X}). \tag{27}$$

The evaluation of the failure probability P_F requires an integration of the joint probability density function $f_{\mathbf{X}}$ over the failure domain

$$P_F = P[g(\mathbf{X}) \leq 0] = \int \cdots \int_{g(\mathbf{X}) \leq 0} f_{\mathbf{X}}(\mathbf{x}) d\mathbf{x}, \tag{28}$$

which might be numerically very demanding for the general case [37].

Within the Design-for-Six-Sigma approach, the optimization constraints are typically formulated in terms of the safety margin of the random output [40]

$$\bar{Y} + \alpha \cdot \sigma_Y \leq Y_{limit}, \tag{29}$$

which is usually defined as the standard deviation times a required sigma level α. Different procedures for this so-called robust design optimization (RDO) are discussed in [41] and more recently in [21]. Typically, a double-loop approach is necessary, where the outer loop performs the optimization and an inner loop estimates the output mean values and standard deviations with respect to the current nominal values of the design variables.

For the optimization of the TMD parameters, we will apply the linearization approach to estimate the mean value and the standard deviation of the individual amplification function values V_{Hi} and V_{Di}. If a certain limit for the dynamic amplification is given, the optimization task can be formulated in terms of a minimization of the nominal TMD mass ratio as follows:

$$\min_{\mathbf{d}} \mu_d, \quad \text{subjected to} \quad \max_i(\bar{V}_{Hi} + \alpha \cdot \sigma_{V_{Hi}}) \leq V_H^{limit}$$
$$\max_i(\bar{V}_{Di} + \alpha \cdot \sigma_{V_{Di}}) \leq V_D^{limit}, \tag{30}$$

where the constraints could be formulated alternatively in terms of the maximum displacements by using Equation (7). If no limit values for the amplification function or displacements are defined, we could either minimize the mass ratio and the maximum amplification values within a multi-objective optimization

$$\min_{\mathbf{d}} \left(\mu_d, \max_i(\bar{V}_{Hi} + \alpha \cdot \sigma_{V_{Hi}}) \right), \tag{31}$$

or the mass ratio could be kept fixed and the maximum amplification values are minimized. The scaling factor α, which defines the safety margin in terms of the standard deviation, is chosen in this study according to the probability levels for the serviceability limit state $\beta \leq 2.9$ and the ultimate limit state $\beta \leq 4.7$ according to the European design code [42].

2.2. *Extension for Multi-Degree-of-Freedom Systems*

2.2.1. Mechanical Model

The equation of motion for a linear multi-degree-of-freedom (MDOF) system reads in matrix-vector notation as follows:

$$\mathbf{M}_H \cdot \ddot{\mathbf{q}}_H(t) + \mathbf{C}_H \cdot \dot{\mathbf{q}}_H(t) + \mathbf{K}_H \cdot \mathbf{q}_H(t) = \mathbf{f}(t), \tag{32}$$

where the mass matrix $\mathbf{M}_H$, the viscous damping matrix $\mathbf{C}_H$ and the stiffness matrix $\mathbf{K}_H$ correspond to an initial main system without a TMD. $\mathbf{f}(t)$ is the external force vector acting on the individual DOFs. The displacement vector $\mathbf{q}_H(t)$ contains the displacements of all degrees of freedom of the original main system. For the free vibration mode of the undamped system, the natural circular frequencies ω_{H_i} and the corresponding mode shapes $\boldsymbol{\phi}_i$ can be obtained by solving the following eigenvalue problem:

$$\left(\mathbf{K}_H - \omega_{H_i}^2 \cdot \mathbf{M}_H\right) \cdot \boldsymbol{\phi}_i = \mathbf{0}. \tag{33}$$

The mode shapes are orthogonal to each other

$$\boldsymbol{\phi}_i^T \cdot \mathbf{M}_H \cdot \boldsymbol{\phi}_j = 0, \quad \boldsymbol{\phi}_i^T \cdot \mathbf{K}_H \cdot \boldsymbol{\phi}_j = 0 \quad \forall \quad i \neq j, \tag{34}$$

and could be normalized arbitrarily. In our study, we assume a modal damping according to [43]

$$\boldsymbol{\phi}_i^T \cdot \mathbf{C}_H \cdot \boldsymbol{\phi}_j = 0 \quad \forall \quad i \neq j, \quad \boldsymbol{\phi}_i^T \cdot \mathbf{C}_H \cdot \boldsymbol{\phi}_i = \tilde{c}_i, \tag{35}$$

which leads to decoupled equations of motion in the modal space

$$\begin{aligned} \boldsymbol{\phi}_i^T \mathbf{M}_H \boldsymbol{\phi}_i \cdot \ddot{\tilde{q}}_{H_i}(t) + \quad \boldsymbol{\phi}_i^T \mathbf{C}_H \boldsymbol{\phi}_i \cdot \dot{\tilde{q}}_{H_i}(t) + \quad \boldsymbol{\phi}_i^T \mathbf{K}_H \boldsymbol{\phi}_i \cdot \tilde{q}_{H_i}(t) &= \quad \boldsymbol{\phi}_i^T \mathbf{f}(t), \\ \tilde{m}_i \cdot \ddot{\tilde{q}}_{H_i}(t) + \quad \tilde{c}_i \cdot \dot{\tilde{q}}_{H_i}(t) + \quad \tilde{k}_i \cdot \tilde{q}_{H_i}(t) &= \quad \tilde{f}_i(t), \end{aligned} \tag{36}$$

where $\tilde{m}_i$, $\tilde{c}_i$ and $\tilde{k}_i$ are the modal mass, modal damping and modal stiffness for the mode shape $\boldsymbol{\phi}_i$, and $\tilde{f}_i(t)$ is the corresponding excitation force. The displacement solution in the n original degrees of freedom can be obtained by a superposition of the decoupled modal solutions as follows:

$$\mathbf{q}_H(t) = \sum_{i=1}^{n} \boldsymbol{\phi}_i \cdot \tilde{q}_{H_i}(t). \tag{37}$$

If we extend the MDOF system by several tuned mass dampers, which are coupled directly to certain degrees of freedom of the initial main system, we get a coupled damping and stiffness matrix in the equation of motion

$$\begin{bmatrix} \mathbf{M}_H & \mathbf{0} \\ \mathbf{0} & \mathbf{M}_D \end{bmatrix} \cdot \begin{bmatrix} \ddot{\mathbf{q}}_H(t) \\ \ddot{\mathbf{q}}_D(t) \end{bmatrix} + \begin{bmatrix} \mathbf{C}_H + \mathbf{C}_{DH} & \mathbf{C}_{DC} \\ \mathbf{C}_{DC}^T & \mathbf{C}_{DD} \end{bmatrix} \cdot \begin{bmatrix} \dot{\mathbf{q}}_H(t) \\ \dot{\mathbf{q}}_D(t) \end{bmatrix} + \begin{bmatrix} \mathbf{K}_H + \mathbf{K}_{DH} & \mathbf{K}_{DC} \\ \mathbf{K}_{DC}^T & \mathbf{K}_{DD} \end{bmatrix} \cdot \begin{bmatrix} \mathbf{q}_H(t) \\ \mathbf{q}_D(t) \end{bmatrix} = \begin{bmatrix} \mathbf{f}(t) \\ \mathbf{0} \end{bmatrix}, \tag{38}$$

where $\mathbf{C}_{DH}$, $\mathbf{C}_{DC}$ and $\mathbf{C}_{DD}$ as well as $\mathbf{K}_{DH}$, $\mathbf{K}_{DC}$ and $\mathbf{K}_{DD}$ are sub-matrices which represent the damping and stiffness coefficients of the tuned mass dampers and their association to the coupled degrees of freedom of the main system. $\mathbf{q}_H(t)$ is the displacement vector of the original DOFs of the main system and $\mathbf{q}_D(t)$ are the displacements of the additional DOFs of the TMDs. The excitation force vector is assumed to act only on the main system DOFs. This extended equation of motion covers arbitrary structures with symmetric mass, damping and stiffness matrices of the original structure, including the TMDs. The corresponding DOFs are ordered with respect to the main system DOFs and additional DOFs of the TMDs. Since the TMDs are coupled to the original structure only by viscous damper and spring elements, the mass coupling between the main system and the TMDs is zero as indicated by the zero off-diagonal mass sub-matrices.

For arbitrary values of the TMD coefficients, Equation (38) does not necessarily fulfill the assumption of modal damping and a decoupled solution could not be obtained in the general case of viscous damping. Therefore, time integration methods such as the Newmark-Beta approach [43] are necessary to solve Equation (38). However, this procedure is very demanding from the computational point of view. Especially, if the maximum amplification function of a certain degree of freedom should be considered in the objective function, a sweep over the excitation frequency has to be considered.

2.2.2. Approximated Decoupled Stationary Solution

In order to decrease the numerical effort, we assume a decoupled influence of each TMD by designing it only for a single vibration mode of the main system. For example, if we would assume a tuning for the first mode shape, we would approximate the displacement solution as the super position of a 2-DOF system for the first mode and several SDOF systems for the higher modes as illustrated in Figure 6.

The stationary displacements of the original DOFs can be assembled for this assumption as follows:

$$\mathbf{q}_{p,H}(t) = \sum_{i=1}^{n} \boldsymbol{\phi}_i \cdot \tilde{q}_{p,H_i}(t), \tag{39}$$

where the decoupled solution of an individual vibration mode reads in case of an application of a TMD

$$\tilde{q}_{p,H_i}(t) = \frac{\hat{\tilde{f}}_i}{\tilde{k}_i} \cdot V_{H,i} \cdot \sin(\Omega \cdot t - \varphi_{H,i}). \tag{40}$$

The modal stiffness $\tilde{k}_i$ corresponds to the main system and the force amplitude $\hat{\tilde{f}}_i$ results from a harmonic excitation as follows:

$$\tilde{f}_i(t) = \boldsymbol{\phi}_i^T \hat{\mathbf{f}} \cdot \sin(\Omega \cdot t) = \hat{\tilde{f}}_i \cdot \sin(\Omega \cdot t). \tag{41}$$

The amplification function $V_{H,i}$ and the phase shift $\varphi_{H,i}$ have to consider the modal mass and modal damping as well as the corresponding undamped circular frequency of the main system for the calculation according to Equation (7)

$$\begin{aligned} V_{H,i} &= V_H(\eta_i, \mu_i, \kappa_i, \zeta_{H_i}, \zeta_{D_i}), \qquad \eta_i = \frac{\Omega}{\omega_{H_i}}, \\ \varphi_{H,i} &= \varphi_H(\eta_i, \mu_i, \kappa_i, \zeta_{H_i}, \zeta_{D_i}), \qquad \kappa_i = \frac{\omega_{D_i}}{\omega_{H_i}}, \quad \mu_i = \frac{m_{D_i}}{\tilde{m}_i}. \end{aligned} \tag{42}$$

The corresponding mode shape $\boldsymbol{\phi}_i$ has to be normalized with respect to the DOF, where the TMD is applied. For the other vibration modes, where no TMD is applied, the decoupled stationary displacement solution reads

$$\tilde{q}_{p,H_j}(t) = \frac{\hat{\tilde{f}}_j}{\tilde{k}_j} \cdot V_{1,j} \cdot \sin(\Omega \cdot t - \varphi_{1,j}), \quad \forall \quad j \neq i, \tag{43}$$

where $V_{1,j}$ and $\varphi_{1,j}$ are the amplification function and the phase shift of a standard SDOF system without a TMD according to Equation (4)

$$V_{1,j} = V_1(\eta_j, \zeta_{H_j}), \quad \varphi_{1,j} = \varphi_1(\eta_j, \zeta_{H_j}). \tag{44}$$

No special normalization is necessary for these mode shapes without TMDs. The maximum stationary displacements of the orginal DOFs can be obtained from the assembled stationary displacements according to Equation (39) for one vibration period while considering the individual phase shifts in Equations (40) and (43). The relative displacements of the TMDs with respect to the coupling DOFs result in the decoupled stationary solution similar to the main system as

$$z_{p,D_i}(t) = \frac{\hat{\tilde{f}}_i}{\tilde{k}_i} \cdot V_{D,i} \cdot \sin(\Omega \cdot t - \varphi_{D,i}), \tag{45}$$

with

$$V_{D,i} = V_D(\eta_i, \mu_i, \kappa_i, \zeta_{H_i}, \zeta_{D_i}), \quad \varphi_{D,i} = \varphi_D(\eta_i, \mu_i, \kappa_i, \zeta_{H_i}, \zeta_{D_i}). \tag{46}$$

In the second example, we will show that the approximated displacement solution agrees very well with the results from a time integration and can be used to estimate the maximum displacements for the full excitation frequency range very efficiently.

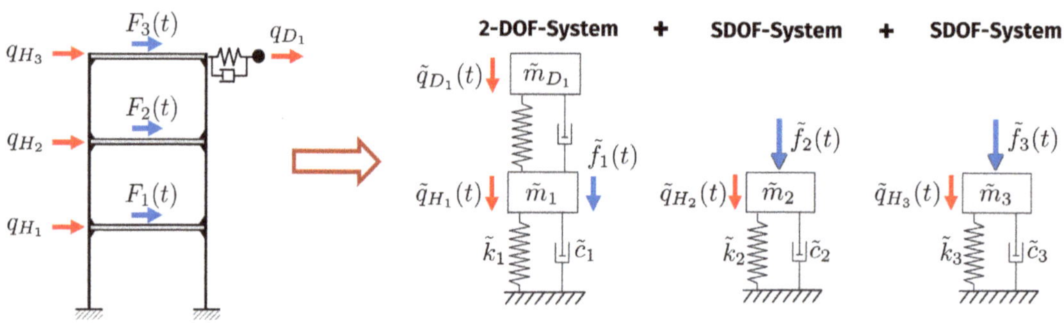

Figure 6. Superposition of the displacements of an MDOF system with a single TMD optimized for the first vibration mode.

2.2.3. Uncertainty Quantification

For the uncertainty quantification, we introduce the maximum stationary displacements of the original DOFs q_{H_n} and maximum relative displacements of the TMDs z_{D_i} for a discrete value of the excitation frequency Ω_k as scalar random outputs

$$Q_{H_{n,k}}(\mathbf{X}) = \max_t(|q_{H_n}(t, \Omega_k, \mathbf{X})|), \quad Z_{D_{i,k}}(\mathbf{X}) = \max_t(|z_{D_i}(t, \Omega_k, \mathbf{X})|), \tag{47}$$

where the linearization approach with respect to the random input vector $\mathbf{X}$ can be applied similarly to the SDOF system

$$Q_{H_{n,k}}(\mathbf{X}) \approx Q^{lin}_{H_{n,k}}(\mathbf{X}) = Q_{H_{n,k}}(\bar{\mathbf{X}}) + \left(\frac{\partial Q_{H_{n,k}}(\mathbf{X})}{\partial \mathbf{X}}\right)^T\Bigg|_{\bar{\mathbf{X}}} (\mathbf{X} - \bar{\mathbf{X}}). \tag{48}$$

The variance of these outputs can be directly estimated using Equation (20). The random input vector contains the random inputs of the main system $\mathbf{X}_H$, which define the scatter of the MDOF system itself, and the random properties of the m TMDs

$$\mathbf{X} = \left[\mathbf{X}_H, m_{D_1}, k_{D_1}, \zeta_{D_1}, \ldots, m_{D_m}, k_{D_m}, \zeta_{D_m}\right]. \tag{49}$$

2.2.4. Optimization Under Uncertainty

As design variables within the robust design optimization, we consider the nominal values of the mass ratio μ_{D_i}, the nominal circular frequencies ω_{D_i} and the nominal damping ratio ζ_{D_i} of each TMD

$$\mathbf{d} = \left[\mu_{D_1,d}, \omega_{D_1,d}, \zeta_{D_1,d}, \ldots, \mu_{D_m,d}, \omega_{D_m,d}, \zeta_{D_m,d}\right]. \tag{50}$$

The choice of design variables is based on the investigations in [44] and is more suitable than the basic mass, stiffness and damping coefficients in the equation of motion.

The objective functions within the robust design optimization are defined similarly to the SDOF system, but the total mass of all TMDs and the safety margin of specific displacements of the main system could be considered for an arbitrary discrete set of excitation frequencies Ω_k

$$\min_{\mathbf{d}}\left(\sum_{i=1}^{m} m_{D_i}, \max_k(\bar{Q}_{H_{n,k}} + \alpha \cdot \sigma_{Q_{H_{n,k}}})\right). \tag{51}$$

Additional constraints with respect to the TMD relative displacements could be formulated in a straightforward manner.

3. Results

3.1. *Single-Degree-of-Freedom Example*

3.1.1. Deterministic Optimization

In the first example we investigated an SDOF system with a TMD as shown in Figure 1. The parameters for the main system have been chosen as examples and are given in Table 1. Additionally, the optimal parameters according to Den Hartog are given for a mass ratio of $\mu = 2.0\%$. In the first step we performed a single-objective optimization with the objective function

$$\min_{\kappa, \zeta_D}\left(\max_\eta(V_H(\eta, \mu = 0.02, \kappa, \zeta_D))\right).$$

The parameter bounds for the optimization variables κ and ζ_D are given in Table 1. In Figure 7 the objective function is shown within the parameter bounds. The figure clearly indicates a unimodal objective function with a single optimum. As the optimization algorithm, the simplex method [31] from the Ansys optiSLang 2025R1 software package [33] was utilized. The optimizer converged within 55 model evaluations. Further details including the full optiSLang projects with the optimization settings and all the results can be found in the accompanying data set. The obtained parameter values of the optimal design agree very well with the optimal values according to Den Hartog as shown in Table 1. The corresponding amplification functions drawn in Figure 8 show very good agreement.

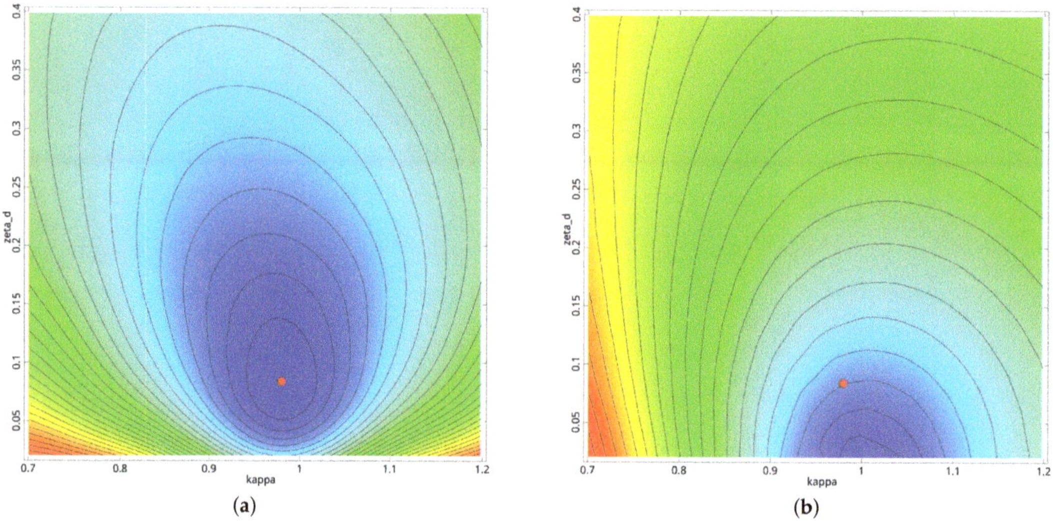

Figure 7. Objective function for the single-objective optimization: (**a**) minimizing $\max(V_H(\eta))$; (**b**) minimizing $V_H(\eta = 1.0)$ with indicated optimal values according to Den Hartog (red dot).

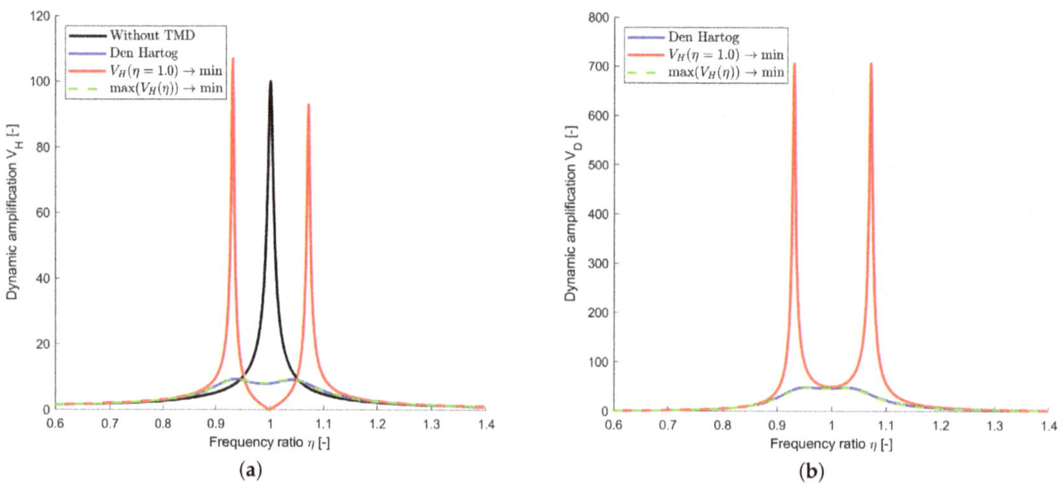

Figure 8. Amplification function for the optimal values from single-objective optimization with different objective functions and according to Den Hartog: (**a**) main system amplification $V_H(\eta)$); (**b**) amplification of the relative displacements $V_D(\eta)$.

Table 1. SDOF system with TMD: reference and optimization bounds and results of the single-objective optimization of $\min(\max(V_H(\eta)))$.

Parameter	Unit	Reference	Den Hartog	Single-Objective Bounds	Single-Objective Optimum	Multi-Objective
m_H	kg	10^4	10^4	10^4	10^4	10^4
k_H	kN/m	10^3	10^3	10^3	10^3	10^3
ζ_H	-	0.005	0.005	0.005	0.005	0.005
μ	-	-	0.020	0.020	0.020	0.00–0.20
κ	-	-	0.980	0.70–1.20	0.979	0.70–1.20
ζ_D	-	-	0.084	0.00–0.40	0.087	0.00–0.40

In the second step, we investigated a different objective function by considering the amplification value at the resonance frequency only

$$\min_{\kappa,\zeta_D}(V_H(\eta = 1.0, \mu = 0.02, \kappa, \zeta_D)).$$

The obtained amplification function is shown additionally in Figures 7 and 8. As indicated in the second figure, the optimized SDOF+TMD system has two significant resonances which have an amplification close to the original SDOF system. Thus, a different excitation frequency would lead to similar displacements as those without a TMD. Even the relative displacements of the TMD with respect to the main system would be very large compared to the values obtained with the Den Hartog parameters. From this finding, we can summarize that the consideration of a single specific excitation frequency might not lead to optimal amplification functions. Therefore, we consider the maximum amplification function as an objective in the following investigations.

In the next step, a multi-objective optimization was performed with two objective functions,

$$\min_{\mu,\kappa,\zeta_D}\left(\mu, \max_{\eta}(V_H(\eta, \mu, \kappa, \zeta_D))\right),$$

by considering the parameter bounds given in Table 1. The NSGA-II method [32] of the Ansys optiSLang 2025R1 software package was used with 50 generations, each having a population size of 50 in order to assure a good convergence. Further details including the optimization settings and results can be found in the accompanying data set.

Figure 9 compares the resulting Pareto front of both objective functions and the results of several single-objective optimization runs with the optimal values according to Den Hartog. The relationship between the optimal mass ratio μ and the corresponding values of the frequency ratio κ and the damping ratio ζ_D is particularly interesting. The scatter in the parameter values results from the stochastic nature of the NSGA-II. The optimal damping ratio according to Den Hartog is visibly lower than that of the optimizations, whereby the maximum values of the magnification functions hardly differ. As a result of the multi-objective optimization, a suitable choice for the mass ratio could be made: the range of about a 2% to 5% mass ratio represents a good compromise between the conflicting objective functions. The relationship between the maximum values of the relative displacement is analogous to that of the main system. However, a limitation of the installation space and thus of the maximum displacements could be considered directly as a constraint in the single-objective and multi-objective optimization.

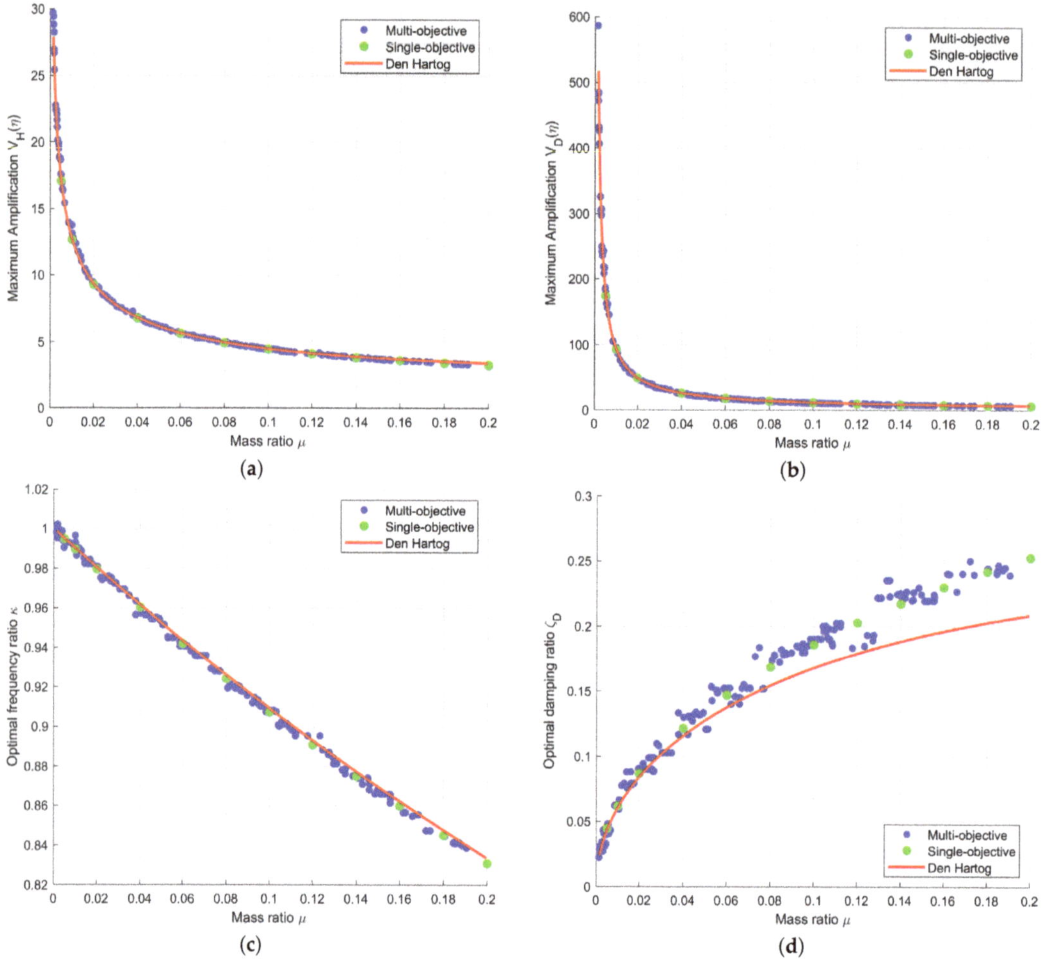

Figure 9. Results of the multi-objective optimization, several single-objective optimization runs with different mass ratios and the corresponding optimal parameters according to Den Hartog for the SDOF example: (**a**) maximum amplification V_H, (**b**) maximum amplification V_D, (**c**) optimal frequency ratio κ and (**d**) optimal damping ratio ζ_D dependent on the mass ratio μ.

3.1.2. Uncertainty Quantification

In the next step, uncertain input parameters were investigated for the given SDOF system. The input parameters were assumed to be independent and normally distributed as given in Table 2. The mean values were assumed as the optimal Den Hartog parameters for a mass ratio of 2% as given in Table 1. First, an improved Latin Hypercube sampling [29] with 1000 samples was generated as the reference with the MATLAB R2024b software package [30] and the corresponding amplification functions for each sample were calculated. The frequency ratio was discretized with 200 equidistant values within the interval $\eta \in [0.8; 1.2]$. The samples and corresponding amplification function values are available in the accompanying data set.

Table 2. Statistical properties of the scattering input parameters for the SDOF example with mean values according to the Den Hartog parameters.

Parameter	Unit	Mean Value	Coefficient of Variation	Distribution Type
m_H	kg	10^4	1.0%	Normal
k_H	kN/m	10^3	2.0%	Normal
ζ_H	-	0.005	5.0%	Normal
m_D	kg	200.0	1.0%	Normal
k_D	kN/m	19.22	2.0%	Normal
ζ_D	-	0.084	5.0%	Normal

Figure 10 shows the first 100 samples which indicate a significant scattering of the magnification function in the range of the two maximum values corresponding to the natural frequencies.

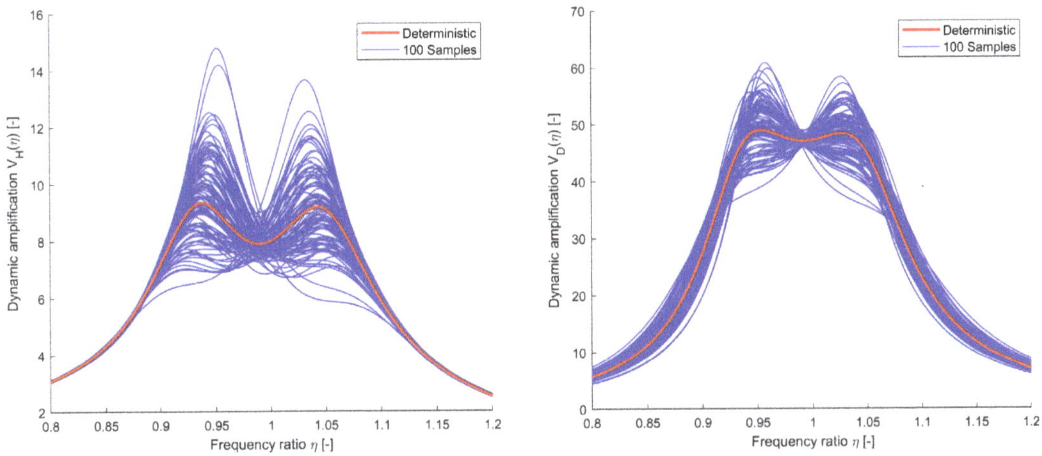

(a) (b)

Figure 10. The 100 Latin Hypercube samples of the amplification functions of the SDOF example compared to the deterministic functions according to Den Hartog for a nominal mass ratio of $\mu = 2.0\%$: (**a**) main system amplification $V_H(\eta)$; (**b**) amplification of the relative displacements $V_D(\eta)$.

In Figure 11 the sample estimates of the mean value and standard deviation for each amplification value V_{Hi} and V_{Di} calculated for every frequency ratio η_i are shown. Additionally, the estimated mean and standard deviation of V_{Hi} and V_{Di} using the analytical linearization approach are plotted. The derivatives required in Equation (18) have been estimated using the central difference method, which required only 12 model calls for the six input parameters. The corresponding derivation interval has been chosen as 1% of the nominal parameter values. The figure indicates very good agreement of both approaches, which indicates a sufficient representation of the scatter of the individual random response values V_{Hi} and V_{Di} by the linearization for the single-degree-of-freedom system.

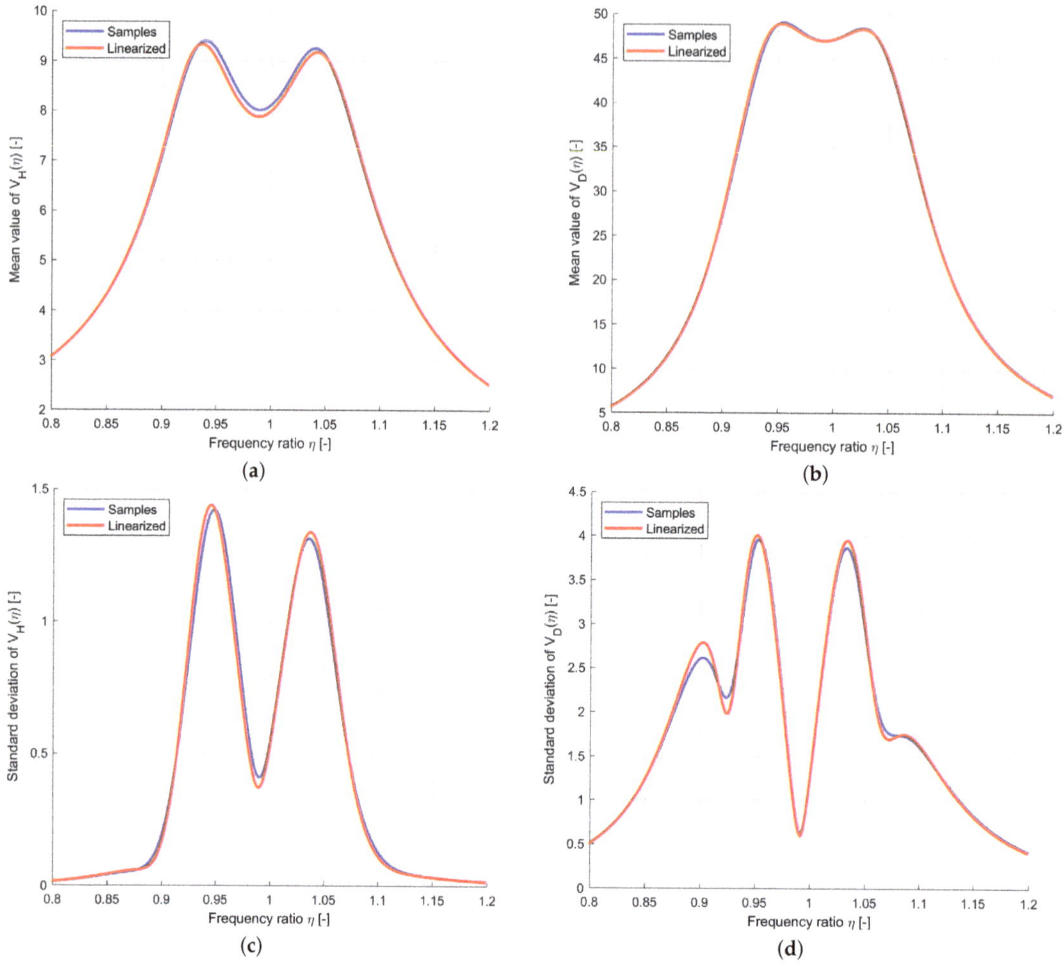

Figure 11. Mean values and standard deviation of the amplification function values for a nominal mass ratio of $\mu = 2.0\%$ estimated from 1000 LHS samples and by using the linearization approach: (**a**) mean values of $V_H(\eta)$, (**b**) mean values of $V_D(\eta)$, (**c**) standard deviation of $V_H(\eta)$ and (**d**) standard deviation of $V_D(\eta)$.

Additionally, the linearized variance contribution $\left(\frac{\partial V_H(\eta_i,\mathbf{X})}{\partial X_k}\right)^2 \sigma_{X_k}^2$ of the input parameters for the individual amplification values was estimated using Equation (20) and is plotted in Figure 12. The figure indicates a significant influence of the stiffness values for the values with the maximum scatter. The influence of the damping of the main system is negligible and the damping of the TMD is mostly significant where the contribution of the mass and stiffness scatter is smallest.

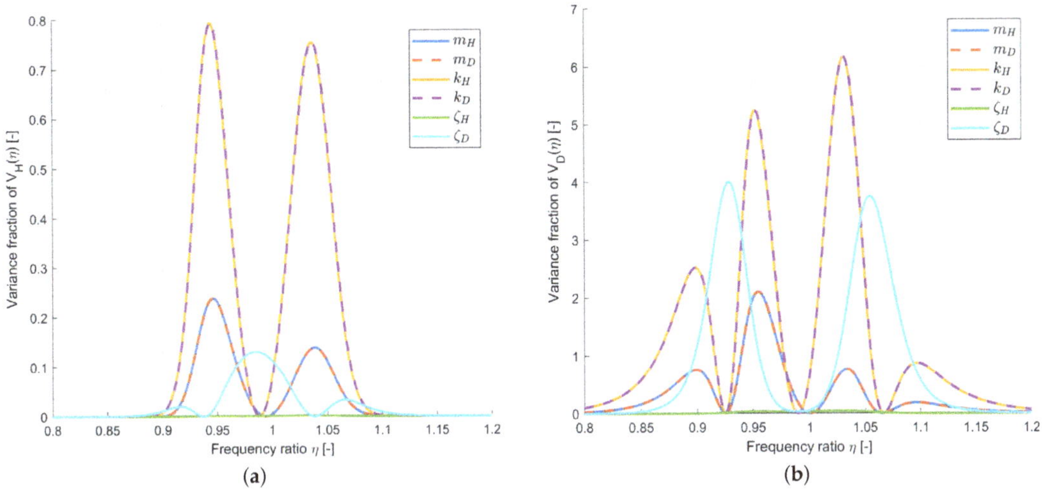

Figure 12. Variance contribution of the input parameters for a nominal mass ratio of $\mu = 2.0\%$ estimated with the linearization approach: (**a**) with respect to main system amplification $V_H(\eta)$; (**b**) with respect to amplification of relative displacements $V_D(\eta)$.

3.1.3. Optimization Under Uncertainty

Finally, the TMD for the SDOF system was optimized considering uncertain input parameters. As the objective function, Equation (31) was considered, whereas the nominal values of κ and ζ_D were taken as the optimization variables and the nominal mass ratio was kept as $\mu = 2.0\%$. The input scatter was defined by the Coefficients of Variation given in Table 2. The mean values and the standard deviation of the individual amplification function values were estimated by the linearization approach. Figure 13 shows the corresponding maximum values of the mean and standard deviation as a function of the nominal values of κ and ζ_D. The figure indicates that a decreasing standard deviation correlates with an increasing mean value. Thus, a combination of both measures forms a compromise between the nominal outputs and the scatter of the results. Similar to the deterministic case, the optimization was performed with the simplex method using Ansys optiSLang 2025R1. Further details including the optimization settings and results can be found in the accompanying data set. The α-factor for the standard deviation in Equation (31) was assumed with $\alpha = 2.9$ and $\alpha = 4.7$. Figure 13 shows the optimal values obtained for both cases in comparison to the deterministic solution. It is clearly recognizable that for a higher weighting of the scattering, the optimal parameters tend towards higher damping values.

The amplification function values from the deterministic optimization and the 4.7σ optimum are compared in Figure 14, where the mean value curves and the mean values + 4.7 times the standard deviation are plotted. The figure clearly indicates an increased maximum mean value of the 4.7σ optimum compared to the deterministic solution, but the scatter is significantly smaller and thus a reduction in the statistical limit is possible.

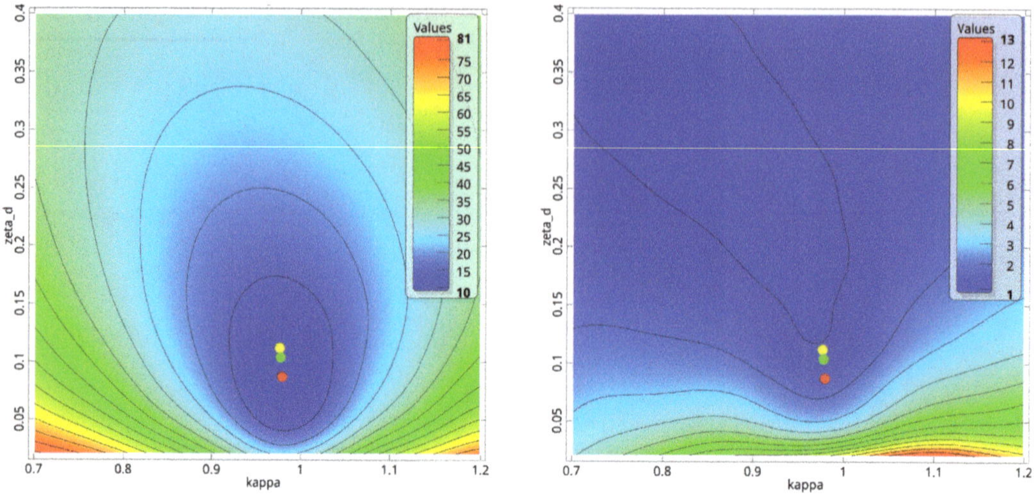

(**a**) (**b**)

Figure 13. (**a**) Maximum mean value and (**b**) standard deviation of $V_H(\eta)$ as a function of the optimization parameters κ and ζ_D, including the deterministic optimum (red dot), the 2.9σ optimum (green dot) and the 4.7σ optimum (yellow dot).

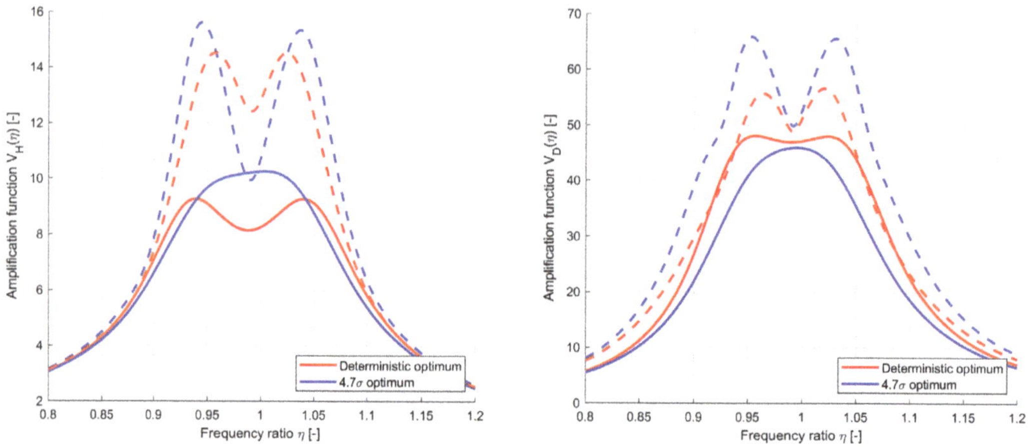

(**a**) (**b**)

Figure 14. Estimated mean values (solid line) and mean values plus 4.7-fold standard deviations (dashed line) for the deterministic optimum and for a safety margin of $\alpha = 4.7$: (**a**) for the main system amplification $V_H(\eta)$; (**b**) for the amplification of relative displacements $V_D(\eta)$.

As the final investigation for the SDOF example, the single-objective robust design optimization was performed for different nominal values of the mass ratio μ and different safety margins α by using again the simplex method. In Figure 15 the obtained maximum values for the mean and standard deviation as well as the corresponding nominal values for κ and ζ_d are shown, whereby, for $\alpha = 0.0$, the nominal values correspond to the deterministic solution presented in Section 3.1.1. The figure indicates that, with an increasing safety margin, the mean values increase, but the standard deviation decreases. Furthermore, the damping ratio increases. This effect is most dominant for small mass ratios, where the influence of the uncertain input parameters is most critical.

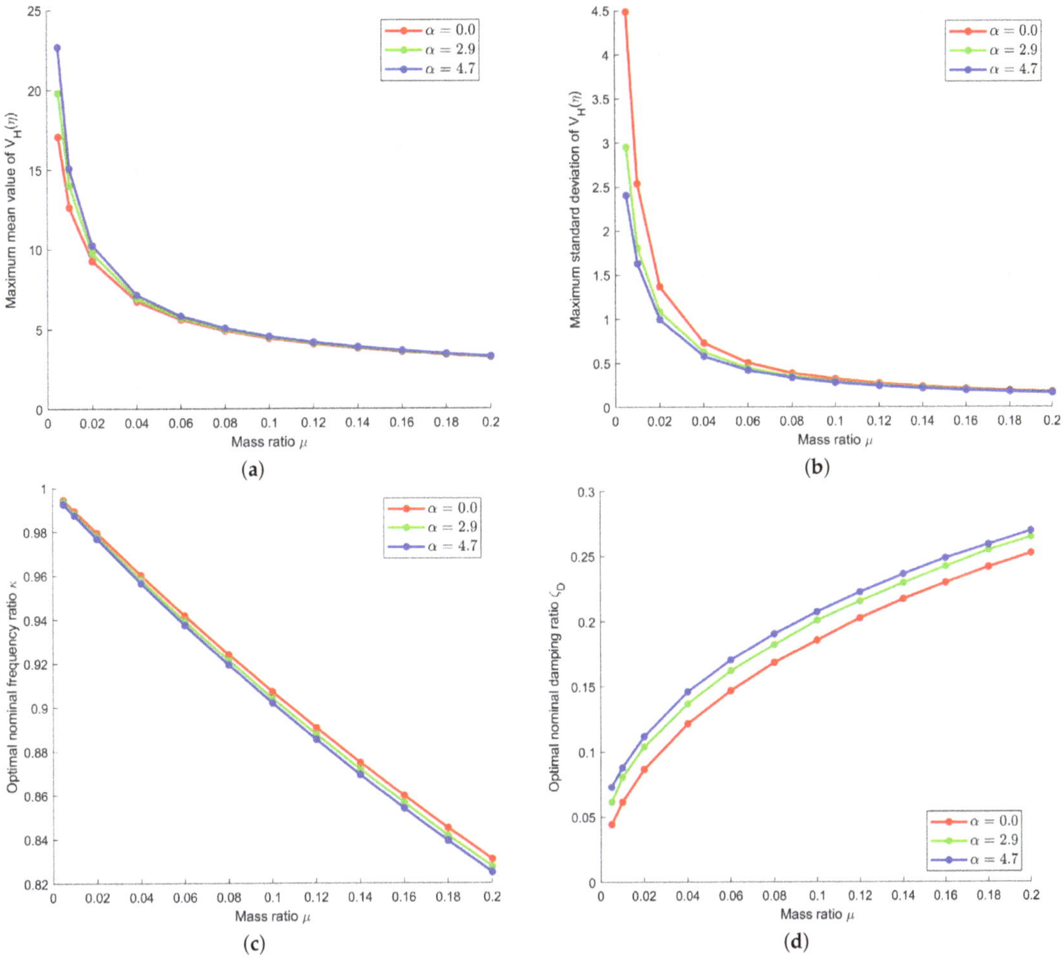

Figure 15. Optimized maximum mean value and standard deviation of the amplification function values for an increasing nominal mass ratio μ and different safety margins α: (**a**) maximum mean values of amplification function $V_H(\eta)$, (**b**) maximum standard deviation of amplification function $V_H(\eta)$, (**c**) optimal frequency ratio κ and (**d**) optimal damping ratio ζ_D.

3.2. Multi-Degree-of-Freedom Example with Multiple Tuned Mass Dampers

3.2.1. Deterministic Analysis

In the second example, a multi-degree-of-freedom system as shown in Figure 16 was investigated. The deterministic values for the stiffnesses, masses and damping ratios have been taken according to the mean values in Table 3. The damping was assumed as modal damping as explained in Section 2.2.1. The three natural circular frequencies of the main system are $\omega_{H_1} = 4.45\frac{1}{s}$, $\omega_{H_2} = 12.47\frac{1}{s}$ and $\omega_{H_3} = 18.02\frac{1}{s}$ and the corresponding mode shape are shown in Figure 17. The amplification functions for a harmonic excitation $F(t)$ are plotted in Figure 18. The peak values decrease for the larger natural frequencies due to the higher damping ratios. The TMDs have been designed for the first and second vibration mode and are coupled to the DOF 3 and 1, respectively, which have the maximum value of the corresponding mode shape. Both mode shapes $\boldsymbol{\phi}_1$ and $\boldsymbol{\phi}_2$ were normalized with respect to the coupling DOFs q_{H_3} and q_{H_1} as shown in Figure 17, which results in the modal masses of

$\tilde{m}_1 = \tilde{m}_2 = 18411.7$ kg. The optimal parameters for the nominal mass of the TMDs are taken according to Den Hartog as $\mu_1 = \mu_2 = 0.027$, $\kappa_1 = \kappa_2 = 0.974$ and $\zeta_{D_1} = \zeta_{D_2} = 0.097$, which results in the nominal stiffnesses given in Table 3. The corresponding amplification functions for the first and second modes including the TMDs are shown additionally in Figure 18.

Table 3. Statistical properties of the input parameters for the MDOF example.

Parameter	Unit	Mean Value	Coefficient of Variation	Distribution Type
m_1,m_2,m_3	kg	10^4	1.0%	Normal
k_1,k_2,k_3	kN/m	10^3	2.0%	Normal
ζ_{H_1}	-	0.010	5.0%	Normal
ζ_{H_2}	-	0.015	5.0%	Normal
ζ_{H_3}	-	0.020	5.0%	Normal
m_{D_1}	kg	500.0	1.0%	Normal
m_{D_2}	kg	500.0	1.0%	Normal
k_{D_1}	kN/m	9.386	2.0%	Normal
k_{D_2}	kN/m	73.69	2.0%	Normal
ζ_{D_1}	-	0.097	5.0%	Normal
ζ_{D_2}	-	0.097	5.0%	Normal

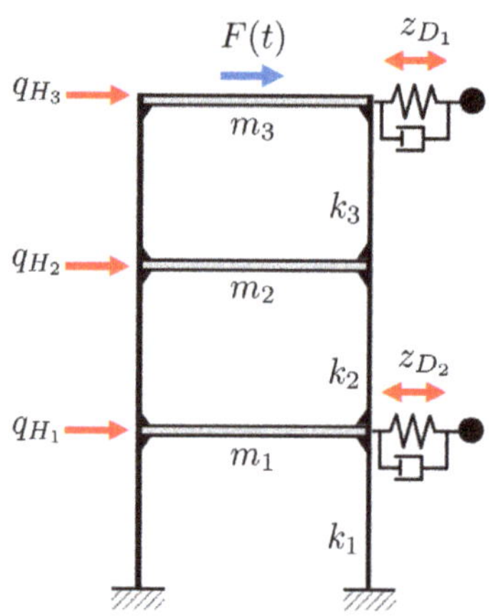

Figure 16. MDOF example with 3 main DOFs and 2 TMDs.

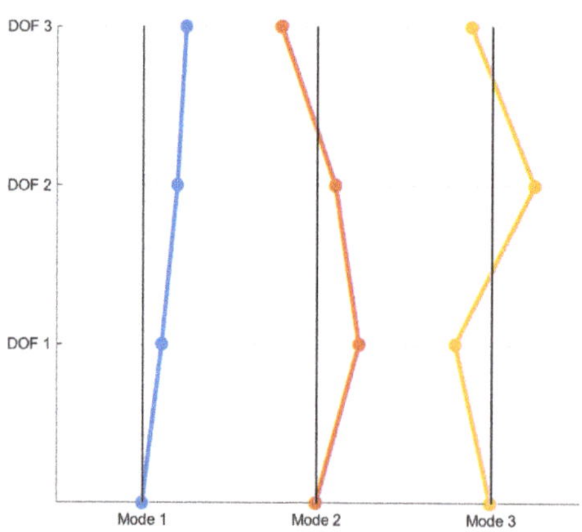

Figure 17. Mode shapes of the initial 3-DOF system with $\omega_{H_1} = 4.45\frac{1}{s}$, $\omega_{H_2} = 12.47\frac{1}{s}$ and $\omega_{H_3} = 18.02\frac{1}{s}$.

In a first step, the accuracy of the approximated displacement solution using the decoupled stationary approach according to Section 2.2.2 was investigated. A harmonic excitation at DOF q_{H_3} was assumed as $F(t) = 1kN \cdot \sin(\Omega \cdot t)$. As the benchmark method, the Newmark time integration was used for the full 5-DOF system including the TMDs within a time range of 100 excitation periods and a time step of $\frac{1}{200}$ of the smallest free vibration period of the initial main system. Figure 18 shows the displacements of DOF q_{H_3} for the first 25 excitation periods with $\Omega = 4.35\frac{1}{s}$. The figure indicates very good agreement of the approximated stationary solution with the Newmark results in the steady state.

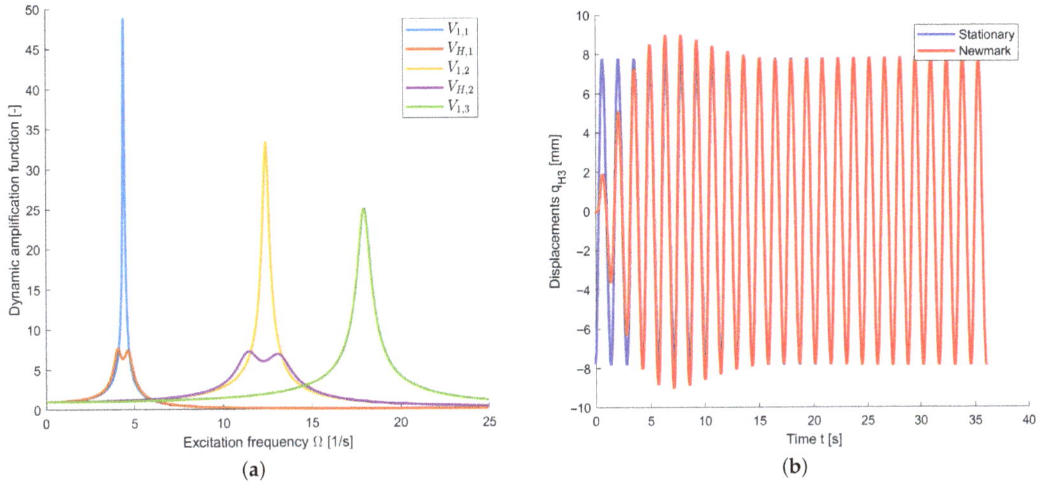

Figure 18. (**a**) Amplification functions for the 3-DOF system with and without TMDs and (**b**) approximated stationary displacements for an excitation frequency of $\Omega = 4.35\frac{1}{s}$.

This analysis was repeated for different excitation frequencies, and the maximum amplitudes of the Newmark results have been extracted from the last 25 out of 100 excitation periods. In Figure 19 the vibration amplitudes in the steady state are compared for the displacements of all DOFs of the main system as well as for the relative displacements z_{D_1} and z_{D_2} of the TMDs. Additionally, the maximum values of the drift displacement at the second floor $q_{H_3} - q_{H_2}$ are plotted. The figure indicates very good agreement for the main DOFs. Small deviations could be observed for the drift displacements. The relative displacements of the TMDs are represented very well for the corresponding first and second vibration resonance, but larger deviations could be observed at the additional peaks for the other resonances. This is the case since no interaction of the TMDs with the other vibration modes was considered in the stationary approach. However, the maximum relative displacements could be approximated sufficiently.

3.2.2. Uncertainty Quantification

In the next step the uncertainty propagation was investigated for the MDOF example by using 1000 Latin Hypercube samples and the linearization approach according to Equation (48). The samples were generated again with the MATLAB R2024b software package [30]. The random input parameters are assumed as normally distributed independent random variables with mean values and the CoV according to Table 3. As model responses, the maximum displacements of the three original DOFs and the maximum drift $q_{H_3} - q_{H_2}$ between the second and third DOF were evaluated. The estimated mean values and standard deviations are compared for both approaches depending on the excitation frequency in Figure 20 for the original DOFs and in Figure 21 for the maximum drift. The figure indicates very good agreement in the range of the first two natural frequencies, where the TMDs are active. Larger deviations could be observed in the range of the third natural frequency. The variance contribution of the inputs indicate a significant influence of the stiffness and mass coefficients of the main system as well as the modal damping ζ_3 on this displacement variation. However, the scatter of the TMD coefficients is significant only for the variation in the drift displacement for an excitation frequency around the first and second natural frequency.

3.2.3. Optimization Under Uncertainty

In the final analysis, a robust design optimization was performed with different displacement measures. First, the total mass of both TMDs was minimized and the mean value and standard deviation of the maximum displacements at the third main DOF $Q_{H_{3,k}}(\mathbf{X}) = \max_t(|q_{H_3}(t, \Omega_k, \mathbf{X})|)$ were considered in the second objective:

$$\min_{\mathbf{d}} \left(m_{D_1} + m_{D_2}, \max_{0<\Omega_k \leq 15\frac{1}{s}} \left(\bar{Q}_{H_{3,k}} + \alpha \cdot \sigma_{Q_{H_{3,k}}} \right) \right).$$

The evaluated discrete excitation frequency values were limited to the range of $0 < \Omega_k \leq 15\frac{1}{s}$ to consider only the first two resonance frequencies in the tuning of the TMD parameters. The maximum displacements for the third resonance frequency remain almost constant if the TMD parameters are modified within the optimization. In Figure 22 the maximum mean values and standard deviations of the displacements are shown, dependent on the total mass of the TMDs for the deterministic optimization with $\alpha = 0$ and for the optimization under uncertainty assuming $\alpha = 4.7$. As in the SDOF example, a significant reduction in the scatter can be observed for the RDO case while the mean values are slightly increased. The TMD tuned for the second vibration mode gets more mass in the RDO case since the scatter of the displacements around the second resonance is larger than for the first resonance frequency as shown in Figure 23.

Second, we considered the maximum drift in the upper floor

$$D_{H_{2-3,k}}(\mathbf{X}) = \max_t(|q_{H_2}(t, \Omega_k, \mathbf{X})) - q_{H_3}(t, \Omega_k, \mathbf{X}))|)$$

in the second objective

$$\min_{\mathbf{d}} \left(m_{D_1} + m_{D_2}, \max_{0<\Omega_k \leq 15\frac{1}{s}} \left(\bar{D}_{H_{2-3,k}} + \alpha \cdot \sigma_{D_{H_{2-3,k}}} \right) \right).$$

In Figure 24 the obtained mass distribution of the TMDs is shown with the mean values and standard deviations depending on the excitation frequency. The figure indicates that the TMD for the second vibration mode obtained much more mass as in the first investigated case. The mean values and the scatter of the maximum displacements are significantly reduced for the considered frequency range while the values around the third resonance remain almost constant. For the robust optimum obtained with $\alpha = 4.7$, the scatter of the maximum drift displacements are significantly reduced compared to the solutions of the deterministic optimization.

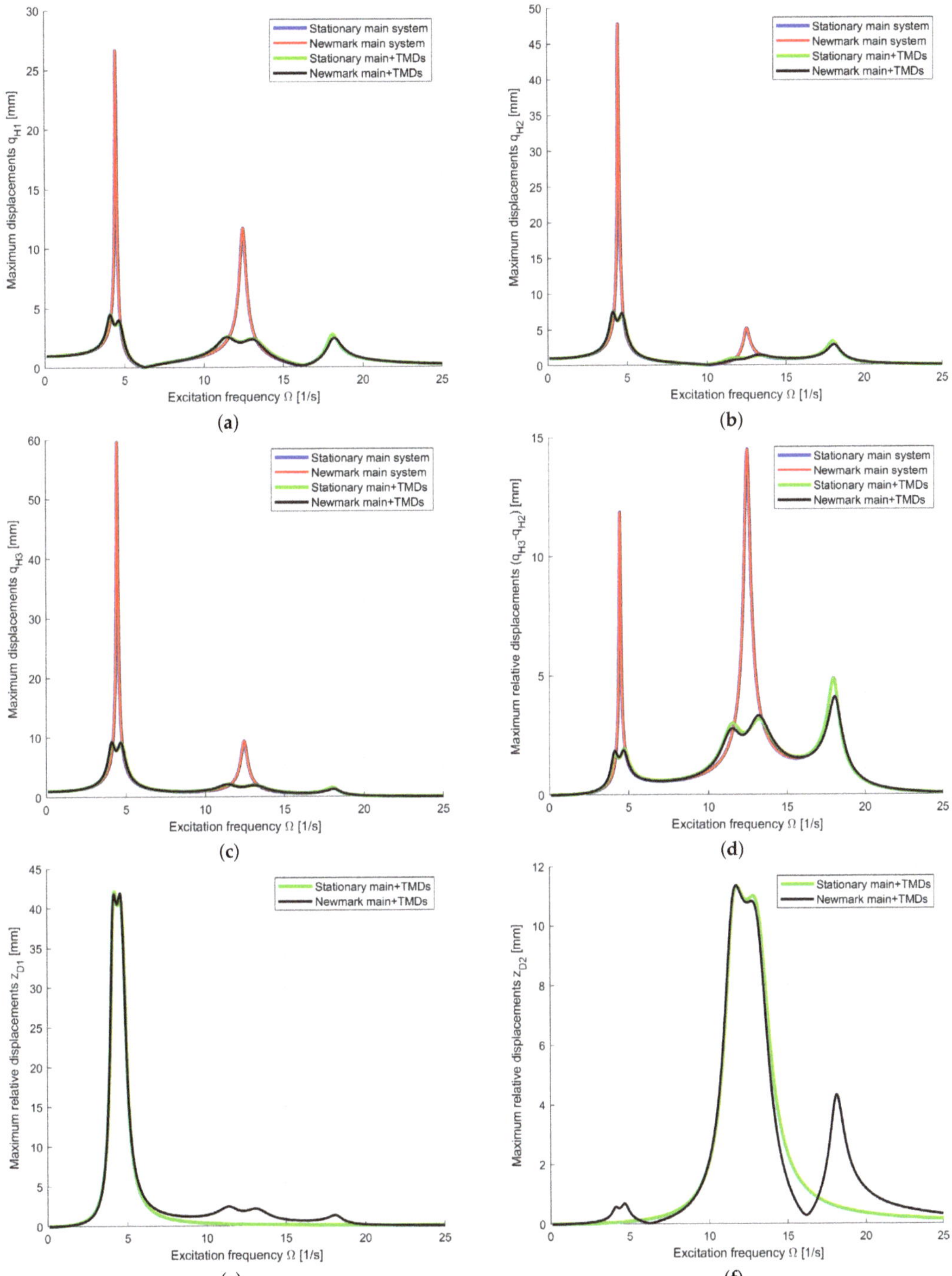

Figure 19. Approximated maximum stationary displacements of the MDOF example compared with the Newmark results: (**a**) for the DOF q_{H_1}, (**b**) for the DOF q_{H_2}, (**c**) for the DOF q_{H_3}, (**d**) for the second-floor drift $q_{H_3} - q_{H_2}$, (**e**) for the TMD relative displacements z_{D_1} and (**f**) for the TMD relative displacements z_{D_2}.

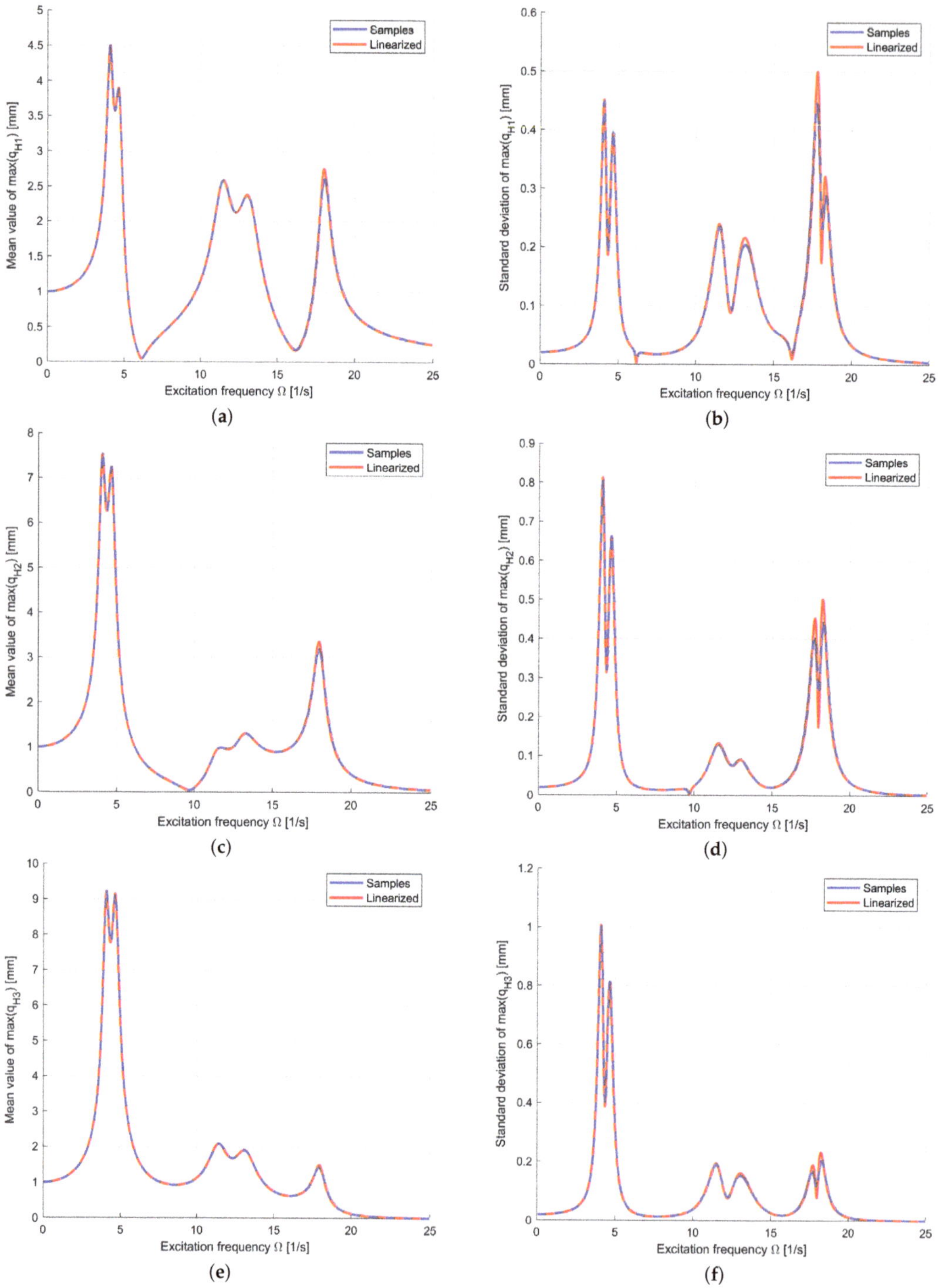

Figure 20. Estimated mean values and standard deviations of the maximum displacements of the three main DOFs estimated from 1000 LHS samples and by using the linearization approach: (**a**) mean value DOF q_{H_1}, (**b**) standard deviation DOF q_{H_1}, (**c**) mean value DOF q_{H_2}, (**d**) standard deviation DOF q_{H_2}, (**e**) mean value DOF q_{H_3} and (**f**) standard deviation DOF q_{H_3}.

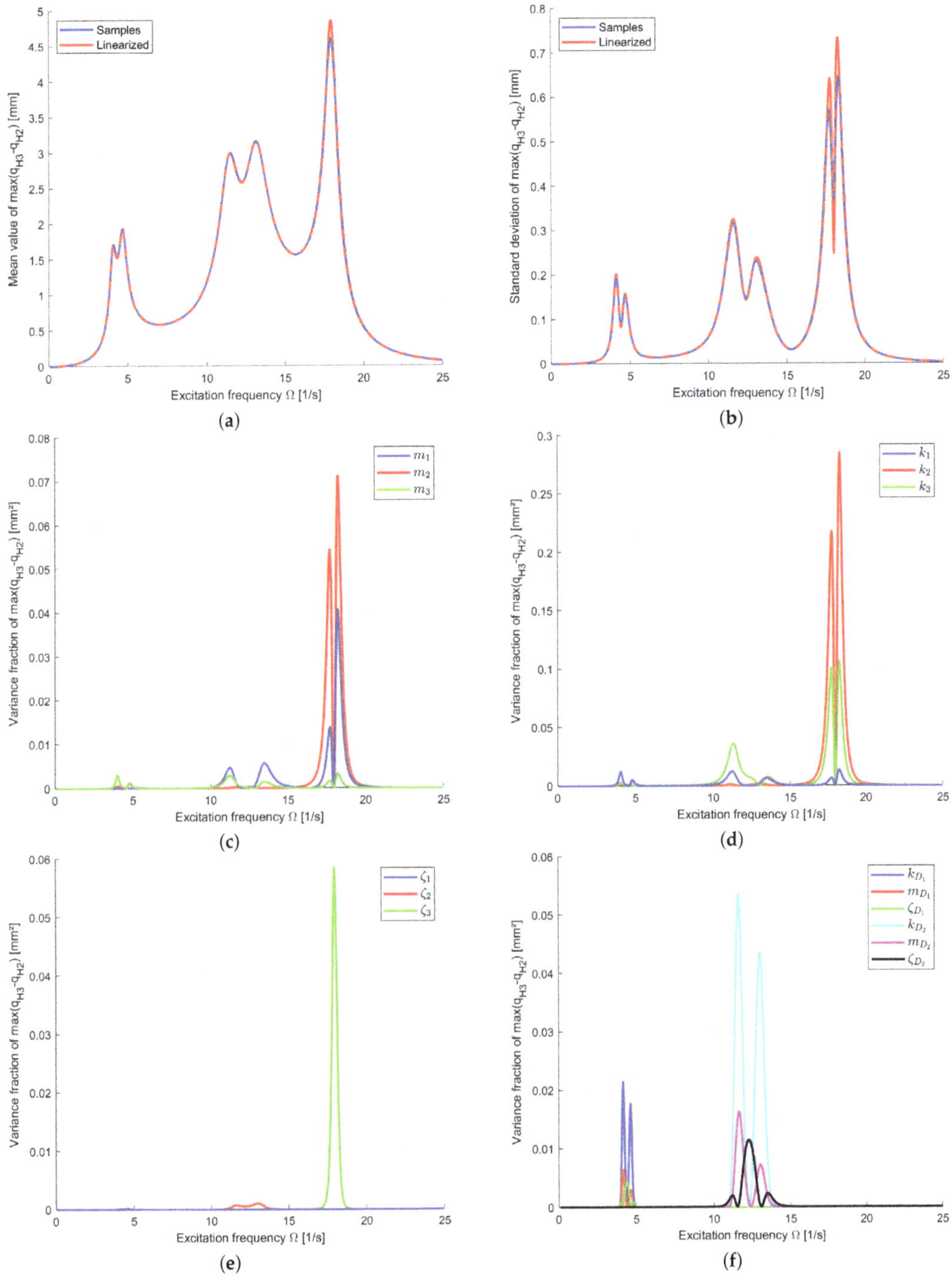

Figure 21. Mean values and standard deviations of the maximum drift $\max(q_{H_3} - q_{H_2})$ estimated from 1000 LHS samples and by using the linearization approach and linearized variance contribution of the input parameters: (**a**) mean values, (**b**) standard deviation, (**c**) variance contribution of masses, (**d**) variance contribution of stiffness, (**e**) variance contribution of damping ratios and (**f**) variance contribution of TMD parameters.

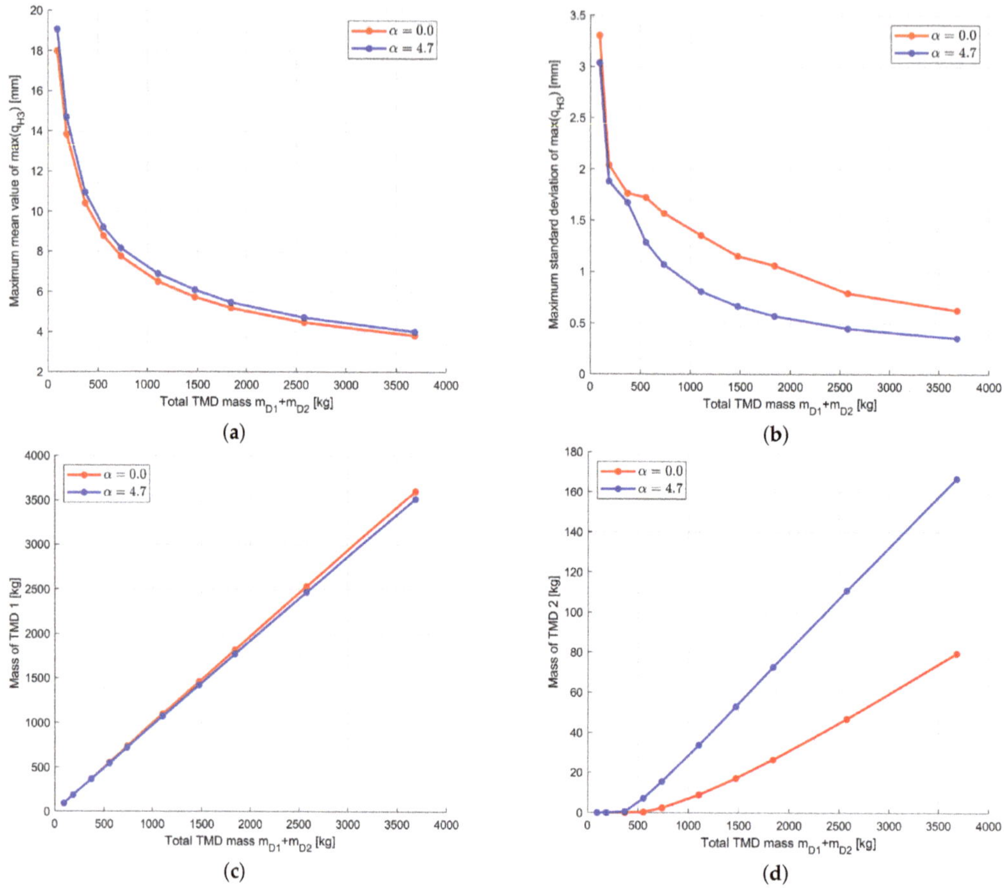

Figure 22. Optimized maximum mean values and standard deviations of the maximum displacement of q_{H_3} for an increasing mass of the TMDs and different safety margins α for the MDOF example: (**a**) optimized maximum mean values, (**b**) optimized maximum standard deviations, (**c**) optimal mass of TMD 1 and (**d**) optimal mass of TMD 2.

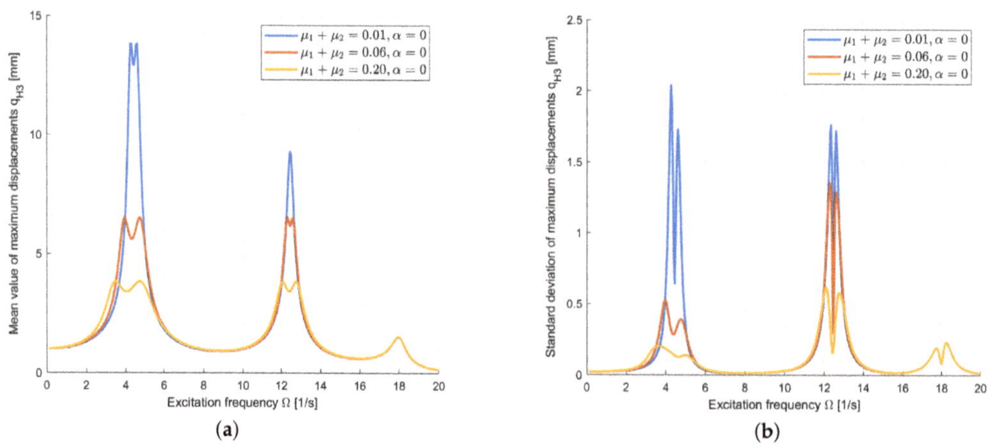

Figure 23. *Cont.*

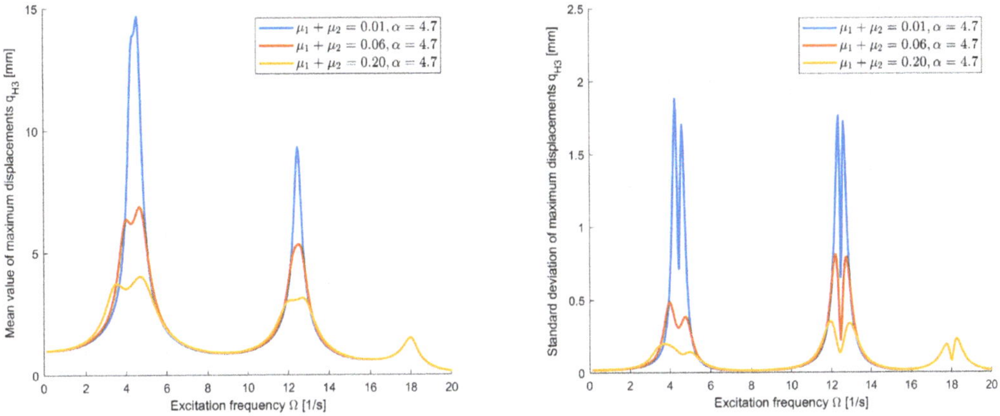

(c) (d)

Figure 23. Mean values and standard deviations of the maximum displacements q_{H_3} for an increasing mass of the TMDs dependent on the excitation frequancy and the safety margin: (**a**) mean values for $\alpha = 0$, (**b**) standard deviation for $\alpha = 0.0$, (**c**) mean values for $\alpha = 4.7$ and (**d**) standard deviation for $\alpha = 4.7$.

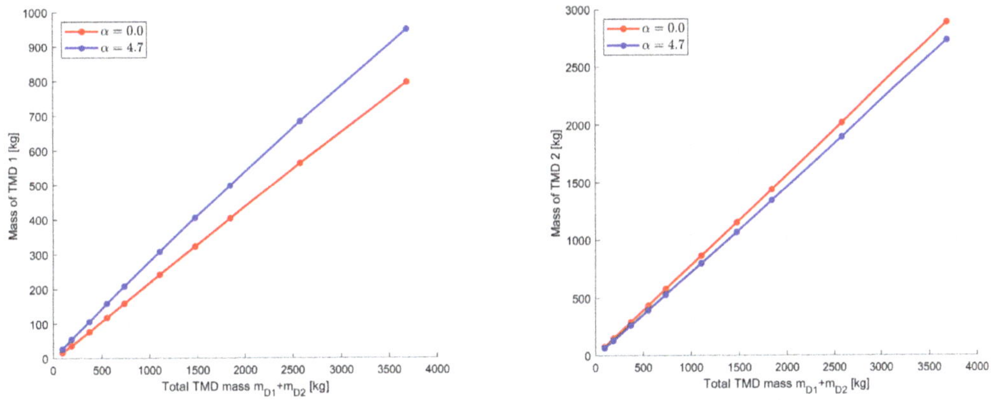

(a) (b)

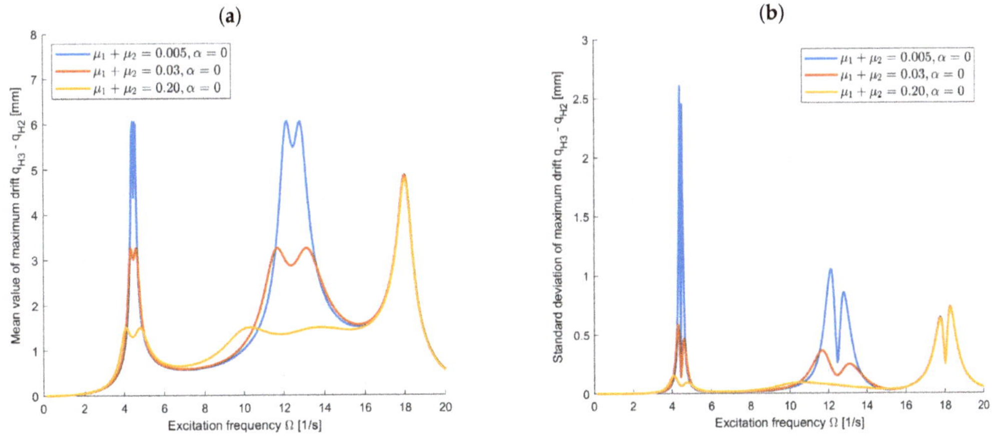

(c) (d)

Figure 24. *Cont.*

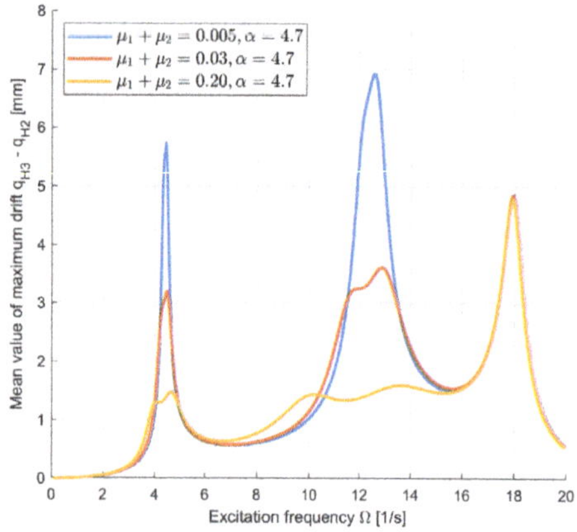

(e)

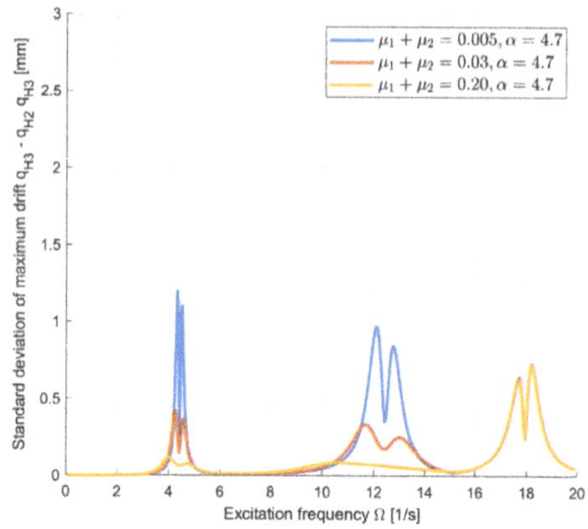

(f)

Figure 24. Optimized maximum mean values and standard deviations of the maximum drift displacement of the third floor $q_{H_3} - q_{H_2}$ for an increasing mass of the TMDs and different safety margins: (**a**) optimal mass of TMD 1, (**b**) optimal mass of TMD 2, (**c**) mean values for $\alpha = 0$, (**d**) standard deviation for $\alpha = 0.0$, (**e**) mean values for $\alpha = 4.7$ and (**f**) standard deviation for $\alpha = 4.7$.

4. Discussion

4.1. Single-Degree-of-Freedom Systems

The deterministic optimization of the TMD parameters is straightforward if the maximum value of the dynamic amplification function for the main system displacements is considered as the objective function. Further constraints could be the maximum relative displacements of the TMD, which would just limit the design space. The single- and multi-objective optimization lead to similar results; therefore, a subsequent single-objective optimization with defined mass ratios seems to be the most promising approach.

The presented analytical uncertainty quantification approach could estimate the mean value and the standard deviation of individual amplification function values with sufficient accuracy for the design process. For the final design, a sampling-based proof of the estimates should be considered. The TMD parameters were assumed within the investigated examples to be independent and normally distributed. Theoretically, the variance estimation is independent of the distribution type as long as the covariance of the random number is available. Here, further investigations for other distribution types and correlated inputs might be necessary.

The definition of the safety margin as the optimization goal or constraint is straightforward and could be directly considered by standard optimization algorithms, such as the simplex Nelder–Mead. However, the calculation of the failure probability itself would require more information, like the definition of a limit state function. Nevertheless, the obtained robust design shows significantly less scatter in the critical values of the amplification function than the pure deterministically optimized design. The optimal nominal parameters changed slightly; especially, the TMD damping is larger than that obtained by the deterministic optimization or the Den Hartog formulas.

4.2. Multi-Degree-of-Freedom Systems

In the MDOF example, we could show that the decoupled stationary solution in the modal space results in similar estimates for the maximum displacements as the time integration approach. For the robust design procedure, this approach seems sufficiently accurate. Nevertheless, the final design should be analyzed with a full time integration for verification purposes, especially since the interaction of the TMDs with the other modes is not represented in the decoupled approach. This may be critical if several natural frequencies of the original system are close to each other.

The implementation of the uncertainty quantification approach is straightforward, whereby the linearization is realized directly for the maximum displacements of the original DOFs and the investigated maximum relative displacements for each discretized excitation frequency. However, the estimated standard deviation shows significant deviations for the higher-resonance frequencies without TMD tuning. Within the design optimization it is necessary to limit the considered frequency range in order to focus on the investigated resonances damped by the TMDs. Within this region, the estimated standard deviation was sufficiently accurate for all the investigated displacement values.

Similarly, as for the SDOF system, the application of the optimization algorithms is straightforward and does not require special attention in the investigated simple example. However, this statement can not be generalized to more complex structures without further investigations. Additionally, the definition of a limit state function and a final reliability analysis might be necessary to proof the robustness of the obtained optimal design. Nevertheless, the results showed a clear difference in the obtained optimal parameters and mass distribution between the robust and the deterministic design. Additionally, the standard deviation of the investigated displacement values could be significantly reduced while the mean values have been increased slightly.

5. Conclusions

The presented uncertainty quantification method seems to be an efficient approach to avoid the typical double loop in sampling-based robust design optimization methods. Since the accuracy of linearization is very sensitive to the investigated structural response values, it should be applied to the individual amplification function or displacement values as presented in the paper. However, an extension for non-normally distributed and dependent input parameters might be the focus of future research. For a large number of random inputs, the central difference approximation of the derivatives may limit its efficiency, and sampling-based methods might be more suitable.

The underlying structural analysis is limited to linear SDOF and MDOF systems with modal damping, which enables the decomposition of the displacement solution in the modal space. For non-linear models and other damping models, the presented decomposition approach might be not suitable anymore and further research would be necessary.

However, the presented approximation could be considered during the robust design optimization to avoid numerically demanding simulation runs. For the final design, a full time integration of the structural system including TMDs should be performed in any case to validate the approximated results. So far, we could show that the procedure itself works for rather simple systems. In further studies, the presented approach should be applied for more sophisticated structures, such as high-rise buildings and pedestrian bridges.

Author Contributions: Conceptualization, T.M., V.Z. and R.R.D.; methodology, T.M., V.Z. and R.R.D.; software, T.M. and A.K.; validation, T.M.; formal analysis, T.M. and A.K.; investigation, T.M. and A.K.; resources, T.M. and V.Z.; data curation, T.M.; writing—original draft preparation, T.M.; writing—review and editing, V.Z. and R.R.D.; visualization, T.M.; supervision, V.Z.; project administration, T.M.; funding acquisition, no funding. All authors have read and agreed to the published version of the manuscript.

Funding: This research received no external funding.

Institutional Review Board Statement: Not applicable.

Informed Consent Statement: Not applicable.

Data Availability Statement: The MATLAB calculation files and Ansys optiSLang project files including all the presented results are available in the publicly accessible repository https://refodat.de/, at the DOI: http://doi.org/10.71758/refodat.48 (accessed on 14 August 2025).

Acknowledgments: We acknowledge support for the publication costs by the Open Access Publication Fund of Bauhaus Universität Weimar and the Deutsche Forschungsgemeinschaft (DFG).

Conflicts of Interest: The authors declare no conflicts of interest.

Abbreviations

The following abbreviations are used in this manuscript:

CoV	Coefficient of Variation
DOF	Degree of Freedom
LHS	Latin Hypercube Sampling
MDOF	Multi-Degree of Freedom
NSGA II	Non-dominated Sort Genetic Algorithm
RDO	Robust Design Optimization
SDOF	Single Degree of Freedom
TMD	Tuned Mass Damper

References

1. Ormondroyd, J.; Den Hartog, J. The theory of the dynamic vibration absorber. *J. Fluids Eng.* **1928**, *49*, 021007. [CrossRef]
2. Gutierrez Soto, M.; Adeli, H. Tuned mass dampers. *Arch. Comput. Methods Eng.* **2013**, *20*, 419–431. [CrossRef]
3. Xu, Y.; Kwok, K.; Samali, B. Control of wind-induced tall building vibration by tuned mass dampers. *J. Wind. Eng. Ind. Aerodyn.* **1992**, *40*, 1–32. [CrossRef]
4. Kwok, K.; Samali, B. Performance of tuned mass dampers under wind loads. *Eng. Struct.* **1995**, *17*, 655–667. [CrossRef]
5. Clark, A. Multiple passive tuned mass dampers for reducing earthquake induced building motion. In Proceedings of the 9th World Conference on Earthquake Engineering, Tokyo, Japan, 2–9 August 1988; Volume 5, pp. 779–784.
6. Sadek, F.; Mohraz, B.; Taylor, A.; Chung, R. A method of estimating the parameters of tuned mass dampers for seismic applications. *Earthq. Eng. Struct. Dyn.* **1997**, *26*, 617–635. [CrossRef]
7. Petersen, C.; Werkle, H. *Dynamik der Baukonstruktionen*, 2nd ed.; Springer Viehweg: Wiesbaden, Germany, 2017.
8. Warburton, G.B. Optimum absorber parameters for various combinations of response and excitation parameters. *Earthq. Eng. Struct. Dyn.* **1982**, *10*, 381–401. [CrossRef]
9. Wang, Y.; Cheng, S. The optimal design of dynamic absorber in the time domain and the frequency domain. *Appl. Acoust.* **1989**, *28*, 67–78. [CrossRef]
10. Rana, R.; Soong, T. Parametric study and simplified design of tuned mass dampers. *Eng. Struct.* **1998**, *20*, 193–204. [CrossRef]
11. De Fazio, N.; Placidi, L.; Tomassi, A.; Fraddosio, A.; Castellano, A.; Paparella, F. Different mechanical models for the study of ultrasonic wave dispersion for mechanical characterization of construction materials. *Int. J. Solids Struct.* **2025**, *315*, 113352. [CrossRef]
12. Lee, C.L.; Chen, Y.T.; Chung, L.L.; Wang, Y.P. Optimal design theories and applications of tuned mass dampers. *Eng. Struct.* **2006**, *28*, 43–53. [CrossRef]
13. Hadi, M.; Arfiadi, Y. Optimum design of absorber for MDOF structures. *J. Struct. Eng.* **1998**, *124*, 1272–1280. [CrossRef]
14. Hoang, N.; Warnitchai, P. Design of multiple tuned mass dampers by using a numerical optimizer. *Earthq. Eng. Struct. Dyn.* **2005**, *34*, 125–144. [CrossRef]
15. Zuo, L.; Nayfeh, S. Optimization of the individual stiffness and damping parameters in multiple-tuned-mass-damper systems. *J. Vib. Acoust.* **2005**, *127*, 77–83. [CrossRef]
16. Leung, A.; Zhang, H. Particle swarm optimization of tuned mass dampers. *Eng. Struct.* **2009**, *31*, 715–728. [CrossRef]
17. Arfiadi, Y. Optimum placement and properties of tuned mass dampers using hybrid genetic algorithms. *Int. J. Optim. Civ. Eng.* **2011**, *1*, 167.

18. Radmard, R.; Könke, C. Seismic control of tall buildings using distributed multiple tuned mass dampers. *Adv. Civ. Eng.* **2019**, *2019*, 6480384. [CrossRef]
19. Yucel, M.; Bekdaş, G.; Nigdeli, S.; Sevgen, S. Estimation of optimum tuned mass damper parameters via machine learning. *J. Build. Eng.* **2019**, *26*, 100847. [CrossRef]
20. Yucel, M.; Bekdaş, G.; Nigdeli, S. Machine learning-based model for prediction of optimum TMD parameters in time-domain history. *J. Braz. Soc. Mech. Sci. Eng.* **2024**, *46*, 192. [CrossRef]
21. Hu, W. *Design Optimization Under Uncertainty*; Springer: Cham, Switzerland, 2023.
22. Papadimitriou, C.; Katafygiotis, L.; Au, S.K. Effects of structural uncertainties on TMD design: A reliability-based approach. *J. Struct. Control* **1997**, *4*, 65–88. [CrossRef]
23. Schmelzer, B.; Oberguggenberger, M.; Adam, C. Efficiency of tuned mass dampers with uncertain parameters on the performance of structures under stochastic excitation. *Proc. Inst. Mech. Eng. Part O J. Risk Reliab.* **2010**, *224*, 297–308. [CrossRef]
24. Moreno, C.; Thomson, P. Design of an optimal tuned mass damper for a system with parametric uncertainty. *Ann. Oper. Res.* **2010**, *181*, 783–793. [CrossRef]
25. Pellizzari, F.; Marano, G.; Palmeri, A.; Greco, R.; Domaneschi, M. Robust optimization of MTMD systems for the control of vibrations. *Probabilistic Eng. Mech.* **2022**, *70*, 103347. [CrossRef]
26. Shamsaddinlou, A.; Shirgir, S.; Hadidi, A.; Azar, B. An efficient reliability-based design of TMD & MTMD in nonlinear structures under uncertainty. *Structures* **2023**, *51*, 258–274. [CrossRef]
27. Li, D.; Tang, H.; Xue, S. Robust design of tuned mass damper with hybrid uncertainty. *Struct. Control Health Monit.* **2021**, *28*, e2803. [CrossRef]
28. Sreeman, D.; Roy, B. Robust design optimization of the friction pendulum system isolated building considering system parameter uncertainties under seismic excitations. *J. Build. Eng.* **2024**, *82*, 108320. [CrossRef]
29. Huntington, D.; Lyrintzis, C. Improvements to and limitations of Latin hypercube sampling. *Probabilistic Eng. Mech.* **1998**, *13*, 245–253. [CrossRef]
30. The MathWorks Inc. *MATLAB*, version R2024b; MATLAB User Documentation; The MathWorks Inc.: Natick, MA, USA, 2024. Available online: https://www.mathworks.com (accessed on 18 June 2025).
31. Luersen, M.A.; Le Riche, R.; Guyon, F. A constrained, globalized, and bounded Nelder–Mead method for engineering optimization. *Struct. Multidiscip. Optim.* **2004**, *27*, 43–54. [CrossRef]
32. Deb, K.; Pratap, A.; Agarwal, S.; Meyarivan, T. A fast and elitist multiobjective genetic algorithm: NSGA-II. *IEEE Trans. Evol. Comput.* **2002**, *6*, 182–197. [CrossRef]
33. Ansys Germany GmbH. *Ansys optiSLang*, version 2025R1; Ansys optiSLang User Documentation; Ansys Germany GmbH: Weimar, Germany, 2025. Available online: https://www.ansys.com/de-de/products/connect/ansys-optislang (accessed on 18 June 2025).
34. Eibl, J.; Häussler-Combe, U. Baudynamik. In *Betonkalender 1997*; Ernst & Sohn: Berlin, Germany 1997; pp. 755–863.
35. Haftka, R.; Gürdal, Z. *Elements of Structural Optimization*; Kluwer Academic Publishers: Dordrecht, The Netherlands, 1992.
36. Nataf, A. Détermination des distributions de probabilités dont les marges sont données. *Comptes Rendus L'academie Des Sci.* **1962**, *225*, 42–43.
37. Bucher, C. *Computational Analysis of Randomness in Structural Mechanics*; CRC Press, Taylor & Francis Group: London, UK, 2009.
38. Rubinstein, R.Y. *Simulation and the Monte Carlo Method*; John Wiley & Sons: New York, NY, USA, 1981.
39. Saltelli, A.; Ratto, M.; Andres, T.; Campolongo, F.; Cariboni, J.; Gatelli, D.; Saisana, M.; Tarantola, S. *Global Sensitivity Analysis. The Primer*; John Wiley & Sons, Ltd: Chichester, UK, 2008.
40. Koch, P.; Yang, R.J.; Gu, L. Design for six sigma through robust optimization. *Struct. Multidiscip. Optim.* **2004**, *26*, 235–248. [CrossRef]
41. Most, T.; Will, J. Robust Design Optimization in industrial virtual product development. In Proceedings of the 5th International Conference on Reliable Engineering Computing (REC), Brno, Czech Republic, 13–15 June 2012; pp. 353–366.
42. *EN 1990:2021-10*; Eurocode—Basis of Structural Design, Regulation 305/2011. The European Union: Brussels, Belgium, 2021.
43. Chopra, A. *Dynamics of Structures*, 5th ed.; Pearson Education, Inc.: Uttar Pradesh, India, 2017.
44. Khadka, A. Optimization of Tuned Mass Dampers in Multi-Degree-of-Freedom Systems. Master's Thesis, Bauhaus-Universität Weimar, Weimar, Germany, 2025.

applied sciences

MDPI

Article

A Unified Modelling Framework Combining FTA, RBD, and BowTie for Reliability Improvement

Mohamad Afiq Amiruddin Parnon [1,2,*], Kassandra A. Papadopoulou [3] and Jyoti K. Sinha [1]

1 Department of Mechanical and Aerospace Engineering, School of Engineering, The University of Manchester, Manchester M13 9PL, UK; jyoti.sinha@manchester.ac.uk
2 Jabatan Teknologi Kejuruteraan Mekanikal, Universiti Teknikal Malaysia Melaka, Hang Tuah Jaya, Durian Tunggal 76100, Malaysia
3 Alliance Manchester Business School, The University of Manchester, Manchester M13 9PL, UK; kassandra.papadopoulou@manchester.ac.uk
* Correspondence: mohamadafiqamiruddin.parnon@postgrad.manchester.ac.uk

Featured Application

The proposed BowTie-based reliability framework provides a structured approach for examining and enhancing wind turbine resilience. It brings together preventive and mitigative pathways, enabling decision-makers, such as asset managers, to pinpoint critical barriers, prioritise maintenance, and evaluate the benefits of redundancy or redesign strategies. The methodology can also be applied to other critical infrastructure, particularly where critical processes are involved and defence-in-depth and quantitative risk justification are crucial. This makes the framework a practical decision-support tool for balancing multiple attributes, including safety, reliability, and cost, within complex engineered systems.

Abstract

Ensuring reliability and safety is essential in complex energy systems such as wind turbines, where failures can trigger unexpected downtimes, severe incidents, and significant costs. This study proposes a hybrid BowTie-based reliability framework that integrates Fault Tree Analysis, Reliability Block Diagrams, and BowTie methodology to quantify risk and evaluate the effectiveness of safety barriers. The framework employs key reliability metrics including availability, probability of failure on demand, and probability of failure per hour, and supports scenario-based sensitivity analyses to explore redesign options. A simulation-based case study of a wind turbine generator subsystem is presented, using parameter values drawn from published reliability data. Results highlight that protective relays and automatic trip systems represent critical single points of defence, while improvements such as enhanced oil analysis and redundant dashboards reduce consequence frequency from 2.912×10^{-17} to 8.257×10^{-19} failures/h (a 97.16% reduction, nearly two orders of magnitude). Compared to conventional models, the proposed framework introduces explicit defence in depth modelling, improves computational compactness, and provides a practical decision support tool for asset managers by balancing safety and reliability. At this stage, the study should be regarded as a proof of concept that demonstrates feasibility and sets a foundation for future research and application to larger, more complex infrastructures.

Keywords: reliability modelling; minimal cut set analysis; defence-in-depth

1. Introduction

Reliability analysis is an essential method in engineering that concentrates on the design, operation, and maintenance of complex systems, guiding maintenance strategies to improve asset performance. Traditional methods, such as Fault Tree Analysis (FTA) and Reliability Block Diagrams (RBDs), remain prevalent, offering structured approaches to identify failure logic and assess system performance. FTA, a top-down approach, uses Boolean logic to estimate failure probability, while RBD offers a complementary view at the system level, illustrating how components and subsystems interact through series and parallel configurations to ensure system functionality. Both techniques have proven effective in safety-critical fields such as nuclear power, aerospace, and energy infrastructure. Despite their long-standing usefulness, each has notable limitations when used alone [1,2]. FTA, although straightforward and rigorous, can become unwieldy in large systems due to combinatorial complexity and is limited by its static nature, which hampers its ability to model repair processes, time-dependent phenomena, or conditional interactions [3]. Conversely, RBD allows quick system-level insights but relies on binary assumptions, making it challenging to represent multi-state, dependent, or sequential behaviours. These constraints are particularly problematic in modern critical infrastructure, such as wind turbines (WT), where systems operate under dynamic conditions and exhibit interdependent failure modes.

Recent studies have attempted to overcome these limitations by hybridising FTA, RBD, and related methods. For example, Rivera et al. [4] suggest an integrated method that combines FTA, RBD, the Analytical Hierarchy Process (AHP), High Reliability Organisation (HRO), and Learning from Failures (LFF) tools to analyse long-term patterns of small to medium chronic failures in an Oil and Gas (O&G) organisation. However, this unified framework increases complexity and is sensitive to expert bias. Conversely, Kabir et al. [5] address individual limitations by assigning large, mainly series or parallel segments to a modularisation technique using Binary Decision Diagram (BDD) and Markov models. This approach can reduce the size and complexity of the embedded FTA, preventing the exponential growth of cut sets. Still, it requires advanced solvers and can make model traceability across modules challenging. Cheng et al. [6] share a similar viewpoint in their study on unified approaches, connecting FTA logic to a Hierarchical Belief Rule Base (BRB). They manage to mitigate issues related to excessive indexes and avoid combinatorial explosion [4], but the framework's scalability remains unvalidated, and it has a limited domain focus.

Recent work has also explored hybrid frameworks that combine the BowTie method with probabilistic reasoning to overcome its static nature and expand its analytical capabilities. For example, Khakzad et al. [7] mapped a BowTie model into a Bayesian network to perform dynamic safety analysis. They noted that the BowTie approach remains popular for accident modelling but is limited by its static representation and inability to represent conditional dependencies. By translating the BowTie to a Bayesian network, probabilities can be updated in response to new information and causal relationships. Similarly, de Barnier et al. [8] introduce a possibility-based BowTie to quantify uncertainties in barrier performance. Another dynamic risk-assessment study [9] emphasises that BowTie is a visual method for depicting event progression from causes to effects, while a Bayesian network captures stochastic relationships. The translation from BowTie to Bayesian network modifies the representation to include intrinsic uncertainties while maintaining logical relationships. Although valuable, these unified methodologies often trade clarity for complexity, a lack of redundancy integration, and risk fragmentation in tool support across different methods [10]. The limitations of hybrid methods could lead to additional design and verification burdens affecting reproducibility and traceability. A comparative taxonomy of recent hybrid methods is summarised in Table 1, which positions the present work relative to state-of-the-art contributions.

Table 1. Comparative summary of hybrid reliability-modelling frameworks (2010–2024).

Study and Year	Hybrid Methods	Application Domain	Main Contribution
Rivera et al. (2021) [4]	FTA + RBD + AHP + HRO + Learning-from-Failures	Oil and Gas organisation	Multi-criteria framework linking organisational reliability factors with classic FTA/RBD models to analyse chronic failures
Kabir et al. (2020) [5]	Modular dynamic fault trees using BDD and Markov models	Safety-critical systems	Hybrid modularisation reduces state-space explosion in dynamic FTA, allowing non-exponential failure distributions
Cheng et al. (2023) [6]	FTA + Hierarchical Belief Rule Base (BRB)	Milling fault detection	Links FTA logic to a BRB to handle uncertainty and avoid combinatorial explosion
Khakzad et al. (2013) [7]	BowTie mapped to Bayesian network	Chemical process safety	Dynamic safety analysis by mapping BowTie to BN; allows update of probabilities and representation of conditional dependencies
de Barnier et al. (2022) [8]	Quantitative BowTie (possibility-based)	Industrial risk assessment	Introduces a possibility-based BowTie to quantify uncertainties in barrier performance
Wu et al. (2023) [9]	BowTie + Bayesian network	Petrochemical risk assessment	Highlights BowTie as a visual method and BN as a probabilistic model; modifies BowTie to incorporate uncertainties and maintain logical relationships

This review highlights clear research limitations within existing hybrid methods. Firstly, conventional models struggle to handle intricate systems due to their inability to reduce combinatorial complexity. Secondly, although hybrid models may capture uncertainty, they neglect redundancy and conditional interactions. Lastly, hybrid models retain domain-specificity that limits generalisation.

To address these limitations, this study introduces a hybrid BowTie-based reliability framework that integrates FTA's top-down logical structure with RBD's system-level quantitative features within a barrier-centric architecture. The framework tackles combinatorial complexity by compressing large FTA subtrees into modular BowTie substructures, where threats and barriers are organised using RBD configurations (series, parallel, or k-out-of-n). This modularisation prevents the exponential growth of minimal cut sets and ensures that barrier interactions remain traceable. Redundancy and conditional dependencies are explicitly represented, with series and parallel redundancy modelled through RBD logic, while barrier sufficiency and dependence are quantified using Risk Achievement Worth (RAW) and Barrier Importance Factor (BIF). Conditional probabilities are assigned to barriers and propagated through the BowTie logic, overcoming the binary constraints of traditional RBD models. Finally, the framework is designed to be generalisable, as the probability inversion and modular substitution techniques are independent of domain specifics and can be applied to any engineered system with barrier data. This makes the approach scalable and transferable beyond a specific case study, supporting wider critical infrastructures that require defence-in-depth and quantitative reliability justification.

Accordingly, this study is guided by the research question: Which failure scenarios and corresponding recovery pathways in the system architecture contribute most to overall risk, and where should redundancy or controls be implemented or redesigned to maximise reliability and safety? To answer this question, this study outlines measurable outcomes: (1) ranking of risk contribution (using Risk Achievement Worth); (2) quantified reliability improvements (reduction in failure frequencies) influenced by implemented strategies; (3) evidence of greater model efficiency, demonstrated through more streamlined pathways compared to conventional reliability model; (4) quantitative justification of barrier sufficiency, using metrics such as availability, average probability of failure on demand (PFD_{avg}), and probability of failure per hour (PFH) under mission time. These outcomes are exemplified through a case study of a WT subsystem in Section 3 to ensure its practicality.

This paper is organised into six sections, including an introduction (Section 1) to promote a more precise understanding. Section 2 introduces the proposed framework architecture underlying the research. Section 3 concentrates on implementing the proposed framework in a case study of WT subsystems. Section 4 examines the results obtained from the BowTie framework computation. Section 5 discusses the outcomes produced by the proposed framework. Finally, Section 6 summarises the paper by reflecting on the significance and contributions of this study within the theoretical context.

2. Proposed BowTie Framework Architecture

This section provides the conceptual and methodological foundation of the study. The unified BowTie framework adapts several core components from traditional BowTie analysis, such as hazard, TE, threat, consequence, barrier, and escalation factor. Standard BowTie analysis employs two main methods: FTA as the threat pathway and Event Tree Analysis (ETA) as the consequence pathway, as shown in Figure 1. Unlike traditional methods discussed in this paper, the BowTie model includes human and socio-technical factors, thus extending beyond purely mechanical systems. Its dual representation allows risks to be quantified and risk-informed in decision-making for safety and performance improvement.

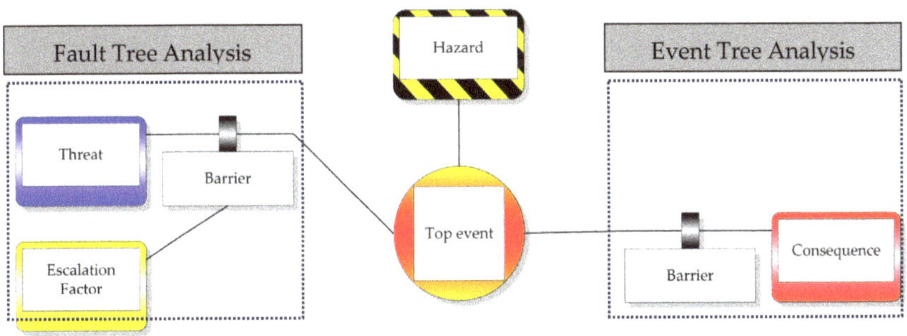

Figure 1. Standard BowTie diagram.

The proposed method enhances traditional BowTie by integrating RBD features, such as series, parallel, and k-out-of-n architectures, to arrange threat, barrier, and consequence positioning and sequencing, thereby offering a more flexible and scalable model. This integration follows a sequential workflow, as shown in Figure 2. First, FTA is initially developed to define threats and their causal structures, identifying logical combinations of basic events that lead to the TE. At this stage, it provides initial threat frequencies, $\lambda_{threat,i}$. Next, the barriers, such as PFD_{avg}, PFH, Human Error Probability (HEP) or conditional probability, $P_{conditional}$, are collected and converted into equivalent failure probability, P_{fail}, using standardised rules as presented in Table 2. This ensures consistent reliability representation and resolves unit mismatches between FTA probabilities and RBD reliability functions. Then, these barriers are organised in RBD, with AND gates represented as series chains and OR gates as parallel structures, preserving the logical relationships from FTA. These outputs are integrated into the BowTie framework, where the prevention side aims to avert threats, and the mitigation side manages the consequences if the TE occurs. This combination calculates the TE frequency, $\lambda_{TE}(t)$, and the associated consequence frequencies, $f_C(t)$. The findings based on $\lambda_{TE}(t)$ and $f_C(t)$ metrics are evaluated to observe the model's effectiveness.

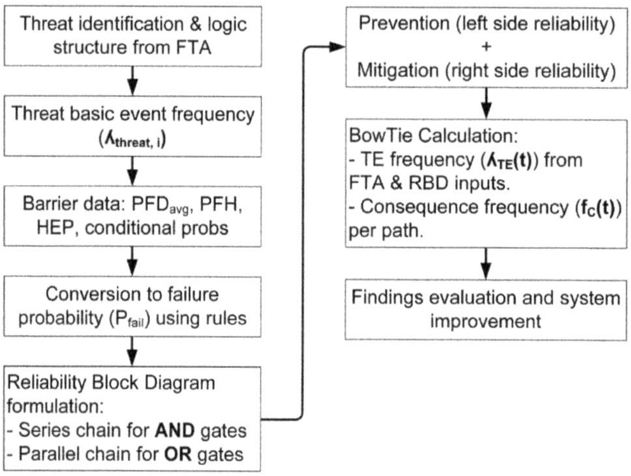

Figure 2. The workflow of FTA, RBD, and the BowTie integration framework.

Table 2. Illustrative of the Barrier Catalogue and conversion rule.

Barrier Input Type	Barrier Expression Type	Unit Dimension	Conversion Rule
PFD_{avg}	Low demand	Probability	$\lambda_b = -\ln(1 - PFD_{avg})/T$ $P_{fail} = 1 - e^{-\lambda_b t}$
PFH	High demand/Continuous	Failure/hour	$\lambda_b = PFH$ $P_{fail} = 1 - e^{-\lambda_b t}$
HEP	Human reliability	Probability	$P_{fail} = 1 - (1 - HEP)^n$
F(t)	Passive device reliability	Probability	$P_{fail} = F(t)$
$P_{conditional}$	Human + machine reliability	Probability	$P_{fail} = P_{conditional}$

Figure 3 shows the detailed process flow in the proposed model to explain how it functions before exploring the following subsections.

2.1. Data Collection and Preprocessing

To advance the process, one should gather the data and preprocess it to achieve a consistent domain. The process starts with a list of threats derived from the basic event of FTA, which must be established beforehand, based on IEC 61025 [11]. The list should include the threat's name, a unique identification code, and a description to facilitate easy tracking and identification. Each threat is then quantified with annual failure rates obtained from an established maintenance record, such as a Computerised Maintenance Management System (CMMS).

Another important tool in the proposed BowTie is the barrier's information, also called the barrier catalogue. This catalogue includes details about preventive actions that stop threats from occurring. It contains the barrier name, barrier input type, expression type, unit dimension, role, input value, data sources and conversion rule, as illustrated in Table 2. Operational data such as annual demands, proof test dates, and repairs must be included in the catalogue.

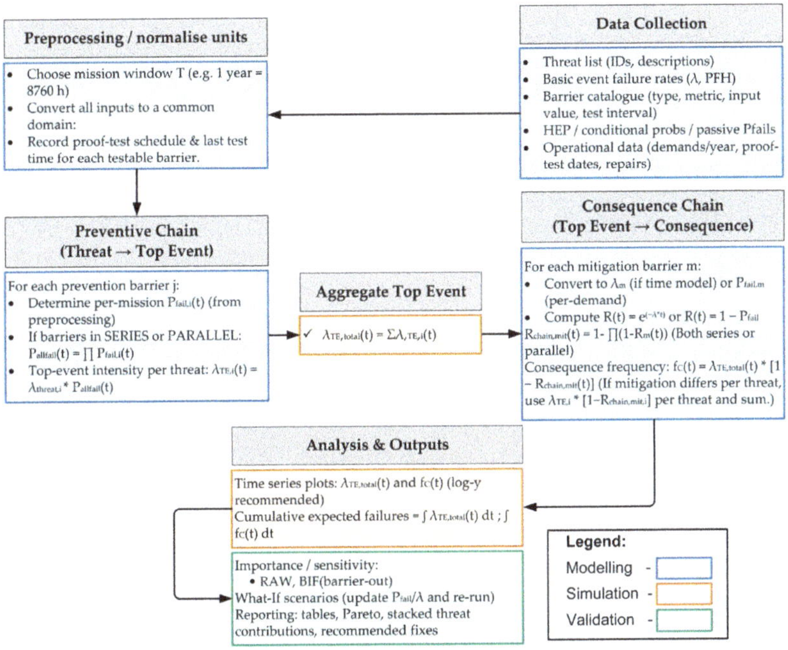

Figure 3. The proposed unified Bowtie framework architecture.

Furthermore, it is important to note that before normalising input values into the same domain, the mission time, T, must be chosen, as each barrier may operate over different time intervals. Once T is selected, all input values of barriers are normalised into the same domain, such as converting failure rates into probabilities; this improves calculation accuracy. If the historical data are inconsistent across reporting years or incomplete, median values are used. Otherwise, if the data are missing, the values are supplemented by peer-reviewed literature or standard reliability databases. After data collection and normalisation are complete, the process moves on to the preventive chain side, as illustrated in Figure 3.

2.2. Preventive Pathway

The proposed framework's preventive chain represents the left-hand side of the BowTie structure, where threats or basic events are systematically determined and controlled by proactive barriers before they escalate into the TE. Each barrier is selected from the barrier catalogue, which specifies its function, type and performance metric as demonstrated in Table 2. These quantitative metrics depend on barrier type, as outlined by the Centre for Chemical Process Safety (CCPS) [12]. Table 3 shows the mapping of barrier types and their corresponding quantitative metrics, along with descriptions.

Table 3. Barrier Types and Their Quantitative Metrics.

Barrier Type	Description	Quantitative Metrics
Behaviour	Human actions, operator compliance	HEP
Socio-technical	Organisational and systemic safeguards	$P_{conditional}$
Active hardware	Components that must act on demand	PFD_{avg}
Continuous hardware	Barriers that operate continuously	PFH
Passive hardware	Inherent design features that require no intervention	$F(t)$

Each barrier is characterised by its attributes, as documented in the barrier catalogue. The proposed framework utilises RBD features by arranging barriers in series when there is a sequential dependence, or in parallel if there is redundancy, depending on the system architecture. For the preventive pathway, all equations are based on failure probability, similar to FTA. However, there is a slight difference when applying equations for series and parallel structures compared to the RBD and FTA theory models.

Within the BowTie model, regardless of whether the configuration is series or parallel, all barriers must fail for a TE to occur. As a result, both configurations have the same equations as follows:

$$P_{\text{all\_fail,pre},i}(t) = \prod_{j \in \mathcal{B}_i} P_{\text{fail,pre},ij}(t) \quad (1)$$

where $j \in \mathcal{B}_i$ means that each preventive barrier j belongs to the set of preventive barriers, $\mathcal{B}_i$. $P_{\text{fail,pre},ij}(t)$ is the failure probability of each preventive barrier j within threat i by time, t, and $P_{\text{all\_fail,pre},i}(t)$ is the prevention path failure probability for threat i when all barriers have failed by time t. The derivation of Equation (1) differs from conventional RBD and FTA methods. In FTA and RBD, a series system requires all components to work, so a single failure causes system failure. Conversely, parallel systems need only one component to operate; for instance, the system fails only if all components fail. In the BowTie approach, these logics do not apply because barriers act as sequential safeguards. For the TE to happen, every barrier in the chain must fail, regardless of how the barriers are arranged. Although the equation appears similar, the model still incorporates redundancy or parallelism to enhance reliability. From $P_{\text{all\_fail,pre},i}(t)$, the TE frequency, $\lambda_{\text{TE}}(t)$, for each threat or basic event can be derived, as explained in Sections 2.3 and 2.4.

Establishing barriers within the BowTie framework follows a few guiding principles, as recommended by CCPS [12]. First, the placement of barriers, whether triggered first or last, depends on the chronological sequence of their effect. This means preventive and mitigative barriers should be positioned logically based on their intended function. Second, active barriers, aside from those mentioned in Table 3, must incorporate elements of Detect-Decide-Action (DDA). Third, barriers' properties should be independent, effective, and auditable. These principles are crucial for ensuring their proper functionality. Finally, to prevent common-mode failure, it is advisable to use more than one type of barrier on each pathway. Having multiple barrier types can compensate for the limitations of others and reduce the risk of simultaneous failure caused by a common cause.

2.3. Aggregate Top Event

At this stage, the goal is to compute the time-dependent aggregated TE frequency, $\lambda_{\text{TE}}(t)$, by summing the contributions of all initiating threats. Each threat's contribution is calculated as the failure rate of the basic event multiplied by the probability that the entire preventive pathway fails, $P_{\text{fail,pre},ij}(t)$. The resulting $\lambda_{\text{TE}}(t)$, is then used to determine the frequencies of potential consequences, $f_C(t)$. Table 4 represents a snapshot of the computation for each threat.

Table 4. Illustrative of each threat computation given barrier value and failure rate.

Threat	Failure Rate ($\lambda_{threat,i}$)	Barrier	Input Value	$P_{fail,pre,ij}(t)$	$P_{all\_fail,pre,i}(t)$	$\lambda_{TE,i}(t)$
Threat *i*	Failure per hour	Behaviour Socio-technical Active hardware Continuous hardware Passive hardware	HEP $P_{Conditional}$ PFD_{avg} PFH F(*t*)	Conversion from input value to P_{fail} for each barrier	$\prod_{j=1}^{n_i} P_{fail,pre,ij(t)}$	$\lambda_{threat,i}$ * $P_{all\_fail,pre,i}(t)$

From Table 4, specific rules of the framework need to be adhered to obtain barrier failure probabilities from the input value. First, if the input value given is PFD_{avg}, it must be inverted numerically to determine an equivalent barrier failure rate, λ_b, by using the following:

$$PFD_{avg} = 1 - \frac{1 - e^{-\lambda_b T}}{\lambda_b T} \tag{2}$$

Equation (2) is solved numerically for λ_b using an initial guess $\lambda_0 \approx 2 \cdot PFD_{avg}/T$. Then

$$R_{b,ij}(t) = e^{-\lambda_b t} \tag{3}$$

$$P_{fail,pre,ij}(t) = 1 - R_{b,ij} = 1 - e^{-\lambda_b t} \tag{4}$$

where $R_{b,ij}$ is the reliability of each barrier *j* within threat *i* by time, *t*. Given that "*t*" is a mission time, *T*. Therefore, $P_{fail,pre,ij}(T) = 1 - e^{-\lambda_b T}$.

Second, if the input value given is PFH (failures per hour), then PFH = λ_b. Therefore, Equation (4) can be used to determine $P_{fail,pre,ij}(t)$. Otherwise, if the input value given is HEP (per-demand), then,

$$P_{fail,pre,ij}(T) = 1 - (1 - HEP)^n \tag{5}$$

In this case, *n* represents the number of demands during mission time, *T*, assuming the HEP remains constant. Lastly, if the input value is a conditional probability, $P_{conditional}$ (socio-technical) or the probability of failure, F(*t*) (passive hardware), they are treated as per-mission $P_{fail,pre,ij}(t)$ (constant),

$$P_{fail,pre,ij}(t) = P_{conditional} \text{ OR } F(t) \tag{6}$$

It is worth noting that for time series plots, the study uses $P_{fail,pre,ij}(t)$ as a vector over time, while HEP, $P_{conditional}$, and F(*t*) remain constant.

Next, in the proposed BowTie logic, a TE occurs only if all preventive barriers for threat *i* fail. When this happens, the individual $P_{fail,pre,ij}(t)$ values for each preventive barrier in the threat are obtained, and Equation (1) is applied to multiply these $P_{fail,pre,ij}(t)$ values together. This product equals $P_{all\_fail,pre,i}$ (*t*). The $P_{all\_fail,pre,i}$ (*t*) is then used in the following formula:

$$\lambda_{TE,i}(t) = \lambda_{threat,i} \times P_{all\_fail,pre,i}(t) \tag{7}$$

$\lambda_{TE,i}(t)$ is the TE frequency of threat *i*, and $\lambda_{threat,i}$ is the failure rate of threat *i*. Equation (7) calculates the individual threat contribution of TE intensity, $\lambda_{TE,i}(t)$, by considering the probability of failure when all barriers are failed within that threat *i*. Each calculated $\lambda_{TE,i}(t)$, based on the individual threat *i* is then aggregated together using the following formula:

$$\lambda_{TE}(t) = \sum_i \lambda_{TE,i}(t) \tag{8}$$

The $\lambda_{TE}(t)$ value is then used in Section 2.4 to determine the consequence frequency, $f_C(t)$, in the mitigation pathway.

2.4. Mitigation Pathway

The mitigation pathway on the right side of the BowTie represents mitigative barriers that reduce the likelihood of a TE leading to a particular consequence. This study computes time-dependent mitigation reliabilities, $R_{mit,ck}(t)$, combines them based on the system's architecture, and determines how frequently consequences occur over time. The analysis process for the mitigation pathway closely resembles that of the prevention pathway, where the barrier input value from the barrier catalogue must be converted into $P_{fail,mit,ck}(t)$, as shown in Table 4. From $P_{fail,mit,ck}(t)$, the mitigation pathway is translated into reliability, $R_{mit,ck}(t)$, reflecting the probability of the system successfully preventing the consequence. Combining these success probabilities in series or parallel yields the overall mitigation chain reliability $R_{chain,mit,c}(t)$.

Apply the same conversion rules used for prevention. If the input value is PFD_{avg}, after solving Equation (2) to find λ_b, use the reliability formula as follows:

$$R_{mit,ck}(t) \;=\; e^{-\lambda_b t}, k \in \mathcal{M}_c \tag{9}$$

where, $k \in \mathcal{M}_c$ means each mitigative barrier, k, belongs to the set of mitigative barriers, $\mathcal{M}_c$. $R_{mit,ck}(t)$ is the mitigative barrier reliability, k, at each consequence, c. If the input value is PFH, it is considered equivalent to λ_b (λ_b = PFH). Subsequently, Equation (9) is used to compute $R_{mit,ck}(t)$. In case of HEP value for behavioural barrier, after solving Equation (5) and obtaining $P_{fail,mit,ck}(t)$, then

$$R_{mit,ck}(t) = 1 - P_{fail,mit,ck}(t) \tag{10}$$

Equation (10) also applies to the input values of $F(t)$ (passive barrier) and $P_{conditional}$ (socio-technical barrier).

This study's mitigation logic topology follows the defence-in-depth principle, where the consequences can be prevented if at least one mitigation measure succeeds. Conversely, the outcome only occurs if all mitigation layers fail. The defence-in-depth principle does not apply to FTA, ETA, and RBD, because, whether in series or parallel arrangements, both topologies require at least one mitigative barrier to prevent the effect. Therefore, to determine $R_{chain,mit,c}(t)$, both cases use the same equation as follows:

$$R_{chain,mit,c}(t) \;=\; 1 - \prod\nolimits_{k \in \mathcal{M}_c} (1 - R_{mit,ck}(t)) \tag{11}$$

Given the aggregated TE frequency, $\lambda_{TE}(t)$, obtained from Equation (8), the time-dependent consequence frequency, $f_{C,c}(t)$, is

$$f_{C,c}(t) = \lambda_{TE}(t) \cdot \Pr\{\text{mitigation chain fails at } t\} \tag{12}$$

Under defence-in-depth,

$$f_{C,c}(t) = \lambda_{TE}(t) \times (1 - R_{chain,mit,c}(t)) \tag{13}$$

If the BowTie framework includes several mitigation pathways resulting in multiple consequences, then Equation (13) can be reformulated as follows:

$$f_C(t) = \sum\nolimits_{c=1} f_{C,c}(t) \tag{14}$$

The computations above are then evaluated, and their outputs can serve as decision-support tools.

2.5. Barrier Parameterisation and Time-Dependent Mapping

In this study, time dependence is obtained directly from the exponential reliability model. The mission time is set to $T = 8760$ h (1 year), consistent with the typical annual operation and maintenance cycle of wind turbines and the reporting horizon of failure statistics in CMMS datasets. This choice ensures that the derived reliability metrics are directly aligned with industry practice and planning intervals. For any barrier characterised by PFD_{avg} (low demand) or PFH, the framework first maps the input to an equivalent constant failure rate λ_b, then computes the time-varying failure probability $P_{fail}(t)$ over the mission horizon [0 until 8760] [0.8760] h with a 1 h step. When PFD_{avg} is specified over the mission time T (here $T = 8760$ h), λ_b is obtained via Equation (2). When PFH is provided, the computation sets $\lambda_b = PFH$. In both cases, once λ_b is known, Equation (4) yields

$$P_{fail}(t) = 1 - e^{-\lambda_b t}, t = 0, 1, 2, \ldots, 8760 \text{ h}.$$

These time series are then propagated through the BowTie logic to produce the top-event rate $\lambda_{TE}(t)$ and consequence frequency $f_C(t)$, making the results explicitly time-dependent rather than single point values. This yields time-dependent results without imposing additional solver assumptions and relies only on the standard exponential form.

2.6. Evaluation, Decision Support and Framework Outputs

This section outlines the comparison framework and decision-support outputs used to assess BowTie scenarios. After calculating time-dependent reliability and availability metrics, as detailed in Sections 2.2–2.4, the study ranks barriers and threats using key measures such as BIF, and RAW. The framework also assesses design modifications like adding redundancy, reducing failure probabilities, and implementing process improvements through scenario comparisons. These evaluation results are summarised in tables and plots and integrated into a decision-making process that connects reliability improvements to practical interventions. This method facilitates transparent prioritisation of prevention and mitigation measures.

2.6.1. Importance Measures

Importance measures determine how each barrier influences overall system risk and aid in guiding design, testing, and investment decisions. This study uses two complementary measures, which are:

- Risk Achievement Worth (RAW) measures the increase in $\lambda_{TE}(t)$ if a particular barrier completely fails ($P_{fail,ij}(t) = 1$). It assesses a barrier's risk significance by comparing the baseline scenario to a scenario where the barrier is entirely failed. The RAW formula for barrier i is expressed as,

$$\text{RAW}\ (\mathcal{B}_i) = \frac{\lambda_{TE(\text{new total when } \mathcal{B}_i \text{ removed})}}{\lambda_{TE(\text{baseline total})}} \quad (15)$$

$\lambda_{TE}(t)$ states that the TE frequency occurs with all barriers modelled as designed. On the other hand, $\lambda_{TE(\text{new total when } \mathcal{B}_i \text{ removed})}$ defines that the TE frequency occurs when barrier i is assumed always to fail ($P_{fail,ij}(t) = 1$). If RAW ≈ 1, this means barrier i has no significant influence on $\lambda_{TE}(t)$, and removing it barely changes them. Otherwise, if RAW ≥ 1, barrier i is highly risk significant and its failure leads to a substantial increase in $\lambda_{TE}(t)$. Therefore, RAW shows how much the risk affects the system if the barrier completely fails.

- The Barrier Importance Factor (BIF) identifies which barriers significantly reduce $\lambda_{TE}(t)$. It measures how much $\lambda_{TE}(t)$ could be lowered if these barriers were perfect. This concept offers a straightforward, quantitative method to prioritise improvements or actions for redundancy. The formula of BIF can be presented as follows:

$$\text{BIF}\ (\mathcal{B}_i) = \frac{\text{TE}_0 - \text{TE}_i}{\text{TE}_0} \times 100\% \tag{16}$$

TE_0 represents the baseline $\lambda_{TE}(t)$ with all barriers modelled as usual. Conversely, TE_i is the $\lambda_{TE}(t)$ if the barrier, $\mathcal{B}_i$, is considered perfect ($P_{fail,ij}(t) = 0$). A high BIF suggests that the barrier is crucial, and making it perfect could significantly lower the overall impact of $\lambda_{TE}(t)$.

All significant measure results are displayed and visualised using a bar chart, clearly illustrating the significance of the findings.

2.6.2. Sensitivity Analysis Workflow

This study adopts a structured, scenario-based sensitivity analysis workflow to evaluate potential design and operational improvements. It begins by establishing a baseline that represents the current system setup. From this baseline, it develops alternative scenarios reflecting proposed barrier upgrades. Each scenario is characterised by explicit modifications to barrier parameters, such as reducing oil system failure probability. Multiple parameter changes can also be combined into scenario sets, allowing the assessment of joint effects.

For each scenario, the reliability parameters are recomputed using the exact inversion, as described in Sections 2.2–2.4. This avoids potential hidden biases that may arise if baseline conversions are reused without modification. Finally, the comparison results are plotted over time, t, using $\lambda_{TE}(t)$ and $f_C(t)$ as performance indicators to analyse the trend. The comparison framework is shown in Figure 4.

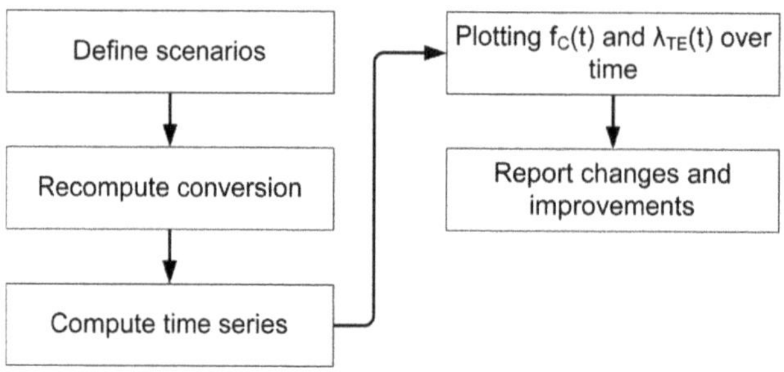

Figure 4. The comparison workflow for comparing baseline and alternative scenarios.

2.6.3. Model Efficiency

Model efficiency assesses computational and representational performance by comparing the BowTie aggregation to explicit ETA/FTA enumeration. This framework reports the minimal cut-set counts and sizes, and the total number of explicit event-tree sequences,

$$N_{ET} = \sum_{i=1}^{N_{threats}} 2^{n_{mit,i}} \tag{17}$$

Each TE with the number of mitigation, n_{mit}, would branch per TE = $2^{n_{mit,i}}$. The number of sequences the BowTie avoids representing explicitly is shown as follows:

$$N_{avoided} = N_{ET} - N_{rep} \tag{18}$$

where N_{rep} indicates the number of sequences explicitly represented by the BowTie. Additionally, to validate the efficiency of the BowTie model, this framework compares the runtimes of calculating the consequence frequency, $f_C(t)$, using the BowTie model versus explicit ETA enumeration. It also performs a numerical comparison of outcomes from both methods, ETA and BowTie, using the following formula:

$$\Delta_f = \| f_{C,bow}(T) - f_{C,ET}(T) \|_\infty \quad (19)$$

This value is then reported to assess whether it lies within tolerance.

3. Case Study

This section demonstrates how the proposed BowTie-reliability framework can be applied through a real case study. The study referenced here is conducted initially by Ozturk [12], who examines WT failures using reliability methods and Machine Learning (ML). Ozturk focused solely on identifying criticality, modelling reliability, and predicting WT failure. However, this paper sees potential to expand the scope further, motivating the application of the proposed BowTie framework using Ozturk's foundational information. Although their study addresses many reliability issues related to WT, it does not aim to reduce failure risks, prevent the TE, or mitigate failure consequences—areas prioritised here. As a result, this work develops two common reliability models, the FTA and RBDs, based on Ozturk's data and additional insights from related studies. From this foundation, the BowTie model is constructed, assessed, and discussed.

3.1. Background of the Case Study

Wind energy is widely regarded as a leading renewable resource among alternative energy sources. Its environmentally friendly and sustainable qualities have driven rapid growth in its deployment. As the failure rate of WT increases significantly, the focus has shifted toward enhancing their reliability and availability to ensure continuous power generation. The study led by Ozturk, part of the Scientific Measurement and Evaluation Programme (WMEP) in Germany, is a key component of a government effort to analyse onshore WTs from 1989 to 2006. It compiled 64,000 maintenance and repair reports from 1500 turbines, representing over 15,000 years of operational data. Ozturk's findings provide valuable insights into WT reliability, aiding the development and validation of a reliability modelling framework.

This study focuses on the generator as a key subcomponent of the WT system, selected based on its criticality in prior research. Both Ozturk [13] and Parnon et al. [14] identify the generators as the highest-ranked subsystem in their criticality analyses, though derived from different frameworks. This finding is also supported by Pulikollu et al. [15], who consistently identify generators as the most vulnerable and costly assets, with failures such as imbalance, cracking, or lubrication deficiencies leading to sudden breakdowns, extended downtime, and high replacement costs. While the generator's annual failure rate falls within the range of 1% to 4%, its downtime may lead to significant production losses and unplanned repair costs, estimated at around \$100,000 to \$225,000 [15]. Owing to this criticality, the generator is selected for developing the FTA and demonstrating the proposed framework. The following section provides further details on the development of both approaches.

3.2. Development of Reliability Modelling

This section illustrates the development of the FTA and RBDs based on a study conducted by Ozturk [13] and other studies [16–18]. It starts with comparing several established FTA diagrams created by Novaković et al. [16], Márquez et al. [17], and Kang et al. [18]. To produce an FTA based on these established studies, a screening criterion is specified as follows:

- Include if:
  - Mentioned in ≥2 literature sources OR
  - Found in the dataset OR
  - Has a significant impact

As stated in Section 3.1, this study focuses on the generator as a subcomponent of the WT that requires evaluation. Consequently, Figure 5 illustrates an FTA diagram of the WT generator, developed based on the referenced studies [16–18] and screening criteria. Table 5, furthermore, showcases the list of TEs, intermediate events, and basic events with sources of references.

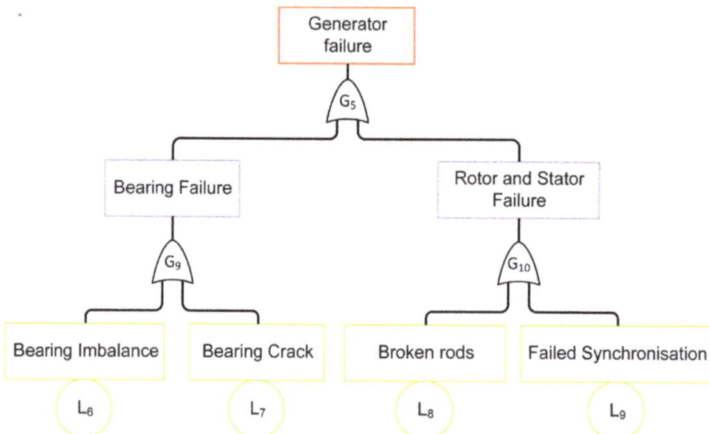

Figure 5. FTA diagram of WT generator for the case study.

Table 5. Logic gates and principal events of the WT generator.

Logic Gates	Codes	Basic Events	Codes
Generator failure	G_5	Bearing Imbalance	L_6
Bearing Failure	G_9	Bearing Crack	L_7
Rotor and Stator Failure	G_{10}	Broken rods	L_8
		Failed Synchronisation	L_9

The RBD model, in Figure 6, is derived from the FTA diagram in Figure 5, based on studies by Ozturk [13], Einarsson et al. [19], and Linsday et al. [20]. Afterwards, the minimal cut set equation is formulated to construct the BowTie model.

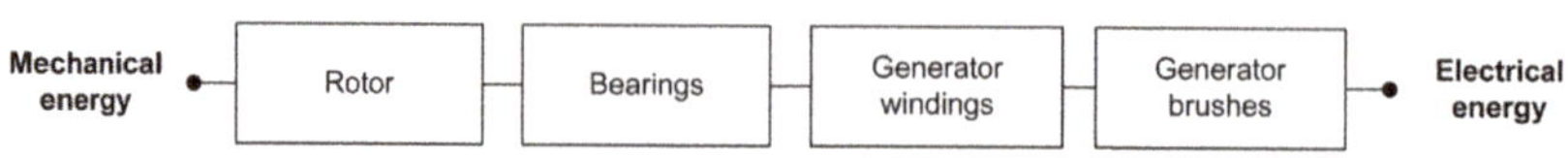

Figure 6. RBD model of WT Generator.

Based on the FTA and RBDs, a minimal cut set can be formulated to determine the smallest set of subcomponent failures needed to cause the entire system to fail. The logic operator in Figure 5 consists solely of OR gates, which correspond to the arithmetic "+" sign in accordance with IEC 61025 [11]. Using this convention, the minimal cut set formula can be derived as follows:

$$Fail\ state = G_5 \tag{20}$$

$$G_5 = G_9 + G_{10} \tag{21}$$

$$G_9 = L_6 + L_7 \tag{22}$$

$$G_{10} = L_8 + L_9 \tag{23}$$

Derivation involving Equations (20)–(23) generates the final minimal cut set equation:

$$Fail\ state = L_6 + L_7 + L_8 + L_9 = \{L_6, L_7, L_8, L_9\} \tag{24}$$

L_6, L_7, L_8, and L_9 constitute the primary first-order minimal cut sets that need to be emphasised in the BowTie model.

3.3. Development of BowTie Model

The BowTie model, clarified in Section 2 with an illustrative diagram in Figure 1 consists of several components, including the hazard, TE, the preventive pathway (left), the mitigation pathway (right), and the associated barriers. This subsection describes how the BowTie model is developed using a case study centred on the minimal cut set identified in Section 3.2.

Section 3.2 of this study reveals that L_6, L_7, L_8, and L_9 are the fundamental events of the developed FTA model in Figure 5. As first-order events, their states are highly susceptible to failure and need focused improvement. To convert information from FTA and RBD into the BowTie model, the minimal cut set is directly identified as threats. The consequences or effects of the TE follow the taxonomy of Ozturk [13]. Barriers, preventive and mitigative, are selected from the literature and industry practice to address each threat and consequence. The full BowTie diagram is presented in Figure 7.

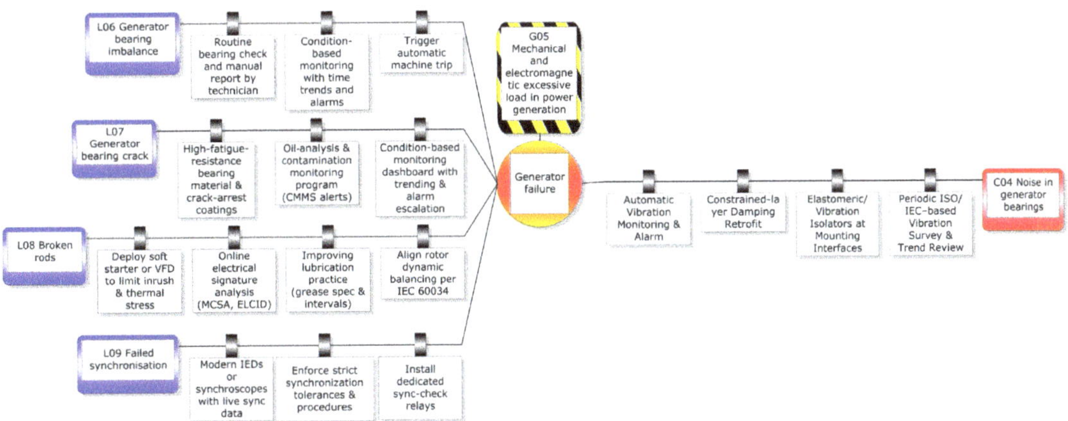

Figure 7. The proposed BowTie model for the WT generator.

To ensure transparency and reproducibility, this study provides supporting materials alongside the BowTie model, including the assumption table listing modelling assumptions and units and quantitative data, such as failure rate and probability. The calculation methods have been presented in Section 2, from Equations (1)–(19). This information is summarised in tabular form, where Table 6 exhibits the modelling assumptions and metrics, Table 7 presents BowTie core elements, while Tables 8–12 detail the preventive and mitigative barriers associated with specific threats and consequences, supported by quantitative evidence from prior studies. The assumptions in Table 6 are adopted for demonstration purposes to keep the case study tractable; they are not rigid requirements of the framework and can be relaxed or adjusted depending on the application context.

Finally, the results and discussion sections include model validation. The validity of key modelling assumptions, as mentioned in Table 6 is implicitly tested through scenario-based sensitivity analysis in Sections 2.6.2 and 3.4. By systematically varying barrier parameters, this analysis provides analytical insights into how deviations from baseline assumptions influence $\lambda_{TE}(t)$ and $f_C(t)$.

Table 6. The modelling assumptions and metrics.

Item	Assumption	Notes
Time basis	Steady-state reliability with constant failure rate.	h
Barrier independence	Barriers are assumed to be independent unless noted; common cause failure is negligible.	–
Threat failure rate	Taken directly from the literature survey.	Failures/h
Mission time, $T_{mission}$	Use 8760 h = 1 year.	h
PFH	Continuous failure rate measure.	Failures/h
PFD_{avg}	Demand frequency is assumed annually.	Probability/demand
P_{fail}	Point probability of the component failure.	dimensionless
$P_{conditional}$	Conditional probability that monitoring fails to detect the threat.	dimensionless
HEP	Derived from human reliability analysis (THERP/SPAR-H)	Probability/event
Maintainability	The repair rate is assumed to be negligible; The renewal function is used (part replaced "as good as new")	–

Table 7. The proposed BowTie core elements.

Element Type	Element Name	Failure Rate (Failure/h)	Sources
Hazard	Mechanical and electromagnetic excessive load in power generation	–	[13]
TE	Generator failure	–	[13]
Threat (L_6)	Generator bearing imbalance	5.85×10^{-6}	[18]
Threat (L_7)	Generator bearing crack	1.17×10^{-6}	[18]
Threat (L_8)	Broken rods	2.10×10^{-7}	[18]
Threat (L_9)	Failed synchronisation	3.61×10^{-6}	[18]

Table 8. Preventive barriers by threat (L_6).

Barrier	Category	Type	Input Value	Sources
Routine bearing check and manual report	Behavioural	HEP	1.10×10^{-1}	[21]
CBM with time trends and alarms	Socio-technical	$P_{conditional}$	3.00×10^{-2}	[21,22]
Trigger automatic machine trip	Active hardware	PFD_{avg}	6.20×10^{-6}	[23]

Table 9. Preventive barriers by threat (L_7).

Barrier	Category	Type	Input Value	Sources
High-fatigue-resistance bearing material and crack-arrest coatings	Passive hardware	P_{fail}	2.40×10^{-2}	[24,25]
Oil-analysis and contamination monitoring program (CMMS alerts)	Socio-technical	$P_{conditional}$	7.20×10^{-1}	[26]
CBM dashboard with trending and alarm escalation	Socio-technical	$P_{conditional}$	1.00×10^{-1}	[27]

Table 10. Preventive barriers by threat (L_8).

Barrier	Category	Type	Input Value	Sources
Deploy soft starter or VFD (Variable Frequency Drive) to limit inrush and thermal stress	Continuous hardware	PFH	3.50×10^{-6}	[28]
Online electrical signature analysis	Active hardware	PFD_{avg}	3.00×10^{-2}/demand	[29]
Improving lubrication practice (grease spec and intervals)	Behavioural	HEP	3.00×10^{-3}	[30]
Align rotor dynamic balancing per IEC 60034	Behavioural	HEP	1.00×10^{-3}	[30]

Table 11. Preventive barriers by threat (L_9).

Barrier	Category	Type	Input Value	Sources
Modern IEDs (Intelligent Electronic Device) or synchroscopes with live sync data	Socio-technical	$P_{conditional}$	1.70×10^{-2}	[31]
Enforce strict synchronisation tolerances and procedures	Behavioural	HEP	3.00×10^{-2}	[21,30]
Install dedicated sync-check relays	Active hardware	PFD_{avg}	1.80×10^{-6}/demand	[32]

Table 12. Mitigative barriers by Consequence (C_{04}).

Barrier	Category	Type	Input Value	Sources
Automatic Vibration Monitoring and Alarm	Active hardware	PFD_{avg}	1.20×10^{-3}	[33]
Constrained-layer Damping Retrofit	Passive hardware	P_{fail}	2.00×10^{-2}	[34]
Elastomeric/Vibration Isolators at Mounting Interfaces	Passive hardware	P_{fail}	3.00×10^{-2}	[35]
Periodic ISO/IEC–based Vibration Survey and Trend Review	Behavioural	HEP	1.00×10^{-2}	[30]

3.4. Sensitivity Analysis

As outlined in Section 2.6.2, a sensitivity analysis is conducted to evaluate the robustness of the model by systematically varying key barrier parameters. Each manipulation represents a distinct scenario to observe the resulting trends in both $\lambda_{TE}(t)$ and $f_C(t)$. In this case study, there are four scenarios, with one baseline for comparison purposes. Table 13 summarises these scenarios, where each represents modifications in barriers such as redundancy, reduction, and frequent proof-test frequency, as well as combined improvements across multiple barriers.

Each scenario is analysed and evaluated using the outlined workflow in Figure 4. The findings are then explained in Section 4.4.

Table 13. Sensitivity analysis scenarios.

Scenario	Barrier Modified	Parameter Adjustment
Baseline	None	Reference value
Oil analysis improved	Oil lab protection	P_{fail} reduced from 0.72 to 0.20
Redundant dashboard	Dashboard	Addition of a 2nd independent dashboard
Faster auto-alarm proof-test	Auto-alarm	PFD_{avg} halved
Combined (Oil + Dashboard)	Oil lab protection, dashboard	Oil analysis and dashboard applied together

4. Results

This section highlights the analysis results from the BowTie model applied to the WT case study. The outcomes emphasise WT's reliability performance and the strategies for preventing and mitigating major failures. Findings are organised according to the research question outlined in Section 1, focusing on four key measurable aspects: risk contribution ranking, quantitative assessment of barrier adequacy, reliability improvements through scenario-based sensitivity analysis, and overall model efficiency.

Using Equation (1) through (19) with the provided information from Tables 6–12, the results are shown in Table 14, which addresses four measurable aspects of outcomes. The inversion of PFD_{avg} values into λ_b is implemented using the built-in fzero function in MATLAB R2025a Update 1 (Version 25.1.0.2973910). Initial guesses for λ_b are taken from $\lambda_0 \approx 2{\cdot}PFD_{avg}/T$, where T = 8760 h represents one year of operation. Convergence is accepted when the absolute residual of the inversion equation, $\epsilon = 1 - \frac{1-e^{-bT}}{\lambda_b T} - PFD_{avg}^{Target}$, is less than 1×10^{-10}, where all results are well within tolerance between 10^{-12} and 10^{-14}, as presented in Table 15. However, if the residual is unable to achieve within the specified tolerance, the λ_b value is taken from the approximated equation, $2{\cdot}PFD_{avg}/T$.

Table 14. Summary of BowTie outcomes.

Preventive Barrier	$P_{all\_fail,pre,i}(t)$	$\lambda_{TE}(t)$ Contributions (Failures/h)
Threat L_{06}: Routine bearing check and manual report Condition-based monitoring with time trends and alarms Trigger automatic machine trip	2.046×10^{-7}	1.197×10^{-12}
Threat L_{07}: High-fatigue-resistance bearing material and crack-arrest coatings Oil-analysis and contamination monitoring program (CMMS alerts) Condition-based monitoring dashboard with trending and alarm escalation	1.728×10^{-3}	2.022×10^{-9}
Threat L_{08}: Deploy soft starter or VFD to limit inrush and thermal stress. Online electrical signature analysis. Improving lubrication practice (grease spec and intervals). Align rotor dynamic balancing per IEC 60034.	5.380×10^{-9}	1.130×10^{-15}
Threat L_{09}: Modern IEDs or synchroscopes with live sync data. Enforce strict synchronisation tolerances and procedures. Install dedicated sync-check relays.	1.836×10^{-9}	6.628×10^{-15}
Total λ_{TE} (sum of contributions)	2.023×10^{-9} failures/h	

Table 14. *Cont.*

Mitigative Barrier	$P_{all\_fail,mit,c}(t)$	f_C (t) Contributions (Failures/h)
Consequence C_{04}: Automatic Vibration Monitoring and Alarm Constrained-layer Damping Retrofit Elastomeric/Vibration Isolators at Mounting Interfaces Periodic ISO/IEC–based Vibration Survey and Trend Review	1.439×10^{-8}	2.912×10^{-17}
Total $f_C(t)$(Product of total $\lambda_{TE}(t)$ and total $P_{all\_fail,mit}(t)$)		2.912×10^{-17} failures/h

Table 15. Validation of PFD_{avg}-to-λ_b inversion for selected barriers (mission time T = 8760 h).

Barrier (Type)	Target PFD_{avg}	λ_b	Residual
L06 (PFD_{avg})	3.100×10^{-5}	7.078×10^{-9}	-5.151×10^{-14}
L08 (PFD_{avg})	3.000×10^{-2}	6.990×10^{-6}	-4.038×10^{-14}
L09 (PFD_{avg})	1.800×10^{-6}	4.110×10^{-10}	2.077×10^{-12}
Mitigation (PFD_{avg})	1.200×10^{-3}	2.742×10^{-7}	-1.823×10^{-14}

4.1. Risk Contribution Ranking

RAW analysis, as described in Section 2.6.1, assesses barriers' contribution to $\lambda_{TE}(t)$ assuming they fail completely. Figure 8 indicates that most barriers have relatively low RAW values, such as the CBM dashboard, routine checks, and oil analysis. This suggests that their absence would only slightly raise $\lambda_{TE}(t)$. However, a few barriers show very high RAW values, notably the sync-check relays and trigger auto trip, which exceed 10^4. Additionally, rotor balancing, lube practice, and modern IED are also highly critical. These barriers act as vital single points of defence, with their effectiveness significantly controlling the likelihood of avoiding a hazardous outcome.

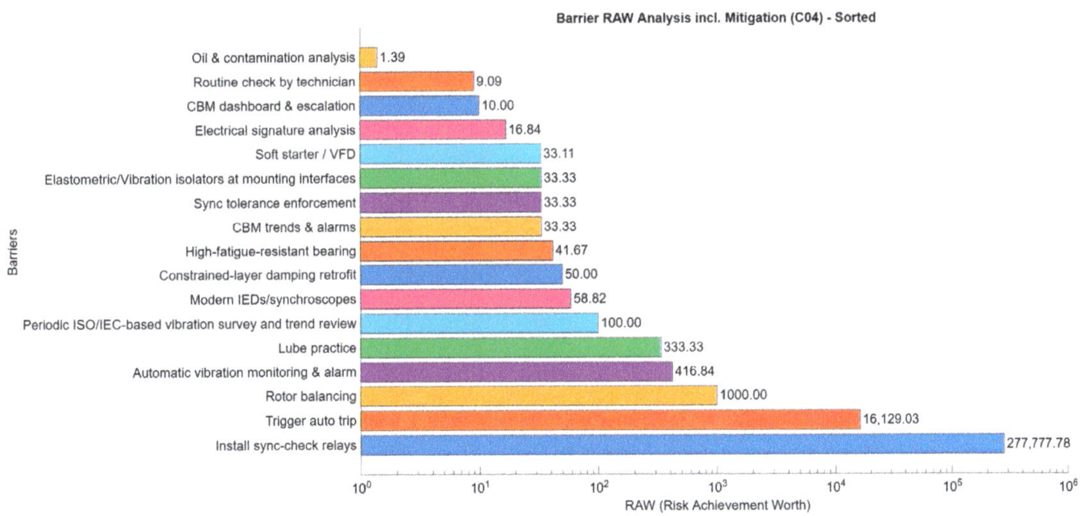

Figure 8. Barrier RAW analysis, including mitigation outcomes.

The pattern also demonstrates the defence-in-depth approach of the BowTie model. Barriers relying solely on human action, like lubrication practices, or static hardware, such as a soft starter, provide additional protection. However, active barriers, such as auto-trip triggers or check relays, play a more significant role in the RAW distribution. This suggests that while barriers not aligned with DDA elements can help mitigate hazards, those fulfilling DDA criteria are primarily responsible for protecting the system.

From a decision perspective, RAW analysis indicates that prioritising investment in maintaining and testing high-RAW barriers is crucial, since their failure could lead to exponential increases in system risk. Conversely, for barriers with lower RAW values, the model suggests they still contribute to system resilience but may need optimisation or streamlining, ensuring safety remains unaffected.

4.2. Barrier Importance Factor

The BIF measures how much each preventive barrier helps prevent $\lambda_{TE}(t)$. A higher BIF shows that improving these barriers can significantly reduce $\lambda_{TE}(t)$. Figure 9 displays the BIF results from the BowTie model.

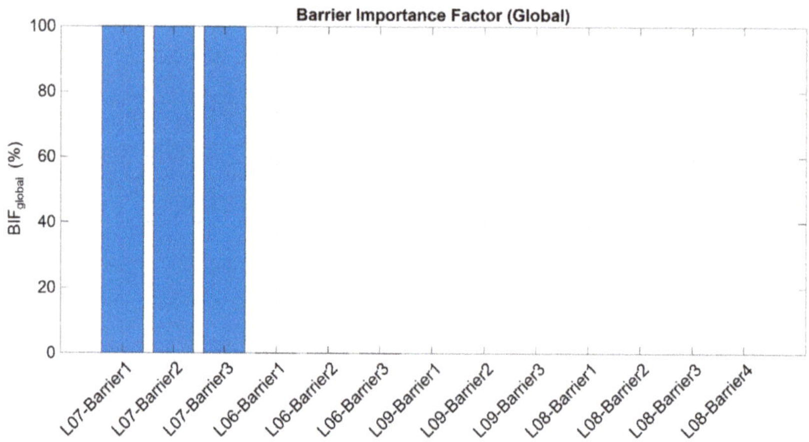

Figure 9. Barrier Importance Factor with respect to $\lambda_{TE}(t)$.

Figure 9 clearly shows that all three preventive barriers under threat L_{07} collectively contribute 100% to the $\lambda_{TE}(t)$. This implies that the effectiveness of preventing the TE from happening is dependent on maintaining these barriers at their proper functioning level. Furthermore, these results also highlight that L_{07} is fully controlled by its associated barriers, meaning any failure or removal of these barriers would directly lead to an increase in $\lambda_{TE}(t)$. Unlike other threats, where barrier contributions barely emerge and are distributed unevenly, L_{07} requires rigorous monitoring and maintenance.

4.3. Barrier Quantitative Assessment over Time

This section shows the trends of $\lambda_{TE}(t)$ and $f_C(t)$ over mission time, T. The goal is to observe how barrier systems behave along with their threats and consequences as time progresses. Figures 10 and 11 display the $\lambda_{TE}(t)$ and $f_C(t)$ over a year, or 8760 h. As explained in Section 2.5, in all cases, barriers parameterised by PFD_{avg} (over $T = 8760$ h) or PFH are mapped to an equivalent constant rate λ_b, and their time-varying failure probability is computed using Equation (4). This gradual increase in $P_{fail}(t)$ over t explains the near-linear growth of $\lambda_{TE}(t)$ under constant λ_{threat} and λ_b from approximately 2.02×10^{-9} to 2.023×10^{-9} failures per hour during the mission, as shown in Figure 10. Although the increase is monotonic, the change remains minimal, indicating that barrier degradation is slow and the preventive configuration sustains a consistently low top-event rate across the operating horizon.

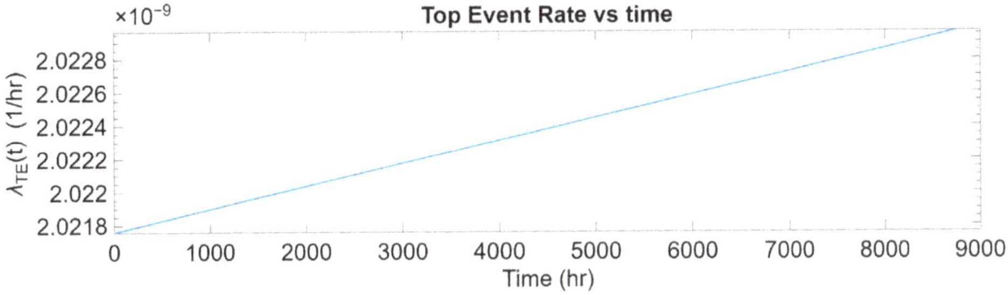

Figure 10. The $\lambda_{TE}(t)$ pattern over mission time, T.

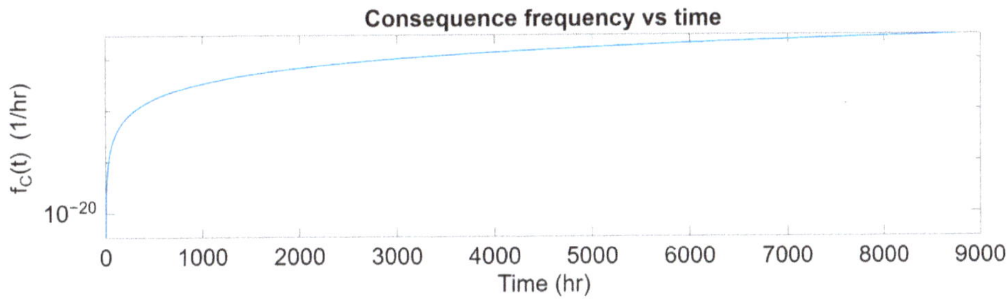

Figure 11. The $f_C(t)$ pattern over mission time, T.

Figure 11 illustrates how $f_C(t)$ evolves over time, showing a rapid increase from 0 to 2000 h during the early mission phase. After this period, it gradually stabilises around 10^{-18} failures per hour. This effectively prevents the escalation of TE into serious consequences. The data suggest that, within a defence-in-depth framework, the occurrence of severe outcomes remains very low.

The combined reliability analysis shows that although $\lambda_{TE}(t)$ continues to rise over time because of barrier wear-out, the mitigation measures keep the impact of TE very low ($\leq 10^{-17}$/h). This indicates the long-term effectiveness of the defence-in-depth strategy.

4.4. Scenario-Based Sensitivity Analysis Outcomes

Section 4.4 assesses the impact of barrier modifications on system reliability using a series of what-if scenarios outlined in Section 3.4. The findings are presented in tabular format in Table 16 and graphically in Figures 12 and 13 to enable a quantitative and visual evaluation of barrier performance.

Table 16. The outcomes of scenario-based sensitivity analysis outcomes.

Scenario	$\lambda_{TE}(t)$	$f_C(t)$	% Reduction vs. Baseline	Notes
Baseline	2.023×10^{-09}	2.912×10^{-17}	–	Reference inputs
Oil -Analysis Improved	5.628×10^{-10}	8.101×10^{-18}	72.18	Oil P_{fail} 0.72 → 0.20
Redundant Dashboard	2.034×10^{-10}	2.034×10^{-18}	89.95	Dashboard P_{fail} 0.10 → 0.01
Faster Auto-Alarm Proof-Test	2.022×10^{-09}	2.911×10^{-17}	0.03	PFD_{avg} halved
Combined (Oil + Dashboard)	5.736×10^{-11}	8.257×10^{-19}	97.16	Both improvements applied

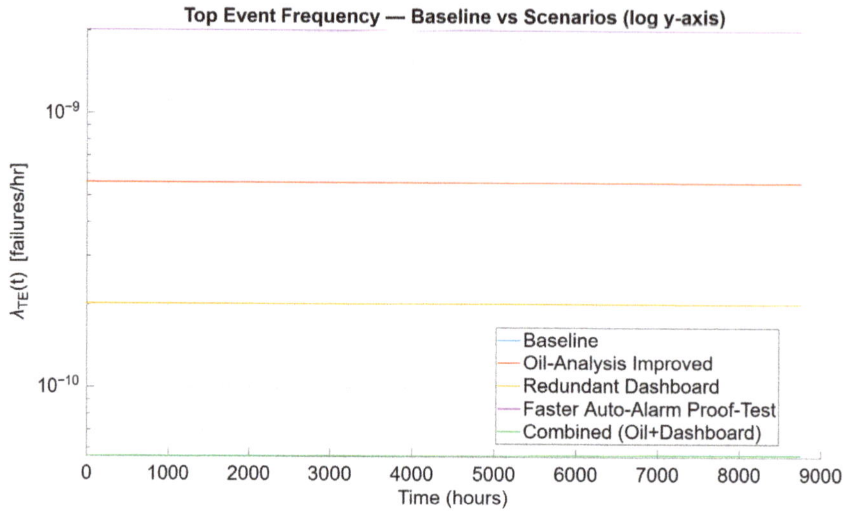

Figure 12. The baseline vs. scenarios of $\lambda_{TE}(t)$.

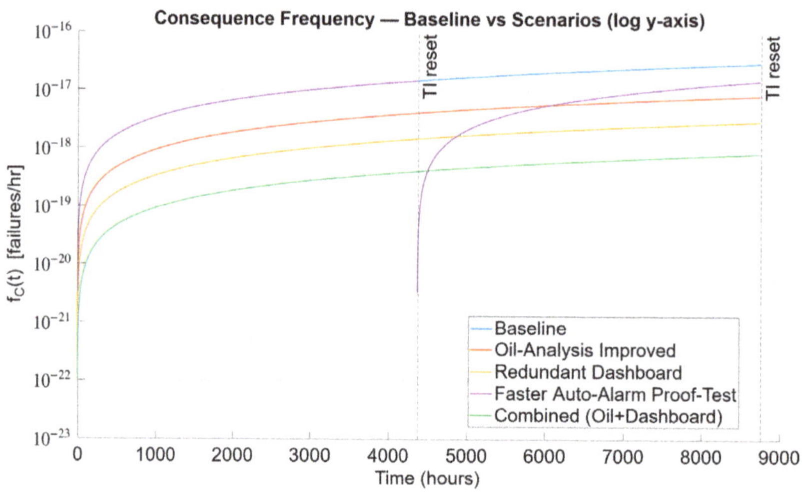

Figure 13. The baseline vs. scenarios of $f_C(t)$.

Table 16 shows that the baseline case results for both $\lambda_{TE}(t)$ and $f_C(t)$ are 2.023×10^{-9} failures per hour and 2.912×10^{-17} failures per hour, which serve as the reference point for comparison. Both tabular and graphical results display this as the highest frequency, emphasising the system's vulnerability under the current barrier effectiveness.

In the initial alternate scenario, the failure probability of the oil analysis barrier decreases from 0.72 to 0.20. As a result, $f_C(t)$ drops sharply to 8.101×10^{-18} failures per hour, marking a 72.18% reduction. This notable decline is clearly shown in Figures 12 and 13, where the curve for this scenario diverges significantly from the original baseline.

Focusing on the redundant dashboard scenario, it significantly benefits by reducing 89.95% of $f_C(t)$, equating to 2.034×10^{-18} failures per hour. Figures 12 and 13 show these findings, with the dashboard curve positioned well below the baseline and oil-analysis curves, highlighting the substantial impact of redundancy in operator interface barriers.

In contrast, cutting the proof-test interval of the auto-alarm in half did not influence $\lambda_{TE}(t)$ because the proof-test modification applies to a mitigation-side barrier, which affects the $f_C(t)$. Therefore, no distinct curve appears in Figure 12. However, this approach slightly reduced failures, $f_C(t)$, to approximately 2.911×10^{-17} per hour, or roughly 0.03%. The results are consistent with the plots in Figure 13, where the new curve overlaps with the baseline, indicating that this scenario does not contribute significantly to overall risk reduction. This behaviour can be explained by the fact that halving PFD_{avg} would reduce the proof-test interval by half. When both parameters are halved, the underlying dangerous failure rate λ_b remains unchanged. Consequently, the function $f_C(t)$ remains unaffected by this change, affecting only the time profile (which resets at each proof test) rather than the entire one-year outcome.

Lastly, the combined improvement between oil analysis and dashboard redundancy achieved the most significant outcome by lowering the $f_C(t)$ to 8.257×10^{-19} failures per hour, approximately 97.16% reduction. Figures 12 and 13 display this curve positioned at the lowest point across the mission time, T.

Table 16 and plotted curves, Figures 12 and 13, consistently show the most effective scenarios for oil analysis and dashboard barriers. Additionally, cutting the proof test period in half does not significantly reduce the auto-alarm proof test interval and has a limited impact. Using multiple approaches together can provide the most significant benefit, with a reduction of nearly two orders of magnitude.

4.5. Model Performance Comparison

This section assesses the effectiveness of the proposed BowTie model, as outlined in Section 2.6.3. It examines this assessment through minimal cut sets, runtime performance, and model compactness. The traditional FTA yields four first-order minimal cut sets, as derived in Equation (24). This equation suggests that these single basic events alone could potentially trigger TE, indicating that their presence makes the system highly fragile. Conversely, the proposed BowTie-based analysis expands the FTA model beyond Boolean logic. It captures the interaction of barriers across both pathways. In this case study, the proposed model identifies four dominant cut-sets of order four or higher, which are 4, 4, 5, and 4, as displayed in Figure 14. This chart illustrates that multiple barriers must fail together for TE to propagate, reflecting more realistically the defence-in-depth principles embedded in engineered systems. While this illustrates the usefulness of BowTie in demonstrating resilience, this study acknowledges that the observed difference is specific to this case, and outcomes may differ for other subsystems or system architectures.

Another key feature of the BowTie model is its computational advantage, which is clear when comparing runtimes. Figure 15 shows that the BowTie aggregation method completes the task in around 0.0009 s, whereas full ETA enumeration takes 0.0729 s. This demonstrates that the framework can achieve efficiency gains while retaining analytical rigour. Although this case study involves a single subsystem, the results highlight the model's potential for scaling to more complex systems where state-space growth poses a challenge.

Ultimately, the proposed BowTie model showcases its capability through its compact representation. As per Equations (18) and (19), the explicit ETA requires 64 distinct states to capture event sequences. In contrast, the BowTie model achieves the same analytical outcome using just 21 states, thereby eliminating 43 redundant states, as illustrated in Figure 16. This simplification of the BowTie model does not compromise analytical rigour, as evidenced by the value of $\Delta_f = 0.0000$. While the state reduction achieved here is a clear advantage, the study notes that efficiency gains must always be balanced with the preservation of critical dependencies. In this case, the framework successfully maintained accuracy while improving compactness, underscoring its potential as a reliable yet efficient alternative to exhaustive enumeration.

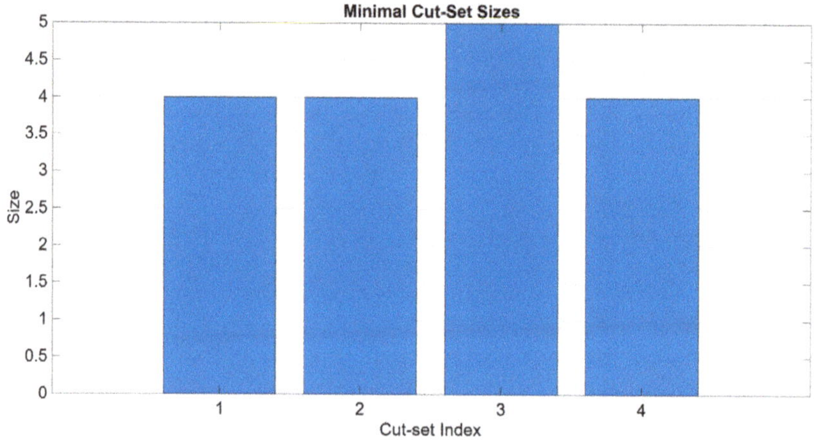

Figure 14. The number and size of the cut-set of the BowTie model.

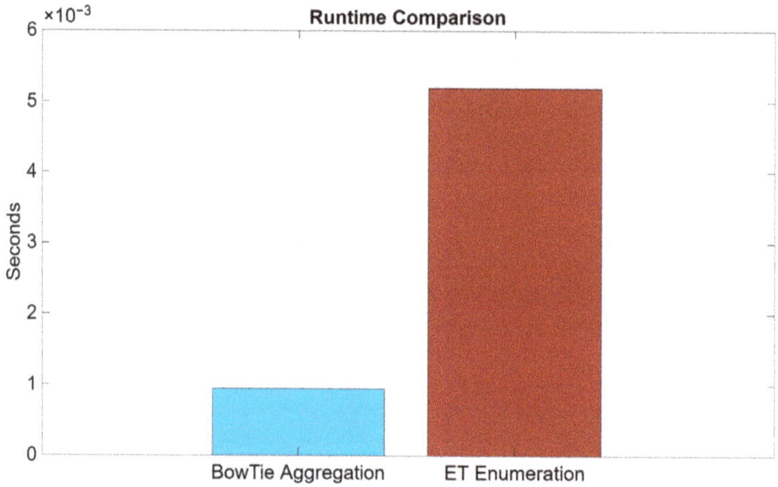

Figure 15. A runtime comparison between the BowTie model and ETA.

Another aspect of model comparison is the performance difference between the TE probability, $P_{TE}(t)$ estimated using the proposed BowTie framework with the conventional FTA (threat-only) approach. Figure shows the FTA curve (orange, dashed) rises sharply during the initial phase and quickly stabilises around 10^{-2}, reflecting the fragility implied when threats are modelled as direct precursors to the top event. In contrast, the BowTie curve (blue, solid) remains several orders of magnitude lower, stabilising in the range of 10^{-6}–10^{-7} over the mission time. This outcome demonstrates the defence-in-depth effect: a top event occurs only when all associated preventive barriers fail, resulting in significantly reduced probabilities compared to the threat-only representation.

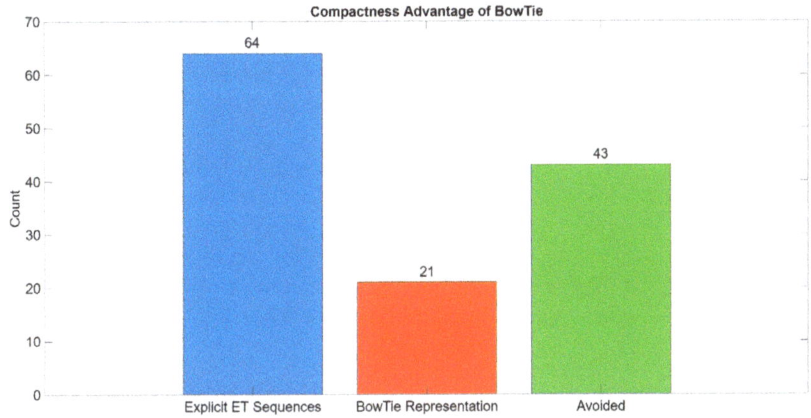

Figure 16. Compactness comparison between BowTie and the ETA model.

5. Discussions

The study answer the research question via four measurable outcomes: (i) risk contribution ranking using RAW and BIF (Figures 8 and 9), (ii) quantified reliability improvements from scenario redesigns (Table 13; Figures 12 and 13), (iii) quantitative evidence of barrier sufficiency over mission time (Figures 10 and 11), and (iv) streamlined analysis relative to ETA/FTA (Figures 14–16). The discussions presented in this section aim to interpret the main findings related to the research question in Section 1 and highlight their implications for system reliability and asset management.

5.1. Criticality of Barriers and Risk Contributions

The RAW findings in Section 4.1 emphasise significant variability in how individual barriers contribute to $\lambda_{TE}(t)$. This inconsistency is clearly evident in key system components, such as sync-check relays and automatic trip triggers, which serve as single points of defence; their failure can directly cause system breakdown. Conversely, barriers like CBM dashboards and routine operator checks show a relatively low risk contribution to $\lambda_{TE}(t)$. This suggests that removing these barriers would not significantly increase the $\lambda_{TE}(t)$. Therefore, high-RAW barriers require strict routine maintenance, regular proof testing, and consideration of redundancy, while low-RAW barriers are optimised or rationalised depending on circumstances, such as operational costs.

Another important aspect of barrier contributions is the findings from BIF. It is essential to highlight that BIF analysis assumes barriers can either be perfected or improved. This implies that the probability of failure is negligible. According to BIF's findings, all barriers at threat level L_{07} could significantly lower $\lambda_{TE}(t)$, contradicting the conclusions reached by RAW. There are several fundamental differences between the BIF and RAW analyses. BIF focuses on the advantages of enhancing a barrier, while RAW evaluates the potential damage that would occur if the barrier fails. A barrier that performs well (with a low P_{fail}) may have a low BIF (indicating limited room for improvement) but a high RAW (suggesting severe consequences if compromised), and the reverse can also be true. In practice, maintenance managers can use RAW to identify which barriers need maximum protection, while BIF helps justify investments by estimating the return on risk reduction [36]. For example, improving the oil analysis programme reduces the probability of failure from 0.72 to 0.20 and lowers $\lambda_{TE}(t)$ by more than 70%, providing a clear business case for enhanced monitoring. By contrast, halving the proof-test interval for the automatic trip barrier has a negligible impact on risk, suggesting that resources would be better spent elsewhere.

These insights bridge the gap between technical analysis and financial decision-making. High-priority barriers (those with a high RAW) should be allocated greater inspection and testing budgets, as their failure would lead to prolonged outages and expensive repairs. In contrast, low-priority barriers may be grouped or de-scoped to save costs. BIF-driven improvements, such as upgrading vibration alarms or adopting more advanced oil-analysis sensors, can be compared against their implementation costs to evaluate the return on investment. This aligns with the broader reliability engineering literature, where risk-based inspection (RBI) and reliability-centred maintenance (RCM) frameworks are utilised to allocate maintenance resources based on both risk and cost considerations [36].

5.2. Reliability Improvements Through Barrier Redesign

The scenario-based sensitivity analysis in Section 4.4 reveals that not all preventive or mitigation strategies yield proportional gains. While redundancy and lowering barrier failure rates lead to significant reductions in system risk, as shown in Table 13, simply increasing proof-test frequency yields minimal additional benefit. In this case study, halving the auto-alarm proof-test interval reduced $f_C(t)$ by $\approx$0.03% (Table 13), indicating minimal marginal benefit versus redundancy or P_{fail} reduction. This outcome is consistent with modern reliability engineering and barrier management principles, which emphasise targeting the most influential safeguards rather than applying uniform maintenance escalation. For example, Yuan et al. [37] present a dynamic risk-informed safety barrier management approach where barrier interventions are prioritised based on current performance and risk contribution rather than scheduled intensification. Their method demonstrates that optimising high-leverage barriers yields more risk reduction per unit cost than blanket increases in inspection or testing intervals. Likewise, in the risk-based maintenance literature, Leoni et al. [38] compare various RBM methods and show that prioritisation based on probability $\times$ consequence offers more efficient resource allocation than across-the-board maintenance increases. Hence, the results of this study, where proof testing offers little marginal gain while barrier redesign and redundancy deliver considerable benefits, align with the principle of maintenance prioritisation by risk importance, reinforcing that strategic intervention wins over intensification. This prioritise-by-influence result mirrors risk-based barrier management and RCM guidance, which favour targeting high-leverage safeguards over blanket escalation of proof-testing or inspection intervals [39–42].

5.3. BowTie Modelling and Scalability

One of the key benefits of the proposed BowTie framework is its ability to simplify reliability modelling without compromising accuracy. This can be seen through the minimal cut set analysis, where the BowTie model can produce higher-order cut sets (size 4–5), requiring multiple barriers to fail to escalate the TE. In contrast, conventional FTA in Figure 5 generates first-order cut sets, suggesting that a single failure of basic events is enough to cause a catastrophic event. This difference indicates that the BowTie framework better reflects the defence-in-depth strategies used in practice, which capture the system's resilience rather than exposing single-point vulnerabilities. Additionally, the BowTie model can reduce computational states compared to ETA enumeration without losing crucial detail. This means the proposed model has a high potential for scalability, making it suitable for larger or more complex systems where combinatorial explosion is a significant concern. This pattern is consistent with defence-in-depth guidance, where layered safeguards raise the minimum failure-combination order required for escalation.

5.4. Barrier Sufficiency over Mission Time

Investigating barrier sufficiency quantitatively shows that $\lambda_{TE}(t)$, increases linearly but remains nearly constant over the mission period due to its small slope effect. The consequence frequency, $f_C(t)$, on the other hand, remains at negligible levels, approximately $10^{-17}/h$. These findings provide strong evidence that the barrier configuration in the case study is sufficient to maintain reliability over the operating horizon. In practice, the defence-in-depth strategy ensures that even with gradual wear-out and ageing effects, the probability of escalation to severe consequences is effectively controlled and minimised, strengthening confidence in current operational practice.

5.5. Practical Implications and Generalisability

From an asset management perspective, this study highlights the benefits of a risk-informed approach to barrier design and maintenance prioritisation. The comparison in Figure 17 reinforces this point, as relying solely on FTA tends to overstate system fragility by overlooking the role of barriers. Incorporating preventive and mitigative defences through the BowTie framework demonstrates that the likelihood of generator failure is drastically reduced by several orders of magnitude over the mission duration. For example, the combined oil-analysis + dashboard scenario reduces $f_C(t)$ by 97.16% to 8.257×10^{-19} h^{-1} (Table 16), illustrating the material impact of diagnostic upgrades on consequence frequency. For asset managers, this distinction is crucial because it clearly shows the value of barrier design and maintenance investments. Measures like trip relays, dashboards, and oil analysis directly lead to lower top-event risks. This approach thus supports risk-informed decisions, helping identify where resources should be focused and illustrating how defence-in-depth strategies offer meaningful improvements in reliability.

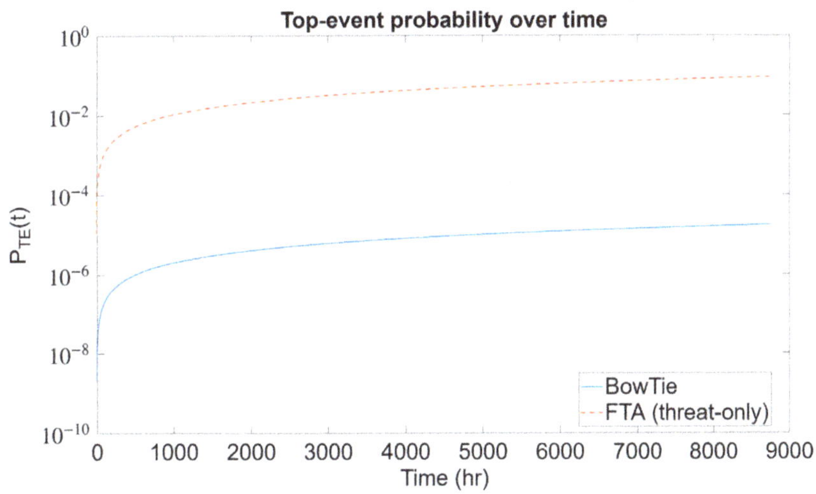

Figure 17. The performance difference of $P_{TE}(t)$ estimated using the proposed model and FTA.

The proposed BowTie model thus offers decision-makers a structured approach to effectively balance safety and reliability by quantitatively ranking barriers, assessing the impacts of redesign, and verifying sufficiency. This approach aligns with risk-based inspection (RBI) by focusing maintenance resources on critical equipment while optimising inspection schedules to reduce operational costs without compromising safety [39,40]. Concerning the WT paradigm, the findings of this study indicate the importance of emphasising diagnostic redundancy and hybridisation, which lead to the most significant decrease in failure rates. Regarding the importance of barriers, critical protective relays, and automatic trip triggers significantly enhance overall system reliability when properly maintained. On the other hand, barriers that have a minimal impact can be optimised or streamlined to prevent unnecessary waste of resources. This viewpoint aligns closely with the principles of reliability engineering, which emphasise the importance of achieving the optimal level of system reliability by balancing risk reduction with economic practicality throughout the asset's lifecycle [41].

Aside from the WT domain, the suggested BowTie methodology can also be applied to various critical infrastructures where a defence-in-depth approach is necessary to prevent a major incident. Yang et al. [42] showcase this approach by combining BowTie analysis with a Bayesian network to improve risk management in oil storage facilities. This enhances emergency response strategies and supports operational decision-making.

5.6. Limitations and Future Research

Several limitations of this study should be noted. First, the case study is limited to the generator subsystem of a WT, which constrains the generalisability of the findings; applying the framework to other subsystems or different infrastructures would provide broader validation. Second, the analysis relies on the assumption of constant failure rates for threats and barriers, without accounting for repair, degradation, or ageing effects. Similarly, HEPs and conditional probabilities are treated as fixed values, overlooking potential variability due to factors such as operator experience, fatigue, or context. Third, the data sources span multiple years and industries, including technical reports and cross-sector studies, which may introduce inconsistencies and reduce direct transferability to wind energy systems. Fourth, barrier independence is assumed, with dependencies and common-cause failures not explicitly considered. Fifth, the framework is presented deterministically, without quantifying uncertainty or providing error bounds for the reported results. Finally, efficiency gains over ETA are demonstrated for a single subsystem only; additional benchmarking on larger and more complex systems is required to substantiate scalability. In this paper, these assumptions are partially probed via scenario-based sensitivity, as explained in Sections 3.4 and 4.4, to observe directional robustness of $\lambda_{\mathrm{TE}}(\mathrm{t})$ and $\mathrm{f_C}(\mathrm{t})$ under barrier redesign.

Future research should build on this work by extending the framework to multiple subsystems and other critical infrastructures, incorporating non-constant failure behaviours such as repair and ageing, and refining the treatment of human and conditional probabilities to capture contextual variability. Explicit modelling of barrier dependencies and common-cause failures, together with validation against field data, would further enhance robustness. In addition, integrating uncertainty propagation techniques, such as Monte Carlo simulation or Bayesian inference, would enable the reporting of confidence intervals or error ranges, thereby strengthening interpretability. Finally, systematic benchmarking across larger case studies is needed to evaluate trade-offs between efficiency and accuracy, and to confirm the framework's scalability for complex engineered systems.

6. Conclusions

This research has illustrated the BowTie framework based on reliability, which delivers a practical and effective method for quantifying system reliability, ranking barriers, and validating the adequacy of defence-in-depth measures. The framework provides a transparent and computationally efficient alternative to traditional reliability models, such as FTA, ETA, and RBD, by incorporating key metrics, including RAW and BIF, alongside cut-set analysis and scenario-based sensitivity evaluations. The findings indicate that high-risk barriers, such as protective relays and automation triggers, play a significant role in escalation pathways. In contrast, low-impact barriers exert only a marginal influence on overall system reliability. Furthermore, the higher-order cut sets generated by the proposed BowTie model provide a more realistic representation of the resilience of WT systems compared to the single-point vulnerabilities identified by conventional FTA.

The study also demonstrates significant efficiency gains: the BowTie model reduces runtime by approximately 40 times compared to ETA enumeration (0.0009 s vs. 0.0729 s) and compresses state requirements from 64 to 21 without loss of fidelity. These improvements highlight its potential for a scalable application. Nevertheless, several limitations remain. The current analysis is restricted to a single subsystem, assumes constant failure rates and barrier independence, and does not explicitly address common-cause failures or repair processes.

Future research should address these limitations by validating the framework across multiple WT subsystems and other infrastructures, incorporating non-constant failure behaviours (repair, ageing, degradation), and explicitly modelling dependencies between barriers. Large-scale benchmarking against conventional reliability techniques is also needed to assess scalability, computational trade-offs, and accuracy in complex engineered systems.

Author Contributions: Conceptualization, M.A.A.P., K.A.P. and J.K.S.; methodology, M.A.A.P.; software, M.A.A.P.; validation, M.A.A.P.; formal analysis, M.A.A.P.; investigation, M.A.A.P.; writing—original draft preparation, M.A.A.P.; writing—review and editing, K.A.P. and J.K.S.; visualization, M.A.A.P.; supervision, K.A.P. and J.K.S. All authors have read and agreed to the published version of the manuscript.

Funding: This research received no external funding.

Data Availability Statement: The original contributions presented in this study are included in the article. Further inquiries can be directed to the corresponding author.

Acknowledgments: The authors would like to acknowledge the University of Manchester for providing facilities for access to journals, conference papers, and other important documents to support this research. During the preparation of this manuscript, the authors used Grammarly (Version v1.2.203.1770) for grammar and language editing purposes only. The authors have reviewed and edited the output and take full responsibility for the content of this publication.

Conflicts of Interest: The authors declare no conflict of interest.

Abbreviations

The following abbreviations are used in this manuscript:

AHP	Analytical Hierarchy Process
BIF	Barrier Importance Factor
BRB	Belief Rule Base
CBM	Condition-based Monitoring
CCPS	Centre for Chemical Process Safety
CMMS	Computerised Maintenance Management System
DDA	Detect-Decide-Action
DFT	Dynamic Fault Tree
ETA	Event Tree Analysis
FTA	Fault Tree Analysis
HEP	Human Error Probability
HRO	High Reliability Organisation
IED	Intelligent Electronic Device
LFF	Learning from Failures

ML	Machine Learning
O&G	Oil and Gas
PAND	Priority-AND
PFD_{avg}	Average Probability of Failure on Demand
PFH	Probability of Failure per Hour
RAW	Risk Achievement Worth
RBD	Reliability Block Diagrams
ROI	Return on Investment
SPAR	Standardised Plant Analysis Risk
TE	Top Event
THERP	Technique for Human Error Rate Prediction
VFD	Variable Frequency Drive
WMEP	Scientific Measurement and Evaluation Programme
WT	Wind Turbine

References

1. Kabir, S. An overview of fault tree analysis and its application in model based dependability analysis. *Expert Syst. Appl.* **2017**, *77*, 114–135. [CrossRef]
2. Ahmad, W.; Hasan, O.; Tahar, S.; Hamdi, M.S. Formal reliability analysis of oil and gas pipelines. *Proc. Inst. Mech. Eng.* **2018**, *232*, 320–334. [CrossRef]
3. Chiacchio, F.; Compagno, L.; D'Urso, D.; Manno, G.; Trapani, N. Dynamic fault trees resolution: A conscious trade-off between analytical and simulative approaches. *Reliab. Eng. Syst. Saf.* **2011**, *96*, 1515–1526. [CrossRef]
4. Rivera, G.; Yunusa-Kaltungo, A.; Labib, A. A hybrid approach for an oil and gas company as a representative of a high reliability organization. *Saf. Reliab.* **2021**, *40*, 130–156. [CrossRef]
5. Kabir, S.; Aslansefat, K.; Sorokos, I.; Papadopoulos, Y.; Konur, S. A hybrid modular approach for dynamic fault tree analysis. *IEEE Access* **2020**, *8*, 97175–97188. [CrossRef]
6. Cheng, X.; Long, M.; He, W.; Zhu, H. Milling fault detection method based on fault tree analysis and hierarchical belief rule base. *Comput. Syst. Sci. Eng.* **2023**, *46*, 2821–2844. [CrossRef]
7. Khakzad, N.; Khan, F.; Amyotte, P. Dynamic safety analysis of process systems by mapping bow-tie into Bayesian network. *Process Saf. Environ. Prot.* **2013**, *91*, 46–53. [CrossRef]
8. de Barnier, T.; Olivier-Maget, N.; Bourgeois, F.; Gabas, N.; Iddir, O. Towards an improved bowtie method for quantifying industrial risks. *Chem. Eng. Trans.* **2022**, *90*, 67–72. [CrossRef]
9. Wu, X.; Huang, H.; Xie, J.; Lu, M.; Wang, S.; Li, W.; Huang, Y.; Yu, W.; Sun, X. A novel dynamic risk assessment method for the petrochemical industry using bow-tie analysis and Bayesian network analysis method based on the methodological framework of ARAMIS project. *Reliab. Eng. Syst. Saf.* **2023**, *237*, 109397. [CrossRef]
10. Zhu, H.-L.; Liu, S.-S.; Qu, Y.-Y.; Han, X.-X.; He, W.; Cao, Y. A new risk assessment method based on belief rule base and fault tree analysis. *Proc. Inst. Mech. Eng. Part O J. Risk Reliab.* **2022**, *236*, 420–438. [CrossRef]
11. International Electrotechnical Commission. *IEC 61025: Fault Tree Analysis (FTA)*; IEC: Geneva, Switzerland, 2006.
12. CCPS; Energy Institute. *Bow Ties in Risk Management*; Wiley: Hoboken, NJ, USA, 2018. [CrossRef]
13. Ozturk, S. Forecasting Wind Turbine Failures and Associated Costs. Master's Thesis, Columbia University, New York, NY, USA, 2019.
14. Parnon, M.A.A.; Papadopoulou, K.A.; Sinha, J.K. A unified framework for asset ranking based on their criticality through case studies. *Int. J. Qual. Reliab. Manag.* **2025**, *42*, 2561–2586. [CrossRef]
15. Pulikollu, R.V.; Dykes, K.; Guo, Y.; Sethuraman, L.; Parsons, T.; King, R.N.; Sethuraman, L.; Veers, P. *Wind Turbine Generator Reliability Analysis to Reduce Operations and Maintenance (O&M) Costs*; Sandia National Laboratories Report; Sandia National Laboratories: Livermore, CA, USA, 2023. [CrossRef]
16. Novaković, B.; Radovanović, L.; Vidaković, D.; Đorđević, L.; Radišić, B. Evaluating wind turbine power plant reliability through fault tree analysis. *Appl. Eng. Lett. J. Eng. Appl. Sci.* **2023**, *8*, 175–182. [CrossRef]
17. Márquez, F.P.G.; Pérez, J.M.P.; Marugán, A.P.; Papaelias, M. Identification of critical components of wind turbines using FTA over the time. *Renew. Energy* **2016**, *87*, 869–883. [CrossRef]
18. Kang, J.; Sun, L.; Soares, C.G. Fault tree analysis of floating offshore wind turbines. *Renew. Energy* **2019**, *133*, 1455–1467. [CrossRef]
19. Einarsson, S. Wind Turbine Reliability Modeling. Master's Thesis, Reykjavík University, Reykjavík, Iceland, 2016.

20. Linsday, J.; Briand, D.; Hill, R.; Stinebaugh, J.; Benjamin, A. *Wind Turbine Reliability: A Database and Analysis Approach*; Sandia National Laboratories: Albuquerque, NM, USA; Lawrence Livermore National Laboratory: Livermore, CA, USA, 2008. [CrossRef]
21. Swain, A.D.; Guttmann, H.E. *Handbook of Human Reliability Analysis with Emphasis on Nuclear Power Plant Applications (NUREG/CR-1278)*; U.S. Nuclear Regulatory Commission: Washington, DC, USA, 1983.
22. May, A.; McMillan, D. Condition based maintenance for offshore wind turbines: The effects of false alarms from condition monitoring systems. In Proceedings of the ESREL 2013—Safety, Reliability and Risk Analysis: Beyond the Horizon, Amsterdam, The Netherlands, 29 September–2 October 2013; CRC Press: Boca Raton, FL, USA, 2013; pp. 783–789.
23. Pepperl+Fuchs GmbH. *Safety Manual SIL KFD0-RSH-1.4S.PS Relay Module*; Pepperl+Fuchs: Mannheim, Germany, 2010.
24. National Renewable Energy Laboratory (NREL). Gearbox reliability database (GRD) statistics. Available online: https://grd.nrel.gov/stats (accessed on 11 September 2025).
25. NSK Ltd. *Advanced Bearing Solutions for the Wind Industry*; NSK Ltd.: Tokyo, Japan, 2018.
26. Brewer, R.; Pelkey, A.; Miller, A. *Limitations of Offline Oil Analysis for Determining Gearbox Health*; Technical Report; Poseidon Systems: Rochester, NY, USA, 2018.
27. Nogueira, W.F.; de Andrade, M.A.H.; de Souza, G.F.M. Wind turbine fault detection through autoencoder-based neural network and FMSA. *Sensors* **2025**, *25*, 4499. [CrossRef]
28. Precision Electric. *Variable Frequency Drives (VFDs)–Comprehensive Technical Overview*; Technical Report; Precision Electric: Livermore, CA, USA, 2025.
29. Kumar, R.R.; Waisale, L.O.; Tamata, J.L.; Tortella, A.; Kia, S.H.; Andriollo, M. Advanced fault detection and severity analysis of broken rotor bars in induction motors: Comparative classification and feature study using dimensionality reduction techniques. *Machines* **2024**, *12*, 890. [CrossRef]
30. Gertman, D.I.; Blackman, H.S.; Marble, J.L.; Byers, J.C.; Smith, C.L. *The SPAR-H Human Reliability Analysis Method (NUREG/CR-6883, INL/EXT-05-00509)*; U.S. Nuclear Regulatory Commission: Washington, DC, USA, 2005.
31. Western Electricity Coordinating Council (WECC). *2023 Protection Systems Performance Report*; WECC: Salt Lake City, UT, USA, 2024.
32. Liebusch, F. *Relays in Safety-Related Control Systems*; White Paper; TE Connectivity: Galway, Ireland, 2020.
33. Generowicz, M.; Martin, R. Introduction to the MoonSIF Spreadsheet Workbook: Failure Probability Model. I&E Systems Pty Ltd., Nov. 2023. Available online: https://creativecommons.org/licenses/by-sa/4.0/legalcode (accessed on 11 September 2025).
34. Lawrance, G.; Paul, P.S.; Shylu, D.S.; Ebenezer Jacob Dhas, D.S.; Dineshkumar, C.; Jeyakumar, P.D.; Muthiya, S.J.; Getachew, N.A. Effect of metallic substrate and rubber elastic materials over passive constrained layer damping on tool vibration during boring process. *J. Low Freq. Noise Vib. Act. Control.* **2024**, *43*, 1139–1157. [CrossRef]
35. U.S. Department of Defense. *MIL-HDBK-217F: Reliability Prediction of Electronic Equipment*; Department of Defense: Washington, DC, USA, 1991.
36. Vairo, T.; Benvenuto, A.C.; Tedeschi, A.; Reverberi, A.; Fabiano, B. Make Bow-Tie Dynamic by Rethinking It as a Hierarchical Bayesian Network: Dynamic Risk Assessment of an LNG Bunkering Operation. *Chem. Eng. Trans.* **2022**, *91*, 277–282. [CrossRef]
37. Yuan, S.; Reniers, G.; Yang, M. Dynamic-Risk-Informed Safety Barrier Management: An Application to Cost-Effective Barrier Optimization Based on Data from Multiple Sources. *J. Loss Prev. Process Ind.* **2023**, *83*, 105034. [CrossRef]
38. Leoni, L.; De Carlo, F.; Paltrinieri, N.; Sgarbossa, F.; BahooToroody, A. On Risk-Based Maintenance: A Comprehensive Review of Three Approaches to Track the Impact of Consequence Modelling for Predicting Maintenance Actions. *J. Loss Prev. Process Ind.* **2021**, *72*, 104555. [CrossRef]
39. Hosseinnia Davatgar, B.; Paltrinieri, N.; Bubbico, R. Safety barrier management: Risk-based approach for the oil and gas sector. *J. Mar. Sci. Eng.* **2021**, *9*, 722. [CrossRef]
40. Kaliszewski, N.; Marian, R.; Chahl, J. A reliability centred maintenance-oriented framework for modelling, evaluating, and optimising complex repairable flow networks. *Complex Intell. Syst.* **2025**, *11*, 223. [CrossRef]
41. Aven, T. Improving the foundation and practice of reliability engineering. *Proc. Inst. Mech. Eng. Part O J. Risk Reliab.* **2017**, *231*, 295–305. [CrossRef]
42. Yang, D.; Xing, K.; Pan, L.; Lu, N.; Yu, J. Sustainable risk management framework for petroleum storage facilities: Integrating bow-tie analysis and dynamic Bayesian networks. *Sustainability* **2025**, *17*, 2642. [CrossRef]

applied sciences

MDPI

Article

Efficient Reliability Block Diagram Evaluation Through Improved Algorithms and Parallel Computing

Gloria Gori *, Marco Papini and Alessandro Fantechi

Department of Information Engineering (DINFO), School of Engineering, University of Florence, Via di S. Marta, 3, 50139 Florence, Italy; marco.papini@unifi.it (M.P.); alessandro.fantechi@unifi.it (A.F.)
* Correspondence: gloria.gori@unifi.it

Abstract

Quantitative reliability evaluation is essential for optimizing control policies and maintenance strategies in complex industrial systems. While Reliability Block Diagrams (RBDs) are a natural formalism for modeling these hierarchical systems, modern applications require highly efficient, online reliability assessment on resource-constrained embedded hardware. This demand presents two fundamental challenges: developing algorithmically efficient RBD evaluation methods that can handle diverse custom distributions while preserving numerical accuracy, and ensuring platform-agnostic performance across diverse multicore architectures. This paper investigates these issues by developing a new version of the *librbd* open-source RBD library. This version includes advances in efficiency of evaluation algorithms, as well as restructured computation sequences, cache-aware data structures to minimize memory overhead, and an adaptive parallelization framework that scales automatically from embedded processors to high-performance systems. Comprehensive validation demonstrates that these advances significantly reduce computational complexity and improve performance over the original implementation, enabling real-time analysis of substantially larger systems.

Keywords: reliability block diagrams (RBD); symmetric multi-processing (SMP); single instruction-multiple data (SIMD); reliability curve; reliability evaluation; software libraries

1. Introduction

Quantitative reliability evaluation enables actionable insights that inform decisions-support systems with different aims, including, e.g., control policy synthesis, system architecture design, and maintenance planning that may involve contrasting measures and trade-offs. As systems become increasingly complex and highly integrated across various industrial sectors, the challenges in reliability assessment intensify. In highly integrated systems, component failures can cascade through interconnected subsystems, causing widespread disruptions that are difficult to isolate and mitigate. This integration trend, while offering significant functional benefits, demands more sophisticated and computationally efficient reliability assessment approaches capable of supporting online evaluation in resource-constrained environments. Reliability is defined as "*the ability of a system or component to perform its required functions under stated conditions for a specified period of time*" [1]. In probabilistic terms, reliability represents the probability that a system successfully performs its required functions during the time interval $[t_0, t)$ given correct operation at time t_0 [2]. This probabilistic definition forms the basis of reliability evaluation theory, enabling systematic quantitative assessment of system dependability.

Modern industrial applications increasingly require these evaluations to be performed online on embedded processors and low-power industrial controllers, where computational efficiency becomes a critical constraint rather than merely a performance optimization goal.

These challenges are widely recognized in the reliability engineering community. For instance, Hollander and Peña [3] introduced non-parametric methods to derive reliability curves for censored data through the survival function, while Xing [4] proposed a framework that simplifies network reliability assessment using Reduced Ordered Binary Decision Diagrams. Green and Vishakha [5] further explored computational efficiency by developing and comparing two parallel computation strategies—batch parallelism and pipeline parallelism—for non-sequential Monte Carlo simulations on multi-core architectures. Additionally, Nelissen et al. [6] proposed the deployment of inline run-time monitoring to realize run-time verification and ensure the required safety properties.

Recent industrial applications have further intensified these challenges, particularly in domains requiring real-time or quasi-runtime reliability assessment:

- Developing efficient and expressive modeling techniques for complex Cyber–Physical Systems and their reliability evaluation [7–9].
- Optimizing software rejuvenation scheduling in distributed environments through the reliability assessment of their software subsystems [10–12].

Our recent industrial case studies on software reliability monitoring and rejuvenation [13] and proactive maintenance [14] have further demonstrated scenarios requiring intensive library usage for quasi-runtime reliability assessment in embedded maintenance systems, revealing critical performance bottlenecks when deployed on resource-constrained hardware. These combined insights from the literature and our practical experience motivated three research questions:

RQ1 *How can the system's reliability be efficiently analyzed in presence of components with arbitrary, non-parametric (i.e., numerical) failure distributions?*

RQ2 *How can computational complexity be reduced for RBD evaluation while preserving numerical accuracy?*

RQ3 *Can modern multicore architectures—from embedded processors to high-performance systems—be effectively leveraged to achieve platform-agnostic performance improvements?*

To address RQ1, we previously released *librbd* [15], an open-source library supporting RBD-based reliability evaluation. *librbd* was specifically created for the numerical computation of reliability curves for all basic RBD blocks with multiplatform compatibility, thereby fully addressing our first research question. This paper presents significant algorithmic and architectural advances addressing all three RQs. We introduce a novel optimized recursive strategy for evaluating complex *KooN* models, combined with parallel computation techniques that provide substantial performance gains across diverse hardware platforms. To the best of our knowledge, no existing open-source implementation provides this level of algorithmic efficiency, platform portability, and computational generality in a lightweight C library suitable for deployment in resource-constrained industrial environments.

The remainder of this paper is organized as follows: Section 2 provides a brief overview of existing methodologies to reliability evaluation; Section 3 presents the mathematical foundations of RBD reliability evaluation and introduces our algorithmic contributions for reducing computational complexity; Section 4 details the enhanced *librbd 2.0* features and validates computed reliability curves through comparison with SHARPE [16,17]; Section 5 describes the experimental methodology and hardware platforms used for performance evaluation; Section 6 presents comprehensive performance results demonstrating the effectiveness of our approach across diverse system configurations.

2. Reliability Evaluation Methodologies

Since reliability evaluation is a fundamental step for the assessment of an industrial system, several modeling formalisms have been developed for reliability assessment, broadly classified into four categories:

- **Combinatorial models** exploit the assumption of statistically independent components, i.e., a failure of one component does not impact the failure rate of other components, to efficiently evaluate the reliability [18,19]. These models include Reliability Block Diagrams (RBDs) [20,21], Fault Trees (FTs) [22,23], Reliability Graphs (RGs) [24,25], and Fault Trees with Repeated Events (FTREs) [23,26].
- **State-space-based models** leverage Markov Processes that allow them to model statistical, temporal, and spatial dependencies among failures at the cost of state-space explosion [18,19]. These include Continuous Time Markov Chains (CTMCs) [27,28], Stochastic Petri Nets (SPNs), and extensions such as Generalized Stochastic Petri Nets (GSPNs) and Stochastic Timed Petri Nets (STPNs) [29–32], Stochastic Reward Nets (SRNs) [33,34], and Stochastic Activity Networks (SANs) [35,36].
- **Hybrid models** augment combinatorial models with time-dependent relationships through state-space-based models [18], including Dynamic RBDs (DRBDs) [37,38] and Dynamic FTs (DFTs) [39–41].
- **Hierarchical models** employ divide-and-conquer strategies by combining different formalisms across system layers [13,18,42–44], exploiting the benefits of both combinatorial and state-space-based approaches while mitigating their respective limitations.

Hierarchical models represent the state-of-the-art approach for reliability evaluation [18], particularly for highly integrated systems where component interactions significantly influence overall system behavior. As system complexity increases, decomposition into subsystems and components becomes necessary for tractable analysis, making RBDs a natural formalism for modeling hierarchical system architectures due to their intuitive graphical representation and computational efficiency.

3. Reliability Block Diagrams

In Section 3.1, we recall the definition of an RBD together with the mathematical formulas used to compute the probability that a block is correctly operating, i.e., its state is equal to *success*. In Section 3.2, we introduce notable tools used for the RBD modeling and evaluation, detailing their core capabilities. In Section 3.3, we analyze the recursive algorithm that is widely used to evaluate the reliability of $KooN$ blocks with generic components, together with its main limitations. In Section 3.4, we propose novel mathematical formulas that can be successfully used to overcome such limitations. Finally, in Section 3.5, we propose novel mathematical formulas also for the computation of Bridge blocks.

3.1. RBD Definitions

An RBD is a graph showing the decomposition of a system into components that may independently fail, plus the logical connections needed for the correct working of the system [15,18–20,45]

Definition 1. *An **RBD model** is a Block Diagram that is used to evaluate the reliability of a system, or sub-system, through the definition of the logical connections needed for its successful operation.*

Definition 2. *A **basic block**, for simplicity also called **block**, is the building unit of an RBD model. A block provides the basic logical connections needed to define an RBD model by aggregating other blocks.*

A block has only two states, i.e., success and failure. An RBD modeling a system represents the success state of the system by means of success paths, that is, connections of the success states of its components.

Basic blocks come in five kinds, recursively defined as follows:

- ***Singleton**. A **component** is a part of the physical system that is considered atomic w.r.t. failures. A physical component is visually represented inside the RBD model by a Singleton block. Singletons can be considered as the elementary bricks of the recursive structure of an RDB model. The state of a Singleton block is equal to success if and only if the associated component is working correctly. An example may be a stand-alone Power Supply.*
- ***Series**. This block is composed by N (sub-)blocks, and its state is equal to success if and only if all its sub-blocks are in success state. Examples are systems in which the single failure of any component produces the failure of the system.*
- ***Parallel**. This block is composed by N blocks, and its state is equal to success if and only if at least one of its sub-blocks is in success state, or, dually, a failure state of a Parallel block occurs if and only if all its blocks are in failure state. An example is a redundant Power Supply system with current sharing.*
- ***K-out-of-N (KooN)**. This block is composed by N blocks, and its state is equal to success if and only if at least K blocks out of N are in success state. An example is a Triple Modular Redundancy computing system (TMR).*
- ***Bridge**. This block is introduced in order to model the possibility of an alternative success path in a parallel/series structure. As shown in Figure 1e, the block in the middle is used as a bridge between two parallel success paths, rerouting success to the other path in case of failure of one block. Typical examples can be found in network infrastructures.*

Definition 3. *We denote as p_i the probability that the state of the i-th component is equal to success, and as $q_i = 1 - p_i$ the probability that the state of the i-th component is equal to failure.*

Definition 4. *A **generic** block uses heterogeneous components, also called **generic** components, i.e., $\exists\, i \neq j$ such that $p_i \neq p_j$. An **identical** block uses homogeneous components, also called **identical** components, i.e., $p_i = p_j = p \;\forall\, i, j$.*

Figure 1 provides the graphical representation of the five kinds of basic blocks, visualizing the success paths among the related sub-blocks, together with the mathematical formulas that are used to perform the reliability analysis of each block according to [18,20]. In particular, Moskowitz [20] defines the formulas for computing the reliability as a probability of series and parallel models, also demonstrating the factoring theorem for simplifying the bridge model, and Trivedi and Bobbio [18] present the formulas for evaluating the reliability curve for series, parallel and *KooN* models, covering both the general case of generic components and the simplified case of identical components. More specifically, the formula in black is the one that can be used to evaluate blocks built with generic components, while the formula in red is the simplified one that can be used to evaluate blocks built with identical components.

For the sake of simplicity, the presented formulas express reliability as a probability. Nevertheless, the same formulations can be straightforwardly generalized to determine the time-varying reliability function. The correct probabilistic analysis of an RBD model requires that a failure probability function is associated to each Singleton block, reflecting the estimated failure probability of the modeled physical component. The Singleton blocks are statistically independent, i.e., given two Singletons A, B such that $A \neq B$, the probability of failure of A, $P(A)$, is not related with the probability of failure of B, $P(B)$, i.e., $P(A|B) = P(A) \;\forall\, A, B \mid A \neq B$.

Throughout the remainder of this paper, for simplicity of description, we consider RBD models composed of N Singleton blocks aggregated into a single system-level block. Also for clarity and simplicity, we refer to the Singleton blocks as *components*, and to the system-level block simply as *block*.

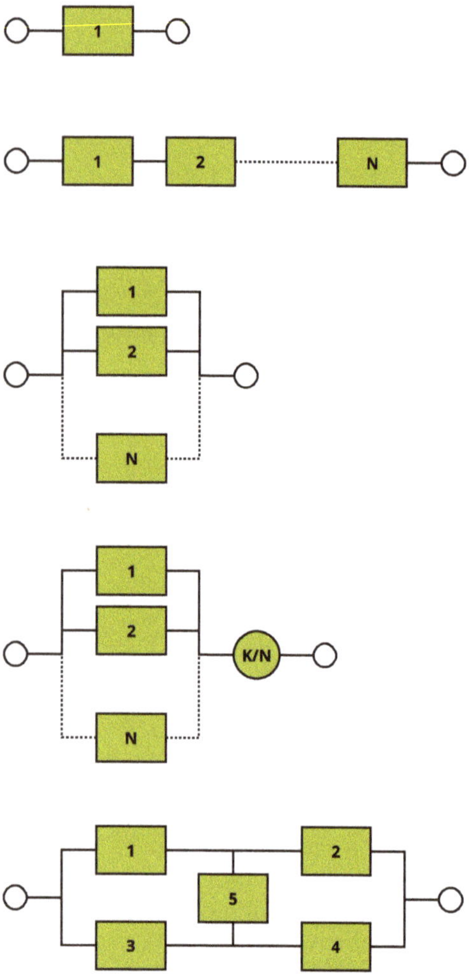

$$p_{singleton} = p_1$$

(**a**)

$$p_{series} = \prod_{i=1}^{N} p_i$$

$$p_{series} = p^N$$

(**b**)

$$p_{parallel} = 1 - \prod_{i=1}^{N} (1 - p_i)$$

$$p_{parallel} = 1 - (1-p)^N$$

(**c**)

$$p_{0ooN} = 1 \quad \forall\, N > 0$$
$$p_{KooN} = 0 \quad \forall\, K > N$$
$$p_{KooN} = p_N \cdot p_{(K-1)oo(N-1)} + q_N \cdot p_{Koo(N-1)} \quad \forall\, 0 < K \leq N$$

$$p_{KooN} = \sum_{J=K}^{N} \binom{N}{J} \cdot p^J \cdot q^{N-J}$$

(**d**)

$$p_{bridge} = p_5 \cdot (1 - q_1 \cdot q_3) \cdot (1 - q_2 \cdot q_4) + \\ + q_5 \cdot (1 - (1 - p_1 \cdot p_2) \cdot (1 - p_3 \cdot p_4))$$

$$p_{bridge} = p \cdot (1 - q^2)^2 + q \cdot (1 - (1 - p^2)^2)$$

(**e**)

Figure 1. Layout of RBD blocks and mathematical formulas used to perform the respective reliability analysis under the generic case (black text) and under the identical case (red text): (**a**) Singleton. (**b**) Series. (**c**) Parallel. (**d**) *KooN*. (**e**) Bridge.

3.2. Existing RBD Modeling and Evaluation Tools

To facilitate the accurate definition and evaluation of RBD models, various supporting software tools have emerged. We can mention:

- **Isograph Reliability Workbench**: this commercial tool supports the hierarchical definition and analysis of scalable RBD models by means of submodels. It also supports the *minimal cut-set* analysis of the RBD model [46]. With respect to RQ1, this tool does not support the computation of the reliability curve.

- **Relyence RBD**: this commercial tool allows the definition and evaluation of series, parallel, and standby configurations, providing the reliability curve using analytical formulas or through Monte Carlo simulation [47]. With respect to RQ1, this tool only supports a set of parametric distributions for its input blocks.
- **ALD RAM Commander—RBD Module**: this commercial tool allows the definition and evaluation of series, parallel, and *KooN* configurations, providing the reliability curve using analytical formulas or through Monte Carlo simulation [48]. With respect to RQ1, this tool only supports a set of parametric distributions for its input blocks.
- **SHARPE**: the Symbolic Hierarchical Automated Reliability and Performance Evaluator (SHARPE) tool supports the definition of hierarchical stochastic models of dependability attributes, including reliability, availability, performance and performability, and the analysis of such models [16,17]. This tool supports several formalisms to define reliability models, including RBDs, and it supports time-dependent reliability analysis. With respect to RQ1, this tool only supports a set of parametric distributions for its input blocks.
- **PyRBD** is an open-source tool particularly effective to model and evaluate communication networks that employs a methodology that generates RBDs from network topologies and decomposes the diagrams for faster processing, with a core focus on utilizing *minimal cut-sets* and Boolean algebra to compute the steady-state availability [49,50]. With respect to RQ1, this tool does not support the computation of the reliability curve and it only supports a limited set of parametric distributions for its input blocks.
- **PyRBD++** is the optimized evolution of PyRBD. While maintaining the same characteristics of its predecessor, it introduces a novel iterative conditional decomposition method to improve scalability performance [51,52]. As its predecessor, this tool does not support the computation of the reliability curve and it only supports a limited set of parametric distributions for its input blocks.

This comparative analysis highlights that no single existing tool fully satisfies the requirements of RQ1: specifically, the capability to compute the reliability curve using arbitrary, non-parametric input distributions. Therefore, we made the strategic decision to further enhance *librbd*, ensuring it uniquely provides the necessary computational foundation for our specialized research projects involving reliability modeling.

3.3. Quantitative Evaluation of KooN Blocks—Traditional Approach

The probability of *success* of a *KooN* block can be computed through the application of a divide and conquer recursive algorithm based on the formulas shown in Figure 1.

The recursive formulation is derived by conditioning on the state of the N-th component [15]. We denote as p_N the probability of the N-th component being in *success* state, and as q_N its probability of being in and *failed* state. If we consider the N-th component as correctly operating: for a *KooN* system to be correctly working, we need at least $K-1$ operating components out of the remaining $N-1$. If, on the other hand, the N-th component is failed, we need at least K operating components out of the remaining $N-1$ to have a correctly operating *KooN* system. The formulas that we report also below include two boundary cases that limit the domain of K and N to significant values and that are used as stop conditions for recursion.

$$
\begin{aligned}
&p_{0ooN} = 1 \quad \forall\, N > 0 \\
&p_{KooN} = 0 \quad \forall\, K > N \\
&p_{KooN} = p_N \cdot p_{(K-1)oo(N-1)} + q_N \cdot p_{Koo(N-1)} \quad \forall\, 0 < K \leq N
\end{aligned}
\tag{1}
$$

This approach can be modeled using an incomplete binary recursion tree since leaf nodes can be found at different depths due to the following two reasons: a node is a leaf node when one of the stop conditions is met, which can occur before the maximum reachable depth N; the initial value K affects the reachability of the stop conditions, with the worst case scenario occurring when $K = N/2$. The theoretical upper limit to the number of recursive calls is given by a complete binary tree, i.e., $\sum_{i=1}^{i \leq N} 2^i = 2^{N+1} - 1$. Figure 2 shows this recursive approach over a 3oo5 block, where the the two numbers in each node represent, respectively, the two parameters K and N, magenta nodes are leaf nodes and cyan nodes are the internal ones.

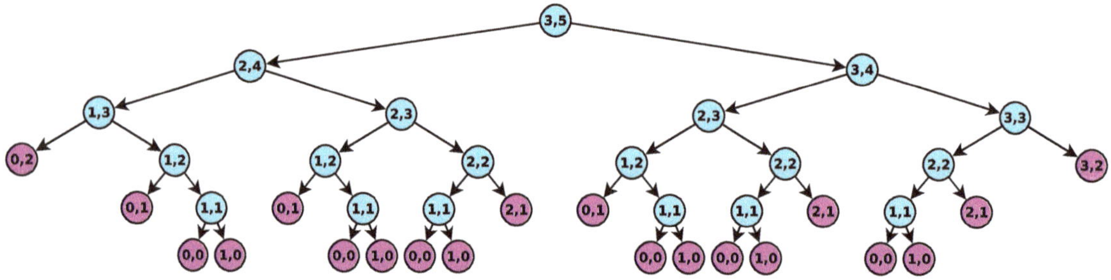

Figure 2. 3oo5 block evaluated through the recursive approach shown in Equation (1). Magenta nodes are the leaf nodes and cyan nodes are the internal ones.

This recursive approach, albeit simple and suitable to analyze $KooN$ blocks with a limited number of components N, soon suffers from exponential blow-up of the number of recursive calls due to the following two reasons:

- The presented approach has trivial stop conditions that exponentially increase the number of leaf nodes, thus increasing the computation time. More specifically, it does not detect trivial $KooN$ configurations such as Series ($NooN$) and Parallel ($1ooN$);
- Several different internal nodes with equivalent input parameters K and N are visited during the recursion tree. More specifically, in the example shown in Figure 2, two different nodes with the same parameters $K = 2$ and $N = 3$ are evaluated.

3.4. Quantitative Evaluation of KooN Blocks—Novel Approach

We propose novel mathematical formulas for the computation of $KooN$ blocks with generic components that allow to reduce the computational complexity and, as a direct consequence, to decrease the computation time. Starting from Equation (1) and keeping in mind its main limitations, we propose two novel sets of mathematical formulas that can be successfully used to reduce the complexity of the already presented recursive algorithm. The first formula reduces the number of recursive calls for the analysis of trivial RBD blocks, hence the reliability of a $KooN$ block can be computed as:

$$\begin{aligned} p_{1ooN} &= p_{parallel} = 1 - \prod_{i=1}^{I} (1 - p_i) \quad \forall N > 0 \\ p_{KooN} &= p_{series} = \prod_{i=1}^{I} p_i \quad \forall K = N \\ p_{KooN} &= p_N \cdot p_{(K-1)oo(N-1)} + q_N \cdot p_{Koo(N-1)} \quad \forall 1 < K < N \end{aligned} \tag{2}$$

Figure 3 shows this recursive approach over the same 3oo5 block using the same notation. It can be easily observed that, by avoiding the analysis of series and parallel blocks, the number of visited nodes decreases drastically with respect to the recursion tree shown in Figure 2.

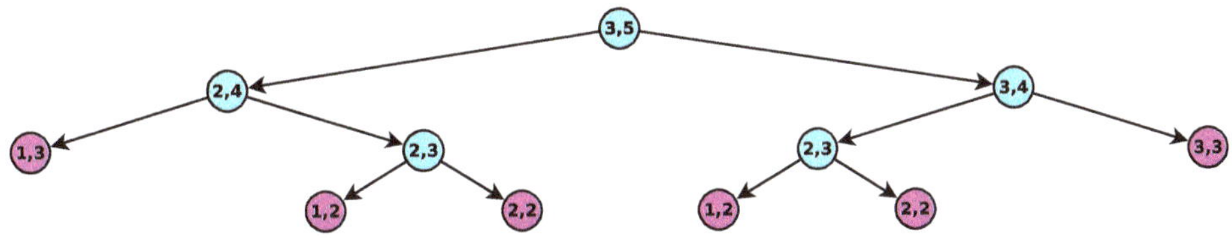

Figure 3. 3oo5 block evaluated through the recursive approach shown in Equation (2). Magenta nodes are the leaf nodes and cyan nodes are the internal ones.

The other novel formula derives from the simplification of Equation (1) under the assumptions that $K > 2$ and $K \leq N-2$. In such case, the reliability of a $KooN$ block can be computed as:

$$\begin{aligned} p_{KooN} &= p_N \cdot p_{N-1} \cdot p_{(K-2)oo(N-2)} + \\ &+ (q_N \cdot p_{N-1} + p_N \cdot q_{N-1}) \cdot p_{(K-1)oo(N-2)} + \\ &+ q_N \cdot q_{N-1} \cdot p_{Koo(N-2)} \end{aligned} \tag{3}$$

Using Equation (3), we can decrease N by 2 instead of 1, thus reducing the number of recursive calls. Furthermore, the two different recursive sub-trees with parameters $(K-1)$ and $(N-2)$, i.e., the first one for which component N is in *failed* state and component $N-1$ is in *success* state and second one for which component N is in *success* state and component $N-1$ is in *failed* state, collapse into a unified sub-tree. By using Equation (3) for all nodes such that $K > 2$ and $K \leq N-2$, the recursion tree ceases to be a binary tree and becomes a ternary tree.

The number of different internal nodes with equivalent input parameters K and N that are visited during the recursion tree achieved through the usage of Equation (3) can be further reduced by iteratively applying the same idea. In particular, let $h \in \mathbb{N}^+$ such that $K-h > 1$ and $N-h \geq K$, then the reliability of a $KooN$ block can be computed as:

$$p_{KooN} = \sum_{i=0}^{h} \left(\sum_{j=1}^{\binom{h}{i}} \left(\prod_{l \in C(h,i,j)} p_l \cdot \prod_{m \notin C(h,i,j)} q_m \right) \right) \cdot p_{(K-i)oo(N-h)} \tag{4}$$

where $\binom{h}{i}$ is the binomial coefficient and $C(h,i,j)$ is the function that returns the j-th unique combination of i working components out of h.

Increasing the value of h results in more identical sub-trees whose computation can be factorized (see the two sub-trees with root $(2,3)$ in Figure 3), leading to a reduction in the total number of recursive calls. Hence, to maximize the benefits of Equation (4), we need the maximum value h such that $K-h > 1$ and $N-h \geq K$:

$$h = min(K-1, N-K) \tag{5}$$

where h is computed for each recursive call. Let H be the maximum of all h values computed. The recursive unfolding of Equation (4) follows a recursion tree with degree $H+1$. Figure 4 shows this recursive approach over the same 3oo5 block using the same notation. It can be easily observed that the number of visited nodes decreases drastically with respect to Figure 2. This reduction is caused by the node $(3,5)$, which is the only one, for this simple RBD model, that benefits from the new algorithm. Furthermore, this mathematical representation allows to reduce the number of visited internal nodes with equivalent input parameters K and N.

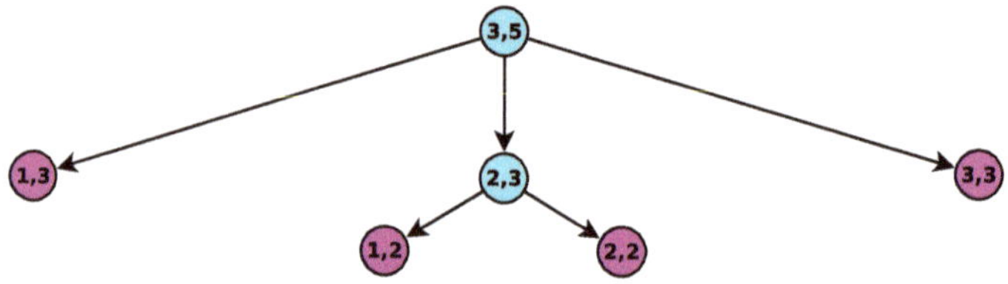

Figure 4. 3oo5 block evaluated through the recursive approach shown in Equation (4). Magenta nodes are the leaf nodes and cyan nodes are the internal ones.

3.5. Quantitative Evaluation of Bridge Blocks—Novel Approach

As already pointed out, we propose novel mathematical formulas for the computation of Bridge blocks with both generic and identical components that allow us to reduce the computational complexity and, as a direct consequence, to decrease the computation time.

To reduce the complexity of Bridge blocks with generic components, we start from the generic equation shown in Figure 1e: after all occurrences of q_i have been replaced with $1 - p_i, i \in [1, 5]$, we obtain the following formula that avoids the explicit computation of unreliability and maximizes the reuse of intermediate values:

$$\begin{aligned} val_1 &= (p_1 + p_3 - p_1 \cdot p_3) \cdot (p_2 + p_4 - p_2 \cdot p_4) \\ val_2 &= p_1 \cdot p_2 + p_3 \cdot p_4 - p_1 \cdot p_2 \cdot p_3 \cdot p_4 \\ p_{bridge} &= p_5 \cdot (val_1 - val_2) + val_2 \end{aligned} \tag{6}$$

To reduce the complexity of Bridge blocks with identical components, we start from the identical equation shown in Figure 1e: after rearranging the order of operations, the computation of the Bridge block can be further simplified to the following formula that avoids a sum, hence it slightly reduces the computational complexity:

$$p_{bridge} = p \cdot (1 + q \cdot (q \cdot (q^2 - 2) + p \cdot (2 - p^2))) \tag{7}$$

4. RBD Computation Library—*librbd 2.0*

As already stated in Section 1, the purpose of this work is to enhance *librbd* [15,53], ensuring that the following original design requirements are fully preserved:

- to support the most common OSes, that is, Windows, MacOS and Linux;
- to support the numerical computation of the reliability curve for all RBD basic blocks; please note that this requirement is necessary to satisfy RQ1;
- to be available as a free software.

Section 4.1 briefly describes the design choices and features that were already integrated in *librbd* and that have been kept, while Section 4.2 provides the new requirements added to address RQs(2) and (3), also providing the detailed description of the new features. Finally, Section 4.3 shows the process used to validate *librbd 2.0* and the obtained results.

4.1. Design Choices Already Present in librbd

The following design choices and features, which were already present in *librdb* [15], have been kept in *librbd 2.0*:

- The implementation of the resolution formulas for series, parallel, $KooN$, and bridge RBD blocks over time, up to 255 components per block.

- The availability of *librbd* on the majority of currently available OSes, i.e., Microsoft Windows, Linux, and MacOS.
- The implementation of *librbd* in C language for higher performance, introducing sporadic conditional compilation when interaction with the OS is deemed necessary [54].
- The availability of *librbd* as both a dynamic and static library.
- In order to minimize the numerical error, all computations use double-precision floating-point format (*double*) compliant with binary64 format [55].
- The implementation of formulas for RBD blocks both with generic components and with identical components.
- The implementation of several optimizations for the *KooN* RBD block computation, e.g., the minimization of computation steps when $N - K > K - 1$ for blocks with identical components.
- The adoption of the Symmetric Multi-Processing (SMP), that can be enabled or disabled at compile time. When disabled, *librbd* is built as a Single Threaded (ST) library.

4.2. New Design Choices Introduced in librbd 2.0

This section details the significant enhancements and new requirements implemented in *librbd 2.0*, which extend the library's capabilities to meet emerging computational challenges. Specifically, our primary contributions focus on performance optimization to address RQs (2) and (3):

- Algorithmic Complexity Reduction: To address RQ2, we implemented novel algorithms aimed at substantially reducing the computational complexity of *KooN* and Bridge blocks (detailed in Section 4.2.1).
- Cross-Platform Parallel Computation Techniques: To Address RQ3, ensuring optimal performance across diverse hardware architectures and OSes, this requirement necessitated several key developments:
  - Vectorization (SIMD): the addition of native support for the Single Instruction, Multiple Data (SIMD) paradigm for computation acceleration (Section 4.2.2).
  - Multi-Core Parallelism (SMP): the optimization of cache utilization within the SMP paradigm (Section 4.2.3), together with the addition of native SMP support for the Windows OS (Section 4.2.4).

4.2.1. Optimization of Algorithms for KooN and Bridge Blocks

The main limitation of *librbd* was its lower performance when computing the reliability of a *KooN* block with generic components when N is high and $K \approx N/2$. The low performance was due to the fact that both algorithms implemented in *librbd*, i.e., the iterative and recursive one, were trivial and not optimized.

We propose a new version of the recursive algorithm, shown in Algorithm 1, that drastically reduces the number of recursive calls, hence decreasing the computation time. The first and second *if* statements are used to detect, respectively, series and parallel blocks and to treat them as such without using recursion, as shown in Equation (2). The innermost statements, i.e., the computation of h variable and the subsequent *if* statement, implement the optimized recursive algorithm as described in Equations (4) and (5). Finally, the last three statements, including the return, implement the classic recursive algorithm shown in Equation (1).

We state that the performance achieved using this new recursive algorithm is comparable to the one achieved with the iterative algorithm that was described in [15] when K is close to either 1 or N. Hence, the iterative algorithm, together with the heuristic used to select the proper algorithm, has been removed from *librbd 2.0* to reduce source code complexity.

```
Algorithm 1: Computation of KooN block with generic components.
Input: Minimum number of working components k
Input: Total number of components n
Input: Array R of reliabilities, where R⟨i⟩ is the reliability of i-th component
Function R_KooN_Recursive(k, n, R)
    if k = n then
        # Evaluate noon block, i.e., Series block
        res = 1;
        for i ∈ [1, n] do
            res = res · R⟨i⟩;
        return res;
    if k = 1 then
        # Evaluate 1oon block, i.e., Parallel block
        res = 1;
        for i ∈ [1, n] do
            res = res · (1 − R⟨i⟩);
        return 1 − res;
    # Compute the best value h (see Equation (5))
    h = min(k − 1, n − k);
    if h > 1 then
        # Compute the reliability of koon (see Equation (4)) and store the result
          over res variable
        res = 1;
        for i ∈ [0, h] do
            # Compute Σ_{j=1}^{(h i)} (Π_{l∈C(h,i,j)} p_l · Π_{m∉C(h,i,j)} q_m) (see Equation (4)) and
              store the result over step variable
            step = 0;
            for j ∈ [1, nCi] do
                # Compute Π_{l∈C(h,i,j)} p_l · Π_{m∉C(h,i,j)} q_m (see Equation (4)) and store
                  the result over tmp variable
                tmp = 1;
                for l ∈ [1, h] do
                    if l ∈ C(n, i, j) then
                        tmp = tmp · R⟨l⟩;
                    else
                        tmp = tmp · (1 − R⟨l⟩);
                step = step + tmp;
            res = res + step · R_KooN_Recursive(k − i, n − h, R);
        return res;
    # The value h is equal to 1, use traditional recursive formula (see Equation (1))
    res = R⟨n⟩ · R_KooN_Recursive(k − 1, n − 1, R);
    res = res + (1 − R⟨n⟩) · R_KooN_Recursive(k, n − 1, R);
    return res;
```

The key indicator of the performance of a recursive algorithm is the number of calls to the recursive function, which, for a *KooN* block, depends on both parameters

K and N and, given N, it reaches its maximum when $K = {}^{N}/_{2}$. Let $F_1(K,N)$, $F_2(K,N)$, and $F_3(K,N)$ be the functions used to compute the recursive calls using, respectively, Equations (1), (2) and (4), defined as follows:

$$F_1(K,N) = \begin{cases} 1 & \text{if } K = 0 \text{ or } K > N \\ 1 + F_1(K-1,N-1) + F_1(K,N-1) & \text{otherwise} \end{cases} \tag{8}$$

$$F_2(K,N) = \begin{cases} 1 & \text{if } K = 1 \text{ or } K = N \\ 1 + F_2(K-1,N-1) + F_2(K,N-1) & \text{otherwise} \end{cases} \tag{9}$$

$$F_3(K,N) = \begin{cases} 1 & \text{if } K = 1 \text{ or } K = N \\ 1 + \sum_{i=0}^{h} F_3(K-i,N-h) & \text{otherwise, with } h = \min(K-1,N-K) \end{cases} \tag{10}$$

Figure 5 shows the trend, in logarithmic scale, of the number of recursive calls for the functions $F_1({}^{N}/_{2},N)$, $F_2({}^{N}/_{2},N)$, and $F_3({}^{N}/_{2},N)$, where ${}^{N}/_{2}$ is the integer division, in particular for (a) N up to 20 to easily observe the benefits of the different algorithms and (b) N up to 255—the maximum number of components that can be grouped into a single block with *librbd 2.0*—to analyze the behavior at a larger scale.

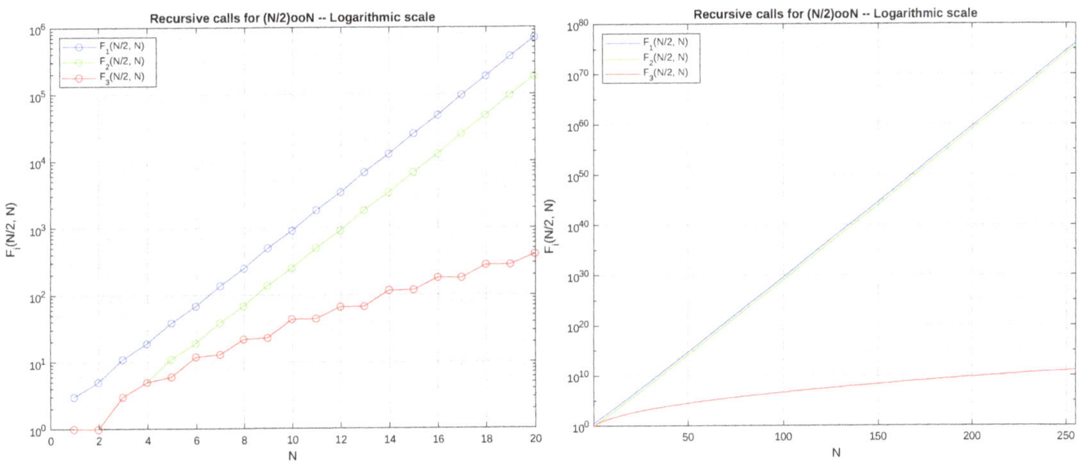

(a) (b)

Figure 5. Number of recursive calls $F_1({}^{N}/_{2},N)$, $F_2({}^{N}/_{2},N)$, and $F_3({}^{N}/_{2},N)$: (**a**) with $N \leq 20$. (**b**) with $N \leq 255$.

We speculate that the observed trend at these two scales can be also extended for increasing values of N. We can observe that, as expected, the recursive algorithm described in Equation (1) is the one that maximizes the number of recursive calls and that this number follows, for the worst case $K = {}^{N}/_{2}$, an exponential function w.r.t. the parameter N. The usage of Equation (2) decreases the number of recursive calls, thus reducing the computational complexity as expected, but the trend is still exponential. Furthermore, for N values up to 20, we can observe that, for a given $n \in \mathbb{N}^+$, the following holds true: $F_1({}^{n}/_{2},n) = F_2({}^{n}/_{2}+1,n+2)$. Finally, the usage of Equation (4) massively decreases the number of recursive calls and the trend is no longer exponential. The number of recursive calls required to analyze a 10oo20 block using Equation (4) is lower than the one needed to analyze a 6oo11 block using Equation (2). The non-exponential trend is due to the different increment patterns that occur for each $N \geq 3$ when N is either even or odd. For odd N, the function exhibits a linear progression, i.e., $F_3(N/2, N) = F_3(N-1/2, N-1) + 1$. Conversely, for even N, the number of recursive calls is marked by a sharp increase, yet the overall number of recursive calls grows sub-exponentially.

For what concerns the computation algorithms of bridge blocks, we propose updated versions both for generic and identical components, which introduce the following benefits:

- they slightly reduce the complexity, hence they decrease the computation time;
- they can be easily implemented with the SIMD paradigm;
- they require a lower number of temporary variables/vectors and hence they decrease the computation time.

Algorithm 2 shows the computation of RBD bridge blocks with generic components, which exploits Equation (6).

Algorithm 2: Computation of Bridge block with generic components.

Input: Array R of reliabilities, where $R\langle i\rangle$ is the reliability of i-th component
Function *R_Bridge_Generic(R)*
  # Compute $val1$ and $val2$ variables (see Equation (6))
  $val1 = (R\langle 1\rangle + R\langle 3\rangle - (R\langle 1\rangle \cdot R\langle 3\rangle) \cdot R\langle 2\rangle + R\langle 4\rangle - (R\langle 2\rangle \cdot R\langle 4\rangle));$
  $val2 = R\langle 1\rangle \cdot R\langle 2\rangle + R\langle 3\rangle \cdot R\langle 4\rangle - R\langle 1\rangle \cdot R\langle 2\rangle \cdot R\langle 3\rangle \cdot R\langle 4\rangle;$
  # Compute reliability of Bridge block (see Equation (6))
  return $R\langle 5\rangle \cdot (val1 - val2) + val2;$

Algorithm 3 shows the computation of RBD bridge blocks with identical components, which exploits Equation (7).

Algorithm 3: Computation of bridge block with identical components.

Input: Reliability R of each component
Function *R_Bridge_Identical(R)*
  # Compute unreliability U of each component
  $U = 1 - R;$
  # Compute reliability of Bridge block (see Equation (7))
  return $R \cdot (1 + U \cdot (U \cdot (U \cdot U - 2) + R \cdot (2 - R \cdot R)));$

4.2.2. Single Instruction, Multiple Data (SIMD)

To further increase performance, the Single Instruction, Multiple Data (SIMD) paradigm has been introduced. SIMD is a type of parallel processing that allows the microprocessor or, in some architectures, a co-processor, to execute the same instruction over multiple data, by means of "SIMD extensions". The SIMD extensions allow for the execution of a single instruction over vectors of data of different size, e.g., 64, 128, 256, and 512 bits. Since a 64-bit vector is able to host a single double-precision floating-point, our interest is placed on SIMD extensions capable of operating on vectors whose size is at least 128 bit. On one hand, the SIMD paradigm offers a parallelism which can greatly decrease the execution time; on the other hand, several considerations that may cause disadvantages must be taken into account:

- Usage of vectors requires large register files that increase the required chip area and the power consumption. Due to the power consumption increase, which causes higher CPU temperatures, Dynamic Frequency Scaling techniques may automatically decrease the CPU frequency.

- The implementation of an algorithm with SIMD instructions requires human effort since compilers typically do not generate SIMD instructions.
- Several SIMD extensions have restrictions on data alignment, thus increasing the complexity of the program. Even worse, different data alignment constraints may apply to different SIMD revisions or different processors of the same manufacturer.
- Due to the increased parallelism, a higher stress is put on the memory bus since a larger data flow is processed. This stress is further increased on multi-threading applications since different threads perform different but "concurrent" memory accesses.
- Specific instructions, like Fused Multiply-Add (FMA), are not available in some SIMD instruction sets.
- SIMD instruction sets are architecture-specific and some architectures lack SIMD instructions entirely, so programmers must provide a generic non-vectorized implementation and a different vectorized implementation for each covered architecture.

Table 1 presents the list of SIMD extensions that have been implemented in *librbd 2.0*, together with their respective vector size and the support to FMA instructions. The SIMD functionality can be enabled or disabled at compile time.

Table 1. SIMD extensions implemented in *librbd 2.0*.

Architecture	SIMD	Vector Size (Bits)	FMA Support
x86	SSE2	128	no
	AVX	256	no
amd64	FMA	256	yes
	AVX512F	512	yes
AArch64	NEON	128	yes

4.2.3. Optimization of Cache Usage with SMP

One of the main difficulties with SMP programming is the subdivision of the entire task, which is responsible to process the whole dataset, into multiple batches, each one responsible for the processing a portion of the original data. *librbd* splits the entire data, i.e., the input reliability curves, into contiguous and non-interleaved batches. This trivial subdivision of the data into batches is affected by a performance issue that has been aggravated with the introduction of newer CPUs with an increased number of cores.

On older CPUs, the limited number of homogeneous cores were clocked at the same frequency. On modern CPUs, the increased number of cores are subdivided into two heterogeneous groups, the Performance Cores (P-Cores) and the Efficient Cores (E-Cores): P-Cores and E-Cores are, in general, clocked with different frequencies.

The performance issue of *librbd* is caused by the cache usage when SMP is used: the inner cache level, which is in general specific of a single core, contains a subset of a specific batch, while the outer cache levels, which are in general shared between multiple cores, contain data from different batches. When multiple threads need to concurrently access new data, i.e., data not available into the outer cache level, the accesses on the memory data bus are serialized, thus creating a bottleneck which reduces the performance. Due to this data organization of batches, and considering the clocking characteristics of both older and modern CPUs, the probability of such an event is non-negligible. This performance loss increases with the number of instantiated threads, thus limiting the benefits of SMP, and modern CPUs with an increased number of cores are then more penalized with respect to older CPUs.

librbd 2.0 splits the entire data into interleaved batches, i.e., given M threads and T time instants to be processed, the m-th thread processes the t time instants for which $t \bmod M \equiv m$. With this subdivision, both the inner and the outer cache levels contain data which are shared between several threads. Also in this case when multiple threads need to concurrently access new data, i.e., data not available into the cache, the accesses on the memory data bus are serialized. The main difference is that, after the first thread has retrieved the data from the external memory, all subsequent threads can access the spatially contiguous data from the outer cache levels, thus reducing the latency on memory data access time and minimizing the bottleneck on the memory data bus.

The introduction of the improved algorithms, already described in Section 4.2.1, and the optimization of cache usage when SMP is enabled allow us to increment the maximum batch size from 10,000 time instants to 20,000.

4.2.4. Native Support to SMP on Windows

librbd provided support to SMP through the usage of *pthreads*, a POSIX-compliant [56] library that implements the management of threads and is natively available on both Linux and MacOS.

Microsoft Windows, on the other hand, does not offer a native support to pthreads, although it is still possible to use it either through pthreads-win32 [57] or through Cygwin [58]. To limit the usage of external libraries, which may be not always available, we decided to refactor the SMP implementation on *librbd 2.0* to support, through adequate conditional compiling, both pthreads and Win32 thread models, thus adding native support to SMP also on Microsoft Windows OSes.

4.3. Validation

To validate the *librbd 2.0* library, we use the same methodology already presented in [15], i.e., we perform a comparison of the output provided by *librbd 2.0* with respect to the ones provided by SHARPE [16,17]. For each RBD basic block, we generated two different RBD models, one using generic components and the other using identical components. During the validation process, we analyzed the 42 RBD models shown in Table 2. The validation process has been performed as follows:

- We used failure rate data from twenty components with constant failure rate, estimated according to Telcordia SR-332 [59] and reported in Table 3.
- We evaluated the RBD models for 100,000 h. By setting the sampling period to 1 h, the resulting reliability curve was determined across 100,000 time instants.

Table 2. RBD models used during validation.

RBD Block	Topology	Components
Series identical	20 components	C1
Series generic	20 components	All
Parallel identical	20 components	C1
Parallel generic	20 components	All
KooN identical	*Koo*20, $K \in [2, 19]$	C1
KooN generic	*Koo*20, $K \in [2, 19]$	All
Bridge identical	5 components C1	
Bridge generic	5 components	From C1 to C5

For each evaluated RBD model, we have produced two text output files, one generated using *librbd 2.0* and the other one with SHARPE, that contain the reliability of each analyzed time instant. Each reliability value has been formatted using scientific notation with eight decimal places. Please note that the uncertainty of the comparison operations is limited by the chosen numerical representation, i.e., scientific notation with 8 decimal digits.

Table 3. Components and their respective failure rate λ.

Component	λ (h^{-1})	Component	λ (h^{-1})
C1	0.0000084019	C11	0.0000047740
C2	0.0000039438	C12	0.0000062887
C3	0.0000078310	C13	0.0000036478
C4	0.0000079844	C14	0.0000051340
C5	0.0000091165	C15	0.0000095223
C6	0.0000019755	C16	0.0000091620
C7	0.0000033522	C17	0.0000063571
C8	0.0000076823	C18	0.0000071730
C9	0.0000027777	C19	0.0000014160
C10	0.0000055397	C20	0.0000060697

To perform this validation, we compiled *librbd 2.0* as an ST, SISD library.

$$
\begin{aligned}
err(t) &= \frac{out_{librbd}(t) - out_{SHARPE}(t)}{out_{SHARPE}(t)} \\
RMSE &= \sqrt{\frac{\sum_{t=0}^{max\_t} (out_{librbd}(t) - out_{SHARPE}(t))^2}{max\_t}} \\
MAE &= \frac{\sum_{t=0}^{max\_t} |out_{librbd}(t) - out_{SHARPE}(t)|}{max\_t}
\end{aligned}
\tag{11}
$$

We decided to employ both RMSE and MAE to evaluate regression model performance. While both metrics quantify the mean discrepancy between values produced by the two tools, i.e., *librbd 2.0* and SHARPE, they diverge in how individual errors are penalized: MAE treats all deviations equally by taking absolute values, whereas RMSE squares the deviations, thereby penalizing larger errors.

We chose this dual-metric approach for the following reasons. First, the error distribution is unknown a priori, and different metrics perform differently depending on the tail behavior of the error distribution [60]. Second, using both metrics allowed us to assess the robustness and stability of the error function: MAE provides an indication of average error magnitude under equal weighting, while RMSE emphasizes large deviations, offering insight into worst-case (or large error) behavior.

Finally, for each evaluated RBD block, we extracted both the minimum and maximum values for the relative error $err(t)$, thus computing the relative error range shown in Figure 6. For some RBD models, e.g., parallel block, SHARPE, and *librbd 2.0* produce the same reliability curve, hence the relative error is 0.0 for all time instants: this condition is visually represented in Figure 6 by not showing the relative error range.

We speculate that the relative error, when present, is due to implementative differences between the two tools, in particular regarding the exact sequencing of floating-point operations performed. This is supported by the observation that the largest error is obtained for series block, i.e., the one with reliability close to 0.0 for high values of t. From the analysis of the obtained results, we can observe that the absolute value of the relative error for all evaluated topologies is lower than $1.0 \cdot 10^{-8}$.

Table 4 reports both the RMSE and MAE values obtained from the validation of *librbd 2.0* w.r.t. SHARPE for both generic and identical component configurations. These results confirm an excellent agreement between *librbd 2.0* and SHARPE across all RBD models. For the parallel, bridge and 2oo20 configurations, both RMSE and MAE values are effectively zero, confirming that *librbd 2.0* reproduces the same reliability curves w.r.t. SHARPE.

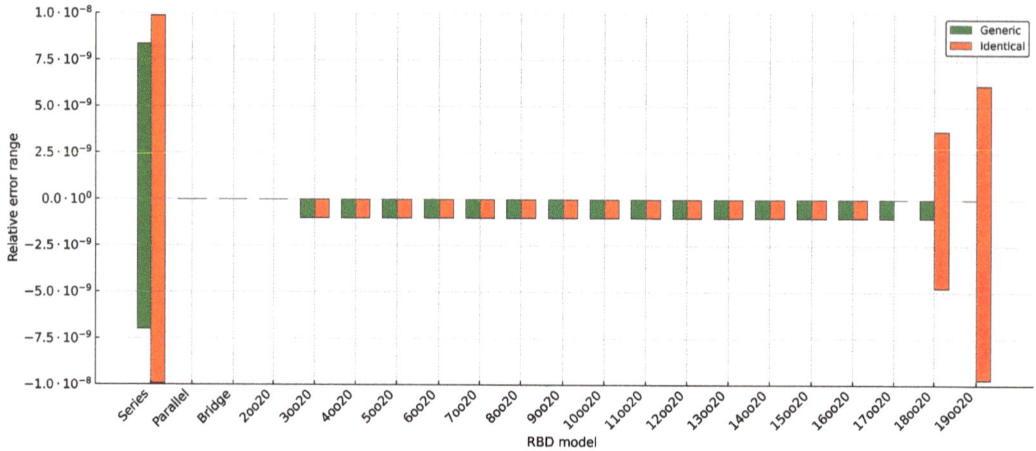

Figure 6. *librbd 2.0* validation against SHARPE—relative error range.

Table 4. *librbd 2.0* validation against SHARPE—RMSE and MAE.

RBD Model	Generic		Identical	
	RMSE	MAE	RMSE	MAE
Series	$3.272 \cdot 10^{-15}$	$1.770 \cdot 10^{-17}$	$5.840 \cdot 10^{-16}$	$1.183 \cdot 10^{-17}$
Parallel	$0.000 \cdot 10^{0}$	$0.000 \cdot 10^{0}$	$0.000 \cdot 10^{0}$	$0.000 \cdot 10^{0}$
Bridge	$0.000 \cdot 10^{0}$	$0.000 \cdot 10^{0}$	$0.000 \cdot 10^{0}$	$0.000 \cdot 10^{0}$
2oo20	$0.000 \cdot 10^{0}$	$0.000 \cdot 10^{0}$	$0.000 \cdot 10^{0}$	$0.000 \cdot 10^{0}$
3oo20	$1.924 \cdot 10^{-11}$	$3.700 \cdot 10^{-13}$	$1.049 \cdot 10^{-11}$	$1.100 \cdot 10^{-13}$
4oo20	$1.517 \cdot 10^{-11}$	$2.300 \cdot 10^{-13}$	$1.225 \cdot 10^{-11}$	$1.500 \cdot 10^{-13}$
5oo20	$1.549 \cdot 10^{-11}$	$2.400 \cdot 10^{-13}$	$1.265 \cdot 10^{-11}$	$1.600 \cdot 10^{-13}$
6oo20	$1.817 \cdot 10^{-11}$	$3.300 \cdot 10^{-13}$	$1.549 \cdot 10^{-11}$	$2.400 \cdot 10^{-13}$
7oo20	$1.761 \cdot 10^{-11}$	$3.100 \cdot 10^{-13}$	$1.225 \cdot 10^{-11}$	$1.500 \cdot 10^{-13}$
8oo20	$1.549 \cdot 10^{-11}$	$2.400 \cdot 10^{-13}$	$1.378 \cdot 10^{-11}$	$1.900 \cdot 10^{-13}$
9oo20	$1.643 \cdot 10^{-11}$	$2.700 \cdot 10^{-13}$	$1.549 \cdot 10^{-11}$	$2.400 \cdot 10^{-13}$
10oo20	$1.673 \cdot 10^{-11}$	$2.800 \cdot 10^{-13}$	$1.225 \cdot 10^{-11}$	$1.500 \cdot 10^{-13}$
11oo20	$1.703 \cdot 10^{-11}$	$2.900 \cdot 10^{-13}$	$1.483 \cdot 10^{-11}$	$2.200 \cdot 10^{-13}$
12oo20	$1.517 \cdot 10^{-11}$	$2.300 \cdot 10^{-13}$	$1.265 \cdot 10^{-11}$	$1.600 \cdot 10^{-13}$
13oo20	$1.225 \cdot 10^{-11}$	$1.500 \cdot 10^{-13}$	$1.140 \cdot 10^{-11}$	$1.300 \cdot 10^{-13}$
14oo20	$1.000 \cdot 10^{-11}$	$1.000 \cdot 10^{-13}$	$1.000 \cdot 10^{-11}$	$1.000 \cdot 10^{-13}$
15oo20	$9.487 \cdot 10^{-12}$	$9.000 \cdot 10^{-14}$	$8.367 \cdot 10^{-12}$	$7.000 \cdot 10^{-14}$
16oo20	$4.472 \cdot 10^{-12}$	$2.000 \cdot 10^{-14}$	$7.071 \cdot 10^{-12}$	$5.000 \cdot 10^{-14}$
17oo20	$5.477 \cdot 10^{-12}$	$3.000 \cdot 10^{-14}$	$0.000 \cdot 10^{0}$	$0.000 \cdot 10^{0}$
18oo20	$3.162 \cdot 10^{-12}$	$1.000 \cdot 10^{-14}$	$3.240 \cdot 10^{-15}$	$1.500 \cdot 10^{-17}$
19oo20	$0.000 \cdot 10^{0}$	$0.000 \cdot 10^{0}$	$6.686 \cdot 10^{-16}$	$8.700 \cdot 10^{-18}$

For the other RBD models, the error remains extremely low—on the order of $1.0 \cdot 10^{-11}$ for RMSE and $1.0 \cdot 10^{-13}$ for MAE in most cases—indicating negligible numerical discrepancies. The slightly higher RMSE compared to MAE across all models is consistent with theoretical expectations, as RMSE magnifies larger deviations due to the squaring of residuals. The stability of both metrics across varying values of K further suggests that *librbd 2.0* maintains numerical robustness independently of the redundancy level within the system.

In conclusion, the joint analysis of relative error, RMSE, and MAE supports the conclusion that *librbd 2.0* achieves numerical equivalence with SHARPE across a wide range of

RBD models, demonstrating high stability, robustness, and consistency irrespective of component heterogeneity. Reporting both metrics provides complementary insights—relative error showing a well-bounded percentage error, MAE confirming the uniform smallness of the errors, and RMSE ensuring that no large outliers are present. Our conclusion is that *librbd 2.0* produces, for each implemented RBD basic block, the correct reliability curve.

5. Performance Evaluation Workbench

In this section, we present the methodology used during the performance evaluation of *librbd 2.0*. In particular, in Section 5.1, we present the materials, i.e., the PCs and their characteristics, used to evaluate the performance, while in Section 5.2, we discuss the actual methodology adopted to evaluate the performance of *librbd 2.0*.

5.1. Materials

The six PCs listed in Table 5 have been used in order to measure the performance of *librbd 2.0* [53]. Please note that (1) PC1 has been tested using three different combinations of OS and Compiler to evaluate the performance on the same hardware equipment; (2) the reported clock frequency corresponds to the maximum frequency: the actual frequency is, in general, set by the OS and/or the CPU itself based on the current CPU load and the number of used cores. To investigate the performance of *librbd 2.0* also on embedded platforms, we have included into the set of used PCs two Raspberry Pi computers.

Both *librbd 2.0* and test binaries have been compiled and linked using the maximum optimization level (−O3).

Table 5. PCs used for performance evaluation.

Name	CPU	Cores/Threads	RAM	OS & Compiler
PC1a	Intel i7-13700K	8/16 @ 5.4GHz + 8/8 @ 4.2GHz	64GB-DDR5 @ 5600MHz	Ubuntu 22.04 GCC 11.4.0
PC1b	Intel i7-13700K	8/16 @ 5.4GHz + 8/8 @ 4.2GHz	64GB-DDR5 @ 5600MHz	Windows 11 Visual Studio 2022
PC1c	Intel i7-13700K	8/16 @ 5.4GHz + 8/8 @ 4.2GHz	64GB-DDR5 @ 5600MHz	Windows 11 GCC 12.4.0
PC2	Apple M3	4/4 @ 4.06GHz + 4/4 @ 2.57GHz	16GB-LPDDR5 @ 3200MHz	Mac OS 14.5 clang 15.0.0
PC3	Intel i7-7700HQ	4/8 @ 3.8GHz	32GB-DDR4 @ 2400MHz	Ubuntu 22.04 GCC 11.4.0
PC4	Intel i7-6700HQ	4/8 @ 3.5GHz	16GB-LPDDR3 @ 2133MHz	Mac OS 10.13.6 clang 10.0.0
PC5	Broadcom BCM2712	4/4 @ 2.4GHz	8GB-LPDDR4X @ 2133MHz	Raspberry Pi OS 12 GCC 12.2.0
PC6	Broadcom BCM2837	4/4 @ 1.2GHz	1GB-LPDDR2 @ 900MHz	Raspberry Pi OS 11 GCC 10.2.1

5.2. Methods

The performance evaluation has been conducted using a test application that has allowed us to measure the execution time needed by *librbd 2.0* to analyze the RBD models shown in Table 6.

We analyzed each RBD model with different time instant configurations, i.e., 1000, 5000, 10,000, 20,000, 50,000, 100,000, and 200,000 time instants. Furthermore, we analyzed each RBD model with 1607 time instants: this number, albeit apparently random, allows us to validate the SIMD computation exploiting hybrid vector lengths of 64 (scalar double), 128, 256, and 512 bits.

Table 6. RBD models used during performance evaluation.

RBD Block	N
Series Generic	$N \in [2,3,4,5,7,10,12,15,20]$
Series Identical	$N \in [2,3,4,5,7,10,12,15,20]$
Parallel Generic	$N \in [2,3,4,5,7,10,12,15,20]$
Parallel Identical	$N \in [2,3,4,5,7,10,12,15,20]$
$N/2$oo$N$ Generic	$N \in [2,3,4,5,7,10,12,15,20]$
$N/2$oo$N$ Identical	$N \in [2,3,4,5,7,10,12,15,20]$
Bridge Generic	$N = 5$
Bridge Identical	$N = 5$
Koo$N$ Generic, $\forall\, 1 < K < N$	$N = 20$
Koo$N$ Identical, $\forall\, 1 < K < N$	$N = 20$

To limit external impacts, e.g., the time consumed by the OS, each RBD model with each time instant configuration has been evaluated 15 times and, after all these 15 experiments were executed, we selected the median time of execution.

The failure rate λ of each component, shown in Table 3, has been chosen using the criteria already described in Section 4.3.

To further investigate the performance improvements of SMP and SIMD, we repeated the experiments four times, once for each combination of SMP and SIMD options, i.e., ST and SISD, ST and SIMD, SMP and SISD, SMP and SIMD.

Please note that, as already stated in Section 4.3, *librbd 2.0* has been validated with respect to SHARPE only when it is compiled as ST and SISD. During this phase, we compared the output for each produced RBD model with all different optimizations to validate that *librbd 2.0* always produces the same results when changing the optimization features. Hence, the usage of all combinations of optimization features on the tested PCs allowed us to almost completely validate all possible versions of *librbd 2.0*. Please note that, since AVX512F ISA is not supported by any CPU used, the support to this SIMD extension is still untested. Nevertheless, the source code used when supporting this ISA is almost equal to the amd64 FMA one, which has been extensively stressed during the tests.

6. Results

To evaluate the performance of *librbd 2.0*, in Section 6.1, we present the performance analysis, while in Section 6.2, we discuss on the obtained results. In Section 6.3, we present the comparison of performance obtained with *librbd 2.0* w.r.t. SHARPE. Finally, in Section 6.4, we discuss the current limitations that have been identified.

6.1. *librbd 2.0 Performance Analysis*

The performance analysis has been performed as described in Section 5.2, and the results are presented in this section as follows:

- For each combination of RBD model, time instants, PC used, and enabled optimizations, the reliability curve produced by the RBD computation has been stored to a file. This allows us to quickly compare the reliability curves both between different architectures and between different enabled optimizations. If the reliability curves for each modeled RBD are the same for all the sets of all enabled optimizations and the PC, then *librbd 2.0* is fully validated.
- For each combination of modeled RBD, time instants, PC used, and enabled optimizations, the *librbd 2.0* minimum, maximum, and median execution time observed on 15 different executions has been stored on a file. This has allowed us to quickly

evaluate the performance both between different architectures and between different enabled optimizations.

6.1.1. Validation with Different Architectures and Enabled Optimizations

Using *librbd 2.0* compiled as ST and SISD, we computed the reliability curves on the eight different architectures already presented in Table 5. We then computed the absolute value of the relative error between the reliability curve obtained with the reference architecture, i.e., PC1b since its results were previously validated with respect to SHARPE, and all other architectures. We observed that only the results obtained with PC5 differ from the expected ones: by further investigating, we observe that this difference is caused by a difference of computation of the input reliability curves, implemented through the usage of `exp` C library function. We suppose that this difference is caused by different versions of the C standard library. Nevertheless, across all the evaluated RBD models, the maximum RMSE and MAE between the reliability curves obtained with PC5 and PC1 are $1.053 \cdot 10^{-6}$ and $1.046 \cdot 10^{-6}$, respectively, indicating that the resulting reliability curves are essentially identical. For all other architectures, the reliability curves correspond to the one obtained with the reference architecture.

For each one of the eight target architectures used, the absolute value of the relative error computed between the reliability curve obtained with the reference library, i.e., the one compiled as ST and SISD, and the other three versions, i.e., SMP and SISD, ST and SIMD, SMP and SIMD, is limited to $1.0 \cdot 10^{-8}$; hence, we can state that the result provided by *librbd 2.0* is the same for all the sets of enabled optimizations and PCs.

6.1.2. Performance Analysis with Different Architectures and Enabled Optimizations

In Figure 7, we show the execution time of *librbd 2.0* on the different PCs with different enabled optimizations.

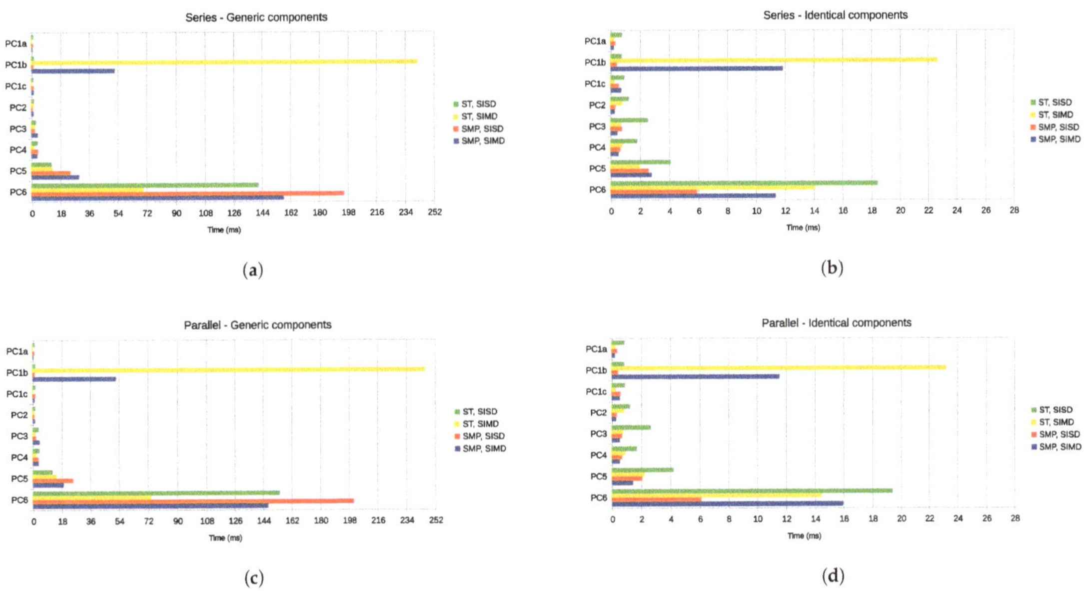

Figure 7. *Cont.*

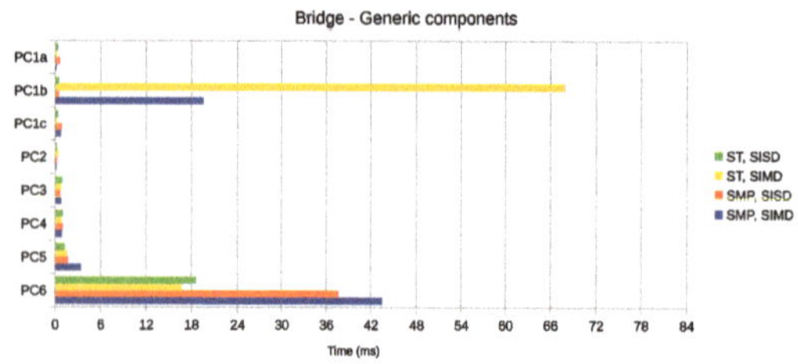

(**e**)

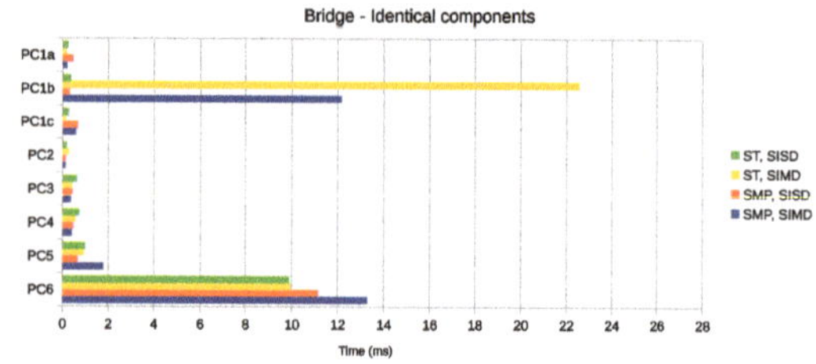

(**f**)

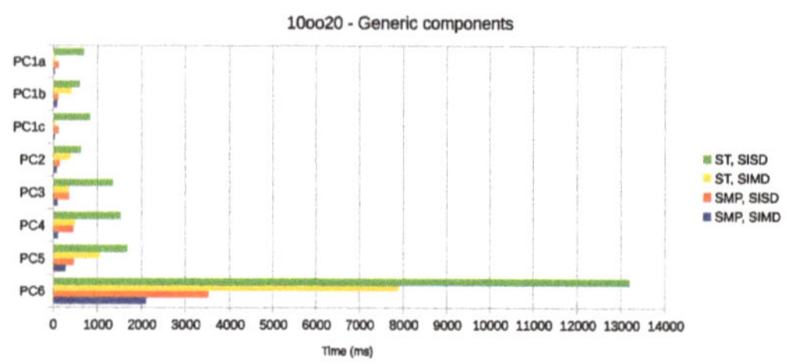

(**g**)

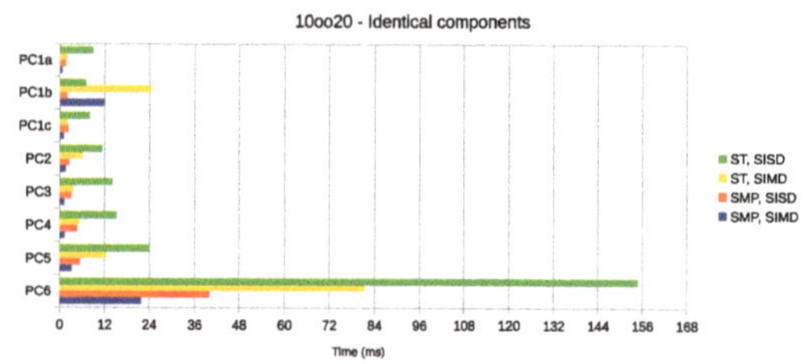

(**h**)

Figure 7. RBD analysis time over different RBD blocks. (**a**) Series with 20 generic components. (**b**) Series with 20 identical components. (**c**) Parallel with 20 generic components. (**d**) Parallel with 20 identical components. (**e**) Bridge with generic components. (**f**) Bridge with identical components. (**g**) 10oo20 with generic components. (**h**) 10oo20 with identical components.

The experimental results shown in Figure 7, which constitute a small subset of all experiments performed and already detailed in Table 6, constitute the worst case execution time of the performed experiments and hence are useful to correctly discuss the improvements introduced by SMP and SIMD.

6.2. Considerations on librbd 2.0 Performance

After the analysis of the execution time of *librbd 2.0* already presented in Section 6.1, we can state that, in general, the adoption of SMP and SIMD considerably improves the performance; however, some exceptions warrant further analysis:

- Considering PC1b, the adoption of SIMD when compiling with Visual Studio 2022 causes a significant loss of performance. This loss of performance may be due to an incorrect setting of file-specific optimization flags or due to compiler issues that limit its capability to effectively optimize source code exploiting SIMD intrinsics.
- Considering PC1 in its three different configurations, i.e., OS and compiler, we observe that, with the exception of the ones involving SIMD and Visual Studio, the variation in the execution time among the different configurations is small, thus showing that *librbd 2.0* can be effectively considered as a multi-platform library.
- Considering PC6, i.e., the PC with the lowest computational power, we observe that, in several occasions, e.g., series, parallel, and bridge with generic components, the introduction of SMP slightly degrades the performance. We suggest that this phenomenon may be caused by the memory latency, since this PC has very limited cache memory with a low bandwidth RAM. Despite PC6 being a low-power and dated embedded platform, its performance is anyway acceptable. Furthermore, its evolution, i.e., PC5, has performance results comparable with the other tested PCs.

- Considering the results obtained on all PCs, excluding the ones for 10oo20 blocks, we observe that, in several occasions, the adoption of SMP slightly degrades the performance. We suggest that this behavior may be caused by a non-optimized choice, for such simple models, of the batch size, i.e., the number of time instants concurrently processed by each thread. As already discussed in Section 4.2.3, the batch size should be properly tuned: if it is too high, a lower number of threads is instantiated, thus limiting the advantages of SMP; on the other side, if it is too low, each thread executes for a negligible amount of time and the overhead introduced by SMP, which includes the threads creation and termination, context switches, and race conditions on memory access, negates the advantages of SMP itself.
- With the exception of PC1b, the usage of SIMD extensions with ST library decreases the execution time. The actual decrease in computation time varies with the complexity of the analyzed block, i.e., more complex blocks with a higher number of computation steps per time instant have the greatest benefits.
- With respect to *librbd* [15], by comparing the results obtained with both *librbd* and *librbd 2.0*, we observe that bridge, parallel, and series blocks *librbd 2.0* have, in general, performance comparable with the one achieved with *librbd*. This may be due to the fact that, for these trivial blocks, the benefits obtained with SIMD are, in general, comparable with the loss of performance due to complexity of the new source code required for the implementation of the new features. For what concerns *KooN* blocks, i.e., blocks characterized by a high complexity and a high number of computations per time instant, we observe huge benefits for both block with generic and identical components. For example, let us consider PC3: using *librbd*, an 8oo15 block with generic components on 200,000 time instants was analyzed in 2278.303 ms, while *librbd 2.0* analyzed the same block in just 22.184 ms.
- Regarding RBDs presenting complex nestings of basic blocks, *librbd 2.0* can be used by applying standard functional composition, that is, if a basic block B is used to compose basic blocks $B_1, \ldots B_n$, if $P_B(p_1, \ldots p_n)$ is the function computing its success probability, the overall success probability is given by $P_B(P_{B_1}(P_{B_{1_1}}, \ldots P_{B_{1_k}}), \ldots P_{B_n}(P_{B_{n_1}}, \ldots P_{B_{nm}}))$. Thus, execution time for each basic block adds up linearly.

6.3. *Performance Comparison w.r.t. SHARPE*

To compare the performance of *librbd 2.0* w.r.t. SHARPE, we created series, parallel, bridge and *K*oo20, with $K \in [2, 19]$, RBD models for the SHARPE tool, both using generic and identical components, and we performed the reliability evaluation for 200,000 time instants using PC1 with Windows OS. We then compared the obtained results with the ones that were already presented in Section 6.1 using PC1c with both SMP and SIMD enabled. Furthermore, we performed some stress tests to understand how the two different tools behave with *KooN* blocks for increasing values of N. In particular, we analyzed 15oo30, 20oo40, and 25oo50 models with both generic and identical components. Table 7 shows the result of such comparison.

In conclusion, the numerical approach of *librbd 2.0* demonstrated a substantial speed advantage over SHARPE's symbolic analytical technique for relatively simple RBD models with exponential distributions, confirming the superior efficiency of numerical solvers. However, for large and complex generic *KooN* models, e.g., 25oo50, *librbd 2.0* experienced a severe performance drop. This behavior reflects an intrinsic limitation of its numerical nature: as detailed in Section 4.2.1, the computational complexity of the numerical evaluation grows sub-exponentially with model size, leading to a significant runtime increase for highly redundant systems.

Table 7. *librbd 2.0* and SHARPE analysis time in milliseconds (ms).

RBD Model	Generic		Identical	
	SHARPE	*librbd 2.0*	SHARPE	*librbd 2.0*
Series	2140.0	0.1	1150.0	0.1
Parallel	2320.0	0.1	1350.0	0.1
Bridge	990.0	0.7	1010.0	0.6
2oo20	2480.0	2.5	1510.0	0.5
3oo20	2590.0	8.1	1640.0	0.7
4oo20	2690.0	16.7	1730.0	0.8
5oo20	2780.0	26.8	1820.0	1.0
6oo20	2910.0	32.6	1900.0	0.9
7oo20	3020.0	35.0	2000.0	1.0
8oo20	3010.0	33.5	2020.0	1.2
9oo20	3110.0	35.8	2070.0	0.9
10oo20	3050.0	33.2	2080.0	1.1
11oo20	3020.0	35.1	2090.0	1.1
12oo20	2970.0	35.0	2030.0	1.0
13oo20	2970.0	32.8	2010.0	1.0
14oo20	2840.0	31.6	1930.0	1.1
15oo20	2810.0	29.9	1900.0	1.0
16oo20	2730.0	25.8	1750.0	0.9
17oo20	2630.0	15.2	1650.0	0.7
18oo20	2410.0	7.3	1490.0	0.6
19oo20	2270.0	2.4	1340.0	0.6
15oo30	4870.0	2024.0	3510.0	1.2
20oo40	7300.0	53,901.3	5450.0	2.1
25oo50	10,150.0	2,277,788.3	7810.0	3.4

Conversely, SHARPE's analytical methods, which rely on closed-form expressions when failure rates follow expolynomial distributions, show a comparatively slower and almost linear growth in computational time as the model size increases. Nevertheless, this advantage rapidly diminishes when SHARPE must handle reliability curves departing from the expolynomial assumption. In such cases, e.g., when components follow a Weibull distribution, SHARPE internally approximates these distributions through expolynomial (phase-type) representations in order to preserve analytical tractability. This approximation process, while avoiding the need for simulation, introduces additional computational overhead and may impact numerical precision.

In contrast, the numerical foundation of *librbd 2.0* ensures that its computational cost and numerical precision remain unaffected by the distributional form of the input reliability functions, providing a unified, accurate, and efficient framework for analyzing systems with non-expolynomial component behaviors.

6.4. Current Limitations

We have identified the following limitations:

- Limited support to compilers: *librbd 2.0*, since it exploits compiler-dependent features, requires the usage of one between GCC, Clang, and MSVC. We consider that this limitation is not important, since MSVC is the primary compiler for Microsoft Windows OSes, while both GCC and Clang are open-source compilers that are widely available over different OSes and architectures.
- Limited support to OSes: *librbd 2.0* exploits both OS-dependent and compiler-dependent features to implement the SMP paradigm. As a result, when SMP is used,

only Microsoft Windows, Linux, and MacOS are supported. We consider this limitation negligible, since these three OSes cover the majority of all OSes used nowadays. Furthermore, since this limitation applies to SMP version only, during compilation *librbd 2.0* automatically detects the target OS and, if it is not supported, it automatically disables SMP.

- Limited support to SIMD: *librbd 2.0* supports only SIMD extensions for x86, amd64, and AArch64 architectures. We believe that the supported SIMD extensions cover the majority of commercially available ones. Nonetheless, in future developments, support to additional SIMD extensions, e.g., PowerPC Vector Scalar Extension (VSX), could be introduced.
- Untested support to amd64 AVX512F SIMD: the support to AVX512F SIMD extension, which has been introduced in *librbd 2.0*, is still untested since the CPUs used during the performance analysis do not support this ISA. We consider this limitation not relevant due to the following two considerations: this ISA is supported by newer server CPUs only; the development of the algorithms exploiting this SIMD extension leverages the porting of FMA extension to AVX512F-specific intrinsics.
- Scalability limitations of *KooN* algorithm: the new algorithm designed and developed for the computation of *KooN* blocks with generic components is a huge improvement with respect to *librbd*, nonetheless its computational complexity is still an issue for big values of N. We consider this limitation negligible, since *librbd 2.0* has shown that the computation of generic 10oo20 blocks is feasible in reasonable time even using low-performance PCs and given that this block can be reasonably considered a worse case scenario in practical RBD usage.

7. Conclusions and Future Developments

This work presented a significant enhancement and validation of *librbd*, an open-source library for RBDs computation, called *librbd 2.0*. *librbd 2.0* successfully addresses the three core RQs posed in this study by introducing novel algorithmic and architectural improvements:

RQ1 Non-parametric reliability analysis. *librbd* was designed to compute the reliability of components with arbitrary, non-parametric failure distributions. Our validation against the established SHARPE tool confirms that *librbd 2.0* maintains this crucial capability, with the absolute maximum relative error between the computed reliability curves remaining consistently below $1.0 \cdot 10^{-8}$ across all models analyzed. This result ensures that the high numerical accuracy required by RQ1 is fully preserved.

RQ2 Reducing computational complexity through novel algorithms. The primary focus for reducing complexity was achieved through the implementation of novel mathematical formulas. Specifically, new algorithms were introduced to efficiently compute the reliability of the *KooN* block with generic components and the Bridge block with both generic and identical components. Performance analysis across various hardware platforms demonstrated the efficacy of these methods, showing a substantial reduction in computational complexity. Crucially, these new algorithms enabled the analysis of complex RBD models—such as large *KooN* configurations—that were previously unfeasible using *librbd*.

RQ3 Leveraging multicore architectures for platform-agnostic performance. To effectively utilize modern multicore architectures, as stipulated by RQ3, we implemented two key architectural improvements: an optimized usage of cache memory within the SMP paradigm, and native support for the SIMD programming paradigm. These techniques successfully increased the throughput of computed data and improved overall performance. Performance tests conducted on diverse PCs, CPUs, and OSes confirmed that these features contribute to achieving the platform-agnostic performance improvements necessary for deployment on a wide range of embedded and high-performance systems.

While the current work successfully addresses the proposed RQs, the following avenues for future research and development are identified to further enhance the utility and robustness of *librbd 2.0*:

- To promote broader adoption and ease of use, we are currently working on the definition of an RBD Description Language (RDL) based on the XML format. This RDL will be leveraged, alongside *librbd 2.0*, to develop an application for generic system-level reliability analysis.
- To address the identified threats to validity concerning SIMD extensions (limited support and untested AVX512F—specific code), we plan to perform a dedicated test session using server CPUs and to add support for additional SIMD extensions, such as PowerPC VSX.
- To resolve the observed performance discrepancy when using SIMD with the MSVC compiler, we plan to investigate the difference in the generated Assembly code between GCC and MSVC and implement necessary tweaks to MSVC-specific compiler options.

Author Contributions: Conceptualization, G.G., M.P. and A.F.; methodology, G.G., M.P. and A.F.; software, M.P.; validation, M.P.; formal analysis, G.G., M.P. and A.F.; investigation, G.G., M.P. and A.F.; resources, M.P. and G.G.; writing—original draft preparation, M.P.; writing—review and editing, G.G., M.P. and A.F.; visualization, G.G., M.P. and A.F.; supervision, G.G. and A.F. All authors have read and agreed to the published version of the manuscript.

Funding: This research received no external funding.

Institutional Review Board Statement: Not applicable.

Informed Consent Statement: Not applicable.

Data Availability Statement: The original contributions presented in the study are included in the article, further inquiries can be directed to the corresponding author.

Acknowledgments: We thank Laura Carnevali and Lorenzo Ciani for their contribution to the initial conceptualization of the *librbd* library.

Conflicts of Interest: The authors declare no conflicts of interest.

References

1. *ISO/IEC/IEEE 24765:2010(E)*; International Standard-Systems and Software Engineering—Vocabulary. IEEE: New York, NY, USA, 2010; pp. 1–418. [CrossRef]
2. *EN 50126-1*; Railway Applications—The Specification and Demonstration of Reliability, Availability, Maintainability and Safety (RAMS)-Part 1: Generic RAMS Process; Technical Report; CENELEC: Brussels, Belgium, 2017.
3. Hollander, M.; Peña, E.A. Nonparametric Methods in Reliability. *Stat. Sci.* **2004**, *19*, 644. [CrossRef] [PubMed]
4. Xing, L. An Efficient Binary-Decision-Diagram-Based Approach for Network Reliability and Sensitivity Analysis. *IEEE Trans. Syst. Man, Cybern.-Part A Syst. Humans* **2008**, *38*, 105–115 . [CrossRef]
5. Green, R.C.; Agrawal, V. A case study in multi-core parallelism for the reliability evaluation of composite power systems. *J. Supercomput.* **2017**, *73*, 5125–5149. [CrossRef]
6. Nelissen, G.; Pereira, D.; Pinho, L.M. A Novel Run-Time Monitoring Architecture for Safe and Efficient Inline Monitoring. In Proceedings of the 2015 International Conference on Reliable Software Technologies (Ada-Europe 2015), Madrid, Spain, 22–26 June 2015; pp. 66–82. [CrossRef]

7. van der Sande, R.; Shekhar, A.; Bauer, P. Reliable DC Shipboard Power Systems—Design, Assessment, and Improvement. *IEEE Open J. Ind. Electron. Soc.* **2025**, *6*, 235–264. [CrossRef]
8. Pan, X.; Chen, H.; Shen, A.; Zhao, D.; Su, X. Reliability Assessment Method for Complex Systems Based on Non-Homogeneous Markov Processes. *Sensors* **2024**, *24*, 3446. [CrossRef] [PubMed]
9. Song, Y.; Wang, X. Reliability Analysis of the Multi-State *k*-out-of-*n*: F Systems with Multiple Operation Mechanisms. *Mathematics* **2022**, *10*, 4615. [CrossRef]
10. Carberry, J.R.; Rahme, J.; Xu, H. Real-Time rejuvenation scheduling for cloud systems with virtualized software spares. *J. Syst. Softw.* **2024**, *217*, 112168. [CrossRef]
11. Nguyen, T.A.; Min, D.; Choi, E.; Tran, T.D. Reliability and Availability Evaluation for Cloud Data Center Networks Using Hierarchical Models. *IEEE Access* **2019**, *7*, 9273–9313. [CrossRef]
12. Dohi, T.; Zheng, J.; Okamura, H.; Trivedi, K.S. Optimal periodic software rejuvenation policies based on interval reliability criteria. *Reliab. Eng. Syst. Saf.* **2018**, *180*, 463–475. [CrossRef]
13. Fantechi, A.; Gori, G.; Papini, M. Software rejuvenation and runtime reliability monitoring. In Proceedings of the 2022 IEEE International Symposium on Software Reliability Engineering Workshops (ISSREW), Charlotte, NC, USA, 31 October–3 November 2022; pp. 162–169. [CrossRef]
14. Carnevali, L.; Fantechi, A.; Gori, G; Vreshtazi, D.; Borselli, A.; Cefaloni, M.R.; Rota, L. Data-Driven Synthesis of Stochastic Fault Trees for Proactive Maintenance of Railway Vehicles. In Proceedings of the 2025 30th International Conference on Formal Methods for Industrial Critical Systems (FMICS), Aarhus, Denmark, 25–30 August 2025; pp. 162–181. [CrossRef]
15. Carnevali, L.; Ciani, L.; Fantechi, A.; Gori, G.; Papini, M. An Efficient Library for Reliability Block Diagram Evaluation. *Appl. Sci.* **2021**, *11*, 4026. [CrossRef]
16. Sahner, R.A.; Trivedi, K.S.; Puliafito, A. *Performance and Reliability Analysis of Computer Systems: An Example-Based Approach Using the SHARPE Software Package*; Kluwer Academic Publishers: Alphen aan den Rijn, The Netherlands, 1996. [CrossRef]
17. SHARPE Portal. Duke University Pratt School of Engineering. Web Page. Available online: https://sharpe.pratt.duke.edu/ (accessed on 7 October 2025).
18. Trivedi, K.S.; Bobbio, A. *Reliability and Availability Engineering*; Cambridge University Press: Cambridge, UK, 2017. [CrossRef]
19. Mahboob, Q.; Zio, E. *Handbook of RAMS in Railway Systems: Theory and Practice*; CRC Press: Boca Raton, FL, USA, 2018. [CrossRef]
20. Moskowitz, F. The analysis of redundancy networks. *Trans. Am. Inst. Electr. Eng. Part I Commun. Electron.* **1958**, *77*, 627–632. [CrossRef]
21. *IEC 61078*; Reliability Block Diagrams. Technical Report; IEC: Geneva, Switzerland, 2016.
22. Hixenbaugh, A.F. *Fault Tree for Safety*; Technical Report; Boeing Aerospace Company: Seattle, WA, USA, 1968.
23. *IEC 61025*; Fault Tree Analysis (FTA). Technical Report; IEC: Geneva, Switzerland, 2006.
24. Rubino, G. Network reliability evaluation. In *State-of-the-Art in Performance Modeling and Simulation*; Bagchi, K., Walrand, J., Eds.; Gordon & Breach Books: London, UK, 1998; pp. 275–301.
25. Bryant, R.E. Graph-Based Algorithms for Boolean Function Manipulation. *IEEE Trans. Comput.* **1986**, *C-35*, 677–691. [CrossRef]
26. Ericson, C.A. Fault Tree Analysis—A History. In Proceedings of the 17th International System Safety Conference, Orlando, FL, USA, 16–21 August 1999; pp. 1–9.
27. Stewart, W. *Introduction to the Numerical Solution of Markov Chains*; Princeton University Press: Princeton, NJ, USA, 1994.
28. *IEC 61165*; Application of Markov Techniques. Technical Report; IEC: Geneva, Switzerland ,2006.
29. Molloy, M. Performance Analysis Using Stochastic Petri Nets. *IEEE Trans. Comput.* **1982**, *31*, 913–917. [CrossRef]
30. Marsan, M.A.; Conte, G.; Balbo, G. A class of generalized stochastic petri nets for the performance evaluation of multiprocessor systems. *ACM Trans. Comput. Syst.* **1983**, *2*, 93–122. [CrossRef]
31. Vicario, E.; Sassoli, L.; Carnevali, L. Using stochastic state classes in quantitative evaluation of dense-time reactive systems. *IEEE Trans. Softw. Eng.* **2009**, *35*, 703–719. [CrossRef]
32. *IEC 62551*; Analysis Techniques for Dependability—Petri Net Techniques. Technical Report; IEC: Geneva, Switzerland, 2012.
33. Ciardo, G.; Blakemore, A.; Chimento, P.F.; Muppala, J.K.; Trivedi, K.S. Automated Generation and Analysis of Markov Reward Models Using Stochastic Reward Nets. In *Linear Algebra, Markov Chains, and Queueing Models*; Meyer, C.D., Plemmons, R.J., Eds.; Springer: New York, NY, USA, 1993; pp. 145–191.
34. Ciardo, G.; Trivedi, K.S. A decomposition approach for stochastic reward net models. *Perform. Eval.* **1993**, *18*, 37–59. [CrossRef]
35. Meyer, J.; Movaghar, A.; Sanders, W. Stochastic Activity Networks: Structure, Behavior, and Application. In Proceedings of the International Workshop on Timed Petri Nets, Torino, Italy, 1–3 July 1985; pp. 106–115.
36. Sanders, W.H.; Meyer, J.F., Stochastic Activity Networks: Formal Definitions and Concepts. In *Lectures on Formal Methods and Performance Analysis: First EEF/Euro Summer School on Trends in Computer Science Bergen Dal, The Netherlands, 3–7 July 2000*; Hermanns, H., Katoen, J.-P., Eds.; Springer: Berlin/Heidelberg, Germany, 2000; pp. 315–343. [CrossRef]

37. Distefano, S.; Puliafito, A. Dynamic reliability block diagrams: Overview of a methodology. In Proceedings of the European Safety and Reliability Conference 2007, ESREL 2007-Risk, Reliability and Societal Safety, Stavanger, Norway, 25–27 June 2007; Volume 2.
38. Distefano, S.; Puliafito, A. Dependability Evaluation with Dynamic Reliability Block Diagrams and Dynamic Fault Trees. *IEEE Trans. Dependable Secur. Comput.* **2009**, *6*, 4–17. [CrossRef]
39. Dugan, J.B.; Bavuso, S.J.; Boyd, M.A. Dynamic fault-tree models for fault-tolerant computer systems. *IEEE Trans. Reliab.* **1992**, *41*, 363–377. [CrossRef]
40. Codetta-Raiteri, D. The Conversion of Dynamic Fault Trees to Stochastic Petri Nets, as a case of Graph Transformation. *Electron. Notes Theor. Comput. Sci.* **2005**, *127*, 45–60. [CrossRef]
41. Volk, M.; Weik, N.; Katoen, J.P.; Nießen, N. A DFT Modeling Approach for Infrastructure Reliability Analysis of Railway Station Areas. In *Formal Methods for Industrial Critical Systems;* Larsen, K.G., Willemse, T., Eds.; Springer International Publishing: Cham, Switzerland, 2019; pp. 40–58. [CrossRef]
42. Dai, Y.-S.; Pan, Y.; Zou, X. A Hierarchical Modeling and Analysis for Grid Service Reliability. *IEEE Trans. Comput.* **2007**, *56*, 681–691. . [CrossRef]
43. Kim, D.S.; Ghosh, R.; Trivedi, K.S. A Hierarchical Model for Reliability Analysis of Sensor Networks. In Proceedings of the 2010 IEEE 16th Pacific Rim International Symposium on Dependable Computing (PRDC), Tokyo, Japan, 13–15 December 2010; pp. 247–248. [CrossRef]
44. Fantechi, A.; Gori, G.; Papini, M. Runtime Reliability Monitoring for Complex Fault-Tolerance Policies. In Proceedings of the 2022 IEEE 6th International Conference on System Reliability and Safety (ICSRS), Venice, Italy, 23–25 November 2022; pp. 110–119. [CrossRef]
45. Siewiorek, D.P.; Swarz, R.S. *Reliable Computer Systems: Design and Evaluation*, 3rd ed.; A. K. Peters, Ltd.: Natick, MA, USA, 1998.
46. Isograph Reliability Workbench. Web Page. Available online: https://www.isograph.com/software/reliability-workbench/rbd-analysis/ (accessed on 7 October 2025).
47. Relyence RBD. Web Page. Available online: https://www.relyence.com/products/rbd/ (accessed on 7 October 2025).
48. ALD RAM Commander—RBD Module. Web Page. Available online: https://aldservice.com/Reliability-Products/reliability-block-diagram-rbd.html (accessed on 7 October 2025).
49. Janardhanan, S.; Badnava, S.: Agarwal, R.; Mas-Machuca, C. PyRBD: An Open-Source Reliability Block Diagram Evaluation Tool. In Proceedings of the 2024 IEEE 38th International Workshop on Communications Quality and Reliability (CQR), Seattle, WA, USA, 9–12 September 2024; pp. 19–24. [CrossRef]
50. PyRBD GitHub Repository. Web Page. Available online: https://github.com/shakthij98/PyRBD (accessed on 7 October 2025).
51. Janardhanan, S.; Chen, Y.; Mas-Machuca, C. PyRBD++: An Open-Source Fast Reliability Block Diagram Evaluation Tool. In Proceedings of the 2025 IEEE 15th International Workshop on Resilient Networks Design and Modeling (RNDM), Trondheim, Norway, 9–11 June 2025; pp. 1–7. [CrossRef]
52. PyRBD++ GitHub Repository. Web Page. Available online: https://github.com/shakthij98/PyRBD_plusplus (accessed on 7 October 2025).
53. librbd GitHub Repository. Web Page. Available online: https://github.com/marcopapini/librbd (accessed on 7 October 2025).
54. Fourment, M.; Gillings, M. A comparison of common programming languages used in bioinformatics. *BMC Bioinform.* **2008**, *9*, 82. [CrossRef] [PubMed]
55. *IEEE Std-754-2019 (Revision IEEE-754-2008)*; IEEE Standard for Floating-Point Arithmetic. IEEE: New York, NY, USA, 2019; pp. 1–84. [CrossRef]
56. *IEEE Std 1003.1-2017 (Revision of IEEE Std 1003.1-2008)*; IEEE Standard for Information Technology–Portable Operating System Interface (POSIX™) Base Specifications, Issue 7. IEEE: New York, NY, USA, 2018; pp. 1–3951. [CrossRef]
57. pthreads-win32—Open Source POSIX Threads for Win32. Web Page. Available online: http://sourceware.org/pthreads-win32/ (accessed on 7 October 2025).
58. Cygwin. Web Page. Available online: https://www.cygwin.com/ (accessed on 7 October 2025).
59. *Telcordia SR-332 Reliability Prediction Procedure for Electronic Equipment;* Technical Report Issue 4; Telcordia Network Infrastructure Solutions (NIS): Bridgewater, NJ, USA, 2016.
60. Hodson, T.O. Root-mean-square error (RMSE) or mean absolute error (MAE): When to use them or not. *Geosci. Model Dev.* **2022**, *15*, 14. [CrossRef]

applied sciences

MDPI

Article

Research on the Effect of Working Memory Training on the Prevention of Situation Awareness Failure in Shearer Monitoring Operations

Xiaofang Yuan [1,2], Ruyi Song [1,*] and Linhui Sun [1,2]

1 School of Management, Xi'an University of Science and Technology, Xi'an 710054, China; yxf@xust.edu.cn (X.Y.); linhuisun@xust.edu.cn (L.S.)
2 Human Factors and Management Ergonomics Research Center, Xi'an University of Science and Technology, Xi'an 710054, China
* Correspondence: songruyi5644@163.com

Featured Application: This study shows that working memory training can improve operators' ability to react quickly and make decisions in complex or emergency situations, preventing situational awareness failure and thus improving workplace safety, especially in industries that require high levels of alertness and accurate judgment, such as mining, aviation, and transportation.

Abstract: The digitization of the instrument control system in monitoring operations makes the problem of the operator's situational awareness failure more prominent. In order to better prevent this occurrence, this paper explores the failure of situational awareness from the perspective of cognitive function. The subjects were randomly divided into two groups: a working memory training group and a control group. Working memory measurements and coal mining machine monitoring simulation system operation tasks were performed before and after training, and the task performance, situational awareness scale, and EEG index data were recorded. The results showed that, after the training, there was a significant improvement in the task performance of the monitoring operation and the scores of the situational awareness scale, and there were different degrees of activation in the θ, α_2, and β_1 frequency bands. It was demonstrated that working memory training could help to improve the rapid reaction and decision-making abilities of operators in complex or emergency situations, thus preventing the failure of situational awareness. This study provides a new direction for research on the prevention of situational awareness failure.

Keywords: working memory training; monitoring; situational awareness failure

1. Introduction

In recent years, with the deepening of coal mining operations, the safe and efficient execution of such operations has attracted widespread attention [1,2]. However, due to the development of smart coal mines, the obvious problems of human error and unsafe behavior have been gradually reduced; moreover, with the continuous improvements in the fault tolerance rates of intelligent systems, safety risks are more likely to arise from the cognitive processing of operators, in which they cannot accurately understand the current situation [3,4]. Therefore, the question of how to effectively prevent the failure of situational awareness is of great significance to ensure the safety of coal mine production.

According to the theoretical model for information processing in situational awareness proposed by Endsley [5], working memory is an important factor that restricts individuals from obtaining and interpreting environmental information to form situational awareness. Limitations in one's working memory capacity make it difficult for individuals to perceive and integrate various types of information in a timely and accurate manner, thus reducing their situational awareness. Individuals with good working memory will also have better situational awareness and operational performance [6]. In complex and changeable operating environments, such as flight driving, limited working memory makes it difficult for pilots to acquire and integrate various types of information in the environment in a timely manner, resulting in SA failure [7]. In driving tasks, individuals with low working memory capacity are prone to task maintenance failure due to driving distractions or external interference [8]. Empirical studies have further confirmed the positive correlation and predictive relationship between working memory capacity and SA [6]. Working memory (WM) is a short-term memory system that stores and manages information and is closely linked to other cognitive processes [9]. Working memory can be evaluated in various ways, including the reading span task, number span task, operation span task, N-back task, etc. The operation span task is a classical measure of working memory. This measure is based on Baddeley's concept of working memory, in which individuals process and store information at the same time, and their final performance on the memory task is used as a measure of their working memory ability [10].

Working memory training is an effective way to improve working memory. Working memory-related tasks are usually used for short-term training to improve individuals' working memory capacity and cognitive abilities, so as to help them to better understand and process complex information [11]. In other words, working memory has high plasticity, and a certain degree of training can increase the capacity of working memory [12] and the performance of training tasks [13], as well as affecting other cognitive abilities related to it [14,15]. The positive effects of working memory training on other higher cognitive functions have been widely recognized, and some studies have proven the close relationship between working memory and situational awareness. However, whether the effects of working memory training can be transferred to practical application scenarios and prevent the failure of situational awareness is still unknown. Therefore, it is of great significance to explore working memory training to prevent situational awareness failure in monitoring workers.

In addition, situational awareness is a complex cognitive process involving multiple cognitive domains, and its measurement methods are complex and diverse. In previous studies, situational awareness was mostly assessed by questionnaires [16,17]. However, this type of measurement has some inaccuracies, as participants may not accurately describe task completion [17]. In contrast, behavioral and physiological measurements can more objectively reflect the responses and emotional states of subjects, making the assessment and analysis of situational awareness more accurate [18]. The recording of brain activity via physiological signals can simultaneously reflect various sensory inputs, so as to conduct an in-depth analysis of participants' physiological responses [19]. Each of the above measurement methods has its own advantages and disadvantages, and a single measurement method is insufficient to achieve the accurate measurement of situational awareness. Therefore, this study combined questionnaire measurements, performance evaluations and physiological measurements to effectively evaluate the level of situational awareness among the subjects.

This study used a randomized controlled trial in which the training group received 15 working days of working memory training and the control group received a blank control. All training and assessments were conducted in the laboratory. This study used multiple measures of situational awareness for pre- and post-training assessments to examine the effects of working memory training on situational awareness in monitoring tasks. This study extends the relevant theory on SA failure prevention to the field of coal mining and provides suggestions to improve the skills of coal mine monitoring personnel.

2. Materials and Methods

2.1. Objects

G* Power 3.1.9.7 was used to calculate the sample size required for this study. According to existing studies, the Cohen effect size f value for working memory training regarding the improvement of working memory function is 0.687, which was taken as the effect size in this study, and the ratio between the training group and the control group was 1:1. It was estimated that the sample size for the training group and control group should be at least 8 cases. A total of 22 participants were randomly selected, including 12 females and 10 males, with an average age of (22 ± 1) years, ranging from 20 to 24 years. All subjects were non-psychology majors, right-handed, with no physical diseases or mental disorders, and had normal visual acuity or corrected vision. The subjects were divided into 2 groups by the random matching method: a training group (n = 10), in which one participant withdrew from the experiment, and a control group (n = 11). There was no significant difference in age [F(1, 19) = 0.149, p = 0.863] or gender [χ^2 = 0.343, p = 0.842] among the two groups, with no statistical significance. All subjects voluntarily participated in the experiment and signed an informed consent form, and they received some remuneration after the experiment.

2.2. Materials

2.2.1. Working Memory Capacity Detection

In this study, operation span tasks were used to assess the working memory capacity of the individuals, and the changes in memory capacity before and after training were investigated. This task was presented by E-prime V.3.0 and was divided into a speech operation span task and space operation span task.

2.2.2. Working Memory Training Procedures

The working memory training tasks were presented by E-prime 3.0. An adaptive two-dimensional N-back task was used. First, two stimuli (visual and auditory) were presented simultaneously for 1000 ms, in which the visual stimulus was a localized picture of the monitoring interface of the coal mining machine, which appeared in the center of the screen. The auditory stimulus was the digits 1–9, which were transmitted through headphones; after an empty screen was presented for 1500 ms, the two stimuli were presented again, and the subjects were asked to judge whether the current stimulus (visual and auditory) was the same as the Nth stimulus (visual and auditory) before the current one. The subjects were also asked to judge whether the current stimuli (visual and auditory) were identical to the previous N stimuli (visual and auditory). When the visual stimuli were identical, "A" was clicked; when the auditory stimuli were identical, "L" was clicked. The interface provided real-time feedback as the subjects responded with keystrokes in order to quickly familiarize them with the rules and keep them motivated. The ratio of consistent to inconsistent trials for both visual and auditory stimuli was 3:7, and the presentation was randomized. The training task was adaptive.

An example diagram for an experiment is shown in Figure 1.

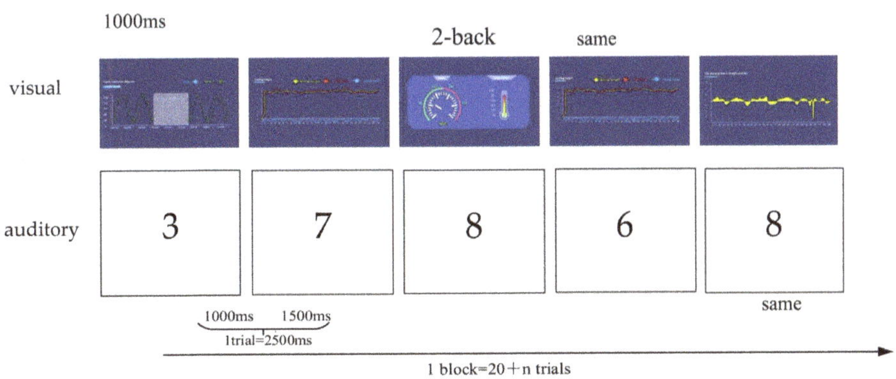

Figure 1. Example working memory training diagram.

2.2.3. Measures of Situational Awareness

The situational awareness task scenarios were rendered in Python 3.9. Prior to the experiment, experts provided all participants with unified operational training to ensure their proficiency in using the coal shearer monitoring simulation system. With the MG500/1180-WD coal shearer as the reference model, Jixi Coal Mine Machinery Co., Ltd., Jixi City, Heilongjiang Province, China. A coal shearer monitoring and operation simulation platform was built based on Python 3.9. The interface was displayed on a 14.1-inch computer screen with a resolution of 1920 × 1080 pixels. The experiment utilized the coal shearer interface of the intelligent coal mine fully mechanized mining control system as the experimental material, where the anomaly information is marked with a yellow rectangle and red font, as shown in Figure 2, indicating abnormalities in the right-arm temperature and gas concentration. During the experiment, the participants were required to monitor various monitoring systems on the coal shearer interface in real time, and, when abnormal data or information appeared on the monitoring interface, they were required to respond to it as accurately and quickly as possible using the mouse. The participants needed to closely monitor the operating status of the coal shearer, accurately and rapidly identifying any abnormal states.

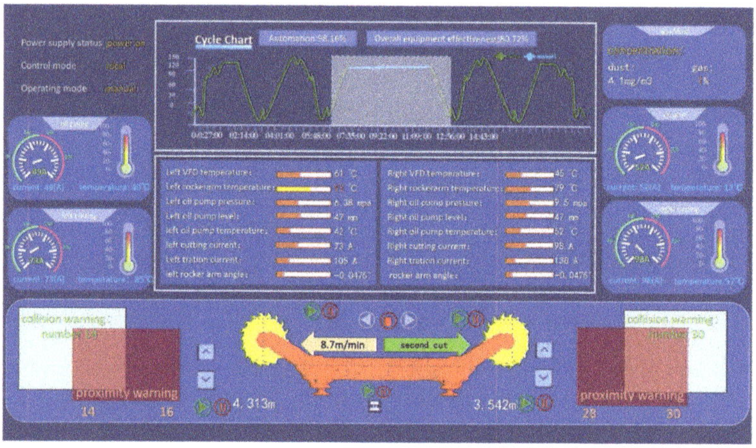

Figure 2. Example of shearer monitoring simulation task interface.

2.3. Experimental Design

The experiment utilized a 2 (training group, control group) × 2 (pre-test, post-test) between-subjects design.

In this study, the task performance indicators, situational awareness scale and EEG indicators were recorded as the evaluation indicators for SA. Among them, the task performance measurement included the reaction time and accuracy rate, which were automatically recorded by the Python 3.9 platform, and the reaction time was the sum of the reaction time for anomaly recognition in the shearer monitoring task. The accuracy rate was the percentage of the number of correct responses to the total amount of abnormal information presented. The situational awareness scale refered to the SAGAT (Situation Awareness Global Assessment Technique) and 3D-SART (Situation Awareness Rating Technique) scores. EEG data were collected by eegoTMmylab in 32 channels: F_7, FT_7, T_3, TP_7, T_5; FP_1, F_3, FC_3, C_3, CP_3, P_3, O_1; F_Z, FC_Z, C_Z, CP_Z, P_Z, O_Z; FP_2, F_4, FC_4, C_4, CP_4, P_4, O_2; F_8, FT_8, T_4, TP_8, T_6. All electrodes were Ag/Agcl, the recording bandwidth was 0~100 Hz, and the sampling rate was 1 000 Hz. The positive reactance of the electrode was kept below 20 kO. GND was set to ground left mastoid A1 as an online reference. We recorded both vertical and horizontal ophthalmograms. The offline EEG signals were re-referenced to the average values of the bilateral mastoid electrodes and were analyzed offline using the MATLAB R2014a EEGLAB141-1B T tool box. The double reference and 1–30 Hz bandpass filtering were performed using bilateral mastoid electrodes. Furthermore, an independent component analysis was conducted, and the ADJUST1.1.1 toolkit was used for artifact removal.

2.4. Experimental Flow

All participants were required to perform the experimental task for 4 weeks. Before the experiment, the principal researcher introduced the content and process of the whole experiment to the subjects, and they conducted sufficient practice experiments to ensure that the subjects were fully familiar with the task. In the first week of the experiment, the subjects were required to perform the first working memory capacity detection task and then the shearer monitoring operation simulation task. In this task, the operation of the subjects was paused randomly to answer the SAGAT questions, and the SART scale was filled in after the end of the experiment to collect the SA scale data. After this, the training was performed for 15 days and lasted about 30 min every day. The control group used a blank control. In the last week, the subjects repeated the working memory capacity detection task and the shearer monitoring simulation task; the specific details were the same as in the first week's task.

2.5. Statistical Methods

SPSS 27.0 was used for statistical analysis. First, an independent-sample *t*-test was used to compare the values of each group before training. A 2 (training group, control group) $\times$ 2 (pre-test, post-test) repeated-measures ANOVA was used to evaluate the effects of working memory training on the working memory capacity and situational awareness in the shearer monitoring tasks, in order to exclude the practice effect and placebo effect. The test level was set at $\alpha = 0.05$.

3. Results

3.1. Training Task Performance

After 15 training sessions, the training group's performance in the dual-dimensional N-back task gradually improved, as shown in Figure 3. In the first 11 sessions, it gradually increased, and then it was relatively stable, with the highest N number reaching 6.10 and the average N number was 3.53. In order to further verify the effect of the dual-dimensional N-back training, the average N value of the training group at the first and 15th instances was measured repeatedly, and the results showed that the average N value increased significantly [$F(1,9) = 18.409$, $p < 0.01$, $\eta^2 = 0.0.672$]. It was found that the working memory ability of the training group was significantly improved.

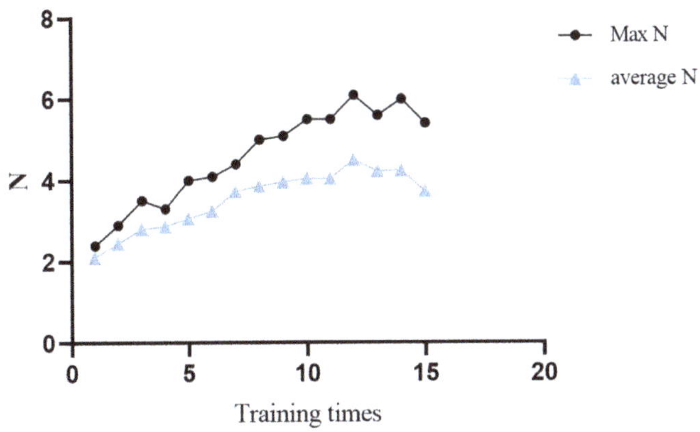

Figure 3. Training process record.

3.2. Working Memory Capacity

The working memory capacity measure was divided into the verbal working memory capacity and spatial working memory capacity; the descriptive results and analysis of variance are shown in Figure 4. The training group showed an improvement in both their verbal and spatial working memory capacity [verbal: $M_1 = 5.70$, $SD_1 = 1.16$, $M_2 = 7.00$, $SD_2 = 1.05$; spatial: $M_1 = 5.20$, $SD_1 = 1.31$; $M_2 = 6.90$, $SD_2 = 0.99$]; the control group did not show significant changes in their verbal and spatial working memory capacity [verbal: $M_1 = 5.55$, $SD_1 = 1.04$, $M_2 = 5.64$, $SD_2 = 0.92$; spatial: $M_1 = 5.46$, $SD_1 = 1.04$, $M_2 = 5.36$, $SD_2 = 0.81$]. Here, M is the mean, SD is the standard deviation, and subscripts 1 and 2 denote the pre-training and post-training values, respectively.

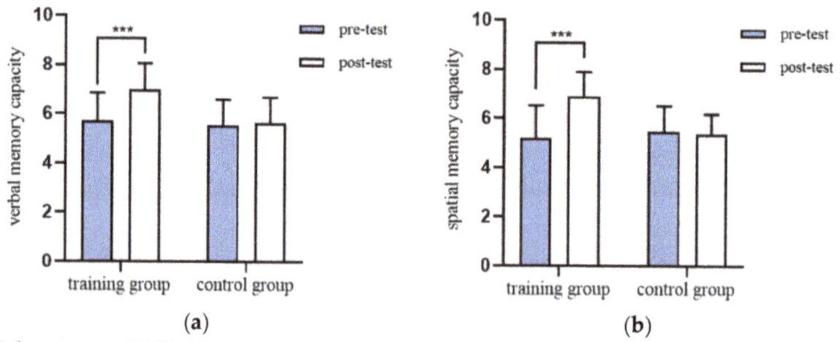

*** denotes $p < 0.001$

Figure 4. Comparison of working memory capacity before and after training. (**a**) Verbal working memory capacity; (**b**) spatial working memory capacity.

Before training, no statistical difference was found in the memory capacity of the training group and the control group in the pre-test in the independent-samples *t*-test [verbal: $T(19) = 0.323$, $p = 0.750$; spatial: $T(19) = -0.332$, $p = 0.743$]. A 2 (pre-test, post-test) × 2 (training group, control group) repeated-measures ANOVA was conducted on the working memory capacity of the two groups, with a non-significant main effect of the grouping [verbal: $F(1, 19) = 3.490$, $p = 0.077$, $\eta^2 = 0.155$; spatial: $F(1, 19) = 2.615$, $p = 0.122$, $\eta^2 = 0.121$], a significant main effect of the time of measurement [verbal: $F(1, 19) = 11.320$, $p = 0.009 < 0.01$, $\eta^2 = 0.373$; spatial: $F(1, 19) = 12.265$, $p = 0.002 < 0.01$, $\eta^2 = 0.392$] and a significant interaction between the grouping and time of measurement [verbal: $F(1, 19) = 8.554$, $p = 0.003 < 0.01$, $\eta^2 = 0.310$; spatial: $F(1, 19) = 15.194$, $p < 0.001$, $\eta^2 = 0.444$]. Further simple effects analyses showed that there was a significant increase in the verbal and spatial working memory capacity in the training group compared to the pre-test period [verbal: $F(1, 19) = 18.878$, $p \leq 0.001$, $\eta^2 = 0.498$; spatial: $F(1, 19) = 26.136$, $p \leq 0.001$,$\eta^2 = 0.579$], and there was no significant difference between the pre-test and post-test in the control group [Verbal: $F(1, 19) = 0.102$, $p = 0.753$, $\eta^2 = 0.005$; Spatial: $F(1, 19) = 0.82$, $p = 0.777$, $\eta^2 = 0.004$]. Bidimensional N-back working memory training can effectively improve the verbal working memory and spatial working memory capacity.

3.3. Situational Awareness

3.3.1. Task Performance

The descriptive statistics and analysis of variance for the task performance measures of the correct rate and response time are shown in Figure 5. Regarding accuracy, the training group remained essentially unchanged [$M_1 = 87.50$, $SD_1 = 5.89$; $M_2 = 88.00$, $SD_2 = 3.44$]; the control group also did not change significantly [$M_1 = 87.73$, $SD_1 = 3.44$, $M_2 = 82.27$, $SD_2 = 5.64$]. Regarding the reaction time, the training group's reaction time decreased after training [$M_1 = 82.27$, $SD_1 = 26.52$; $M_2 = 72.92$, $SD_2 = 19.72$], and there was no significant change in the control group [$M_1 = 79.12$, $SD_1 = 24.57$, $M_2 = 83.46$, $SD_2 = 12.89$].

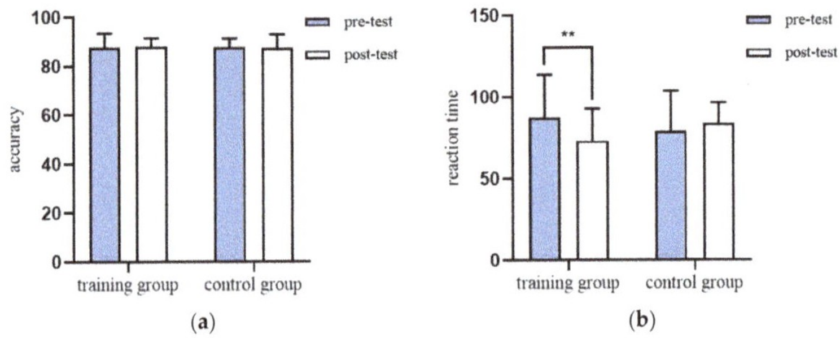

** denotes $p < 0.01$

Figure 5. Comparison of task performance differences before and after training. (**a**) Accuracy; (**b**) reaction time.

Before training, the independent-samples *t*-test found no statistical difference between the training group and the control group in terms of the correctness and response time before training [correctness: $T(19) = -0.109$, $p = 0.914$; response time: $T(19) = 0.731$, $p = 0.474$]. A 2 (pre-test, post-test) × 2 (training group, control group) repeated-measures ANOVA was conducted for all task performance indicators; for the indicator of correctness, the main effects of the grouping, the time of measurement and the interaction between the grouping and time were not significant [$p > 0.05$]. Regarding the reaction time, the main effect of the grouping was not significant [$F(1, 19) = 0.018$, $p = 0.895$, $\eta^2 = 0.001$], the main effect of the time of measurement was not significant [$F(1, 19) = 2.830$, $p = 0.109$, $\eta^2 = 0.130$], and the interaction between the grouping and time was significant [$F(1, 19) = 9.846$, $p = 0.005 < 0.01$, $\eta^2 = 0.341$]. Further simple effects analyses showed that there was no significant difference between the training group and the control group before and after the training. There was also no significant difference in correctness [$p > 0.05$]; however, the reaction time of the training group was significantly reduced after training compared to that before training [$F(1, 19) = 11.088$, $p = 0.004 < 0.01$, $\eta^2 = 0.369$], while the reaction time of the control group was not significantly different before and after training [$p > 0.05$].

3.3.2. SA Scale Data

The measurement index of the situational awareness scale included two aspects, the objective measurement of the SAGAT scale and the subjective measurement of the 3D-SART scale. The descriptive results and analysis of variance are shown in Figure 6. Both the SAGAT and SART scores in the training group were improved before and after training [SAGAT: $M_1 = 6.00$, $SD_1 = 1.10$, $M_2 = 8.10$, $SD_2 = 1.10$; SART: $M_1 = 17.60$, $SD_1 = 5.98$; $M_2 = 23.70$, $SD_2 = 4.56$], while there was no significant change in the control group before and after training [SAGAT: $M_1 = 5.06$, $SD_1 = 1.59$, $M_2 = 5.36$, $SD_2 = 2.01$; SART: $M_1 = 14.09$, $SD_1 = 8.99$, $M_2 = 16.09$, $SD_2 = 5.45$].

Before training, the independent-sample *t*-test found that there was no statistical difference in the objective and subjective situational awareness scales between the training group and the control group before training [SAGAT: $T(19) = 1.558$, $p = 0.136$; SART: $T(19) = 1.042$, $p = 0.311$]. A 2 (pre-test, post-test) × 2 (training group, control group) repeated-measures ANOVA was performed on the situational awareness scale scores of the two groups, and the main effect of the group was significant [SAGAT: $F(1, 19) = 10.254$, $p = 0.005 < 0.01$, $\eta^2 = 0.351$; SART: $F(1, 19) = 5.222$, $p = 0.034 < 0.05$, $\eta^2 = 0.216$], the main effect of the measurement time was significant [SAGAT: $F(1, 19) = 13.028$, $p = 0.002 < 0.01$, $\eta^2 = 0.407$; SART: $F(1, 19) = 7.515$, $p = 0.013 < 0.05$, $\eta^2 = 0.283$], the SAGAT score group had a significant interaction with the measurement time [$F(1, 19) = 7.285$, $p = 0.014 < 0.05$, $\eta^2 = 0.277$] and the SART group had no significant interaction with the measurement time [$p > 0.05$]. Further simple effects analysis of the SAGAT scale data showed that the training group exhibited a significant improvement in both context awareness scale scores after training compared to before training [F(1, 19) = 18.995, $p \leq 0.001$, $\eta^2 = 0.500$], while there was no significant difference in the control group before and after training [F(1, 19) = 0.435, $p = 0.517$, $\eta^2 = 0.022$]. Bidimensional N-back working memory training can effectively improve the SAGAT and SART scores within the situational awareness scale.

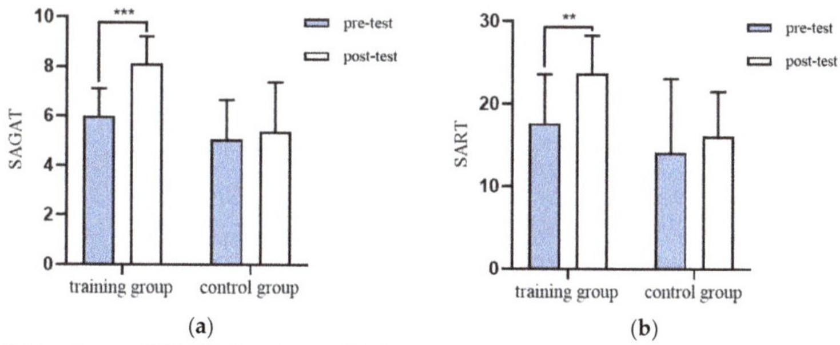

** denotes $p < 0.01$; *** denotes $p < 0.001$

Figure 6. Comparison of differences in situational awareness before and after training. (**a**) SAGAT scale; (**b**) SART scale.

3.3.3. EEG

EEG is usually recorded simultaneously at multiple electrode sites in different locations on the scalp, which are spread across the frontal, temporal, parietal, and occipital regions of the brain. All EEG data were segmented and subjected to a fast Fourier transform, averaging them into six different frequency divisions: δ (1~4 Hz), θ (4~8 Hz), α_1 (8~10 Hz), α_2 (10~13 Hz), β_1 (13~18 Hz), and β_2 (18~30 Hz). Many non-neuroanatomical factors have adverse effects on the absolute power. In order to facilitate the comparison of the resting spontaneous EEG power among different categories of subjects and eliminate the influence of individual differences on the relative power, the ratio of each rhythm to the full frequency band, i.e., the relative power of each frequency band, was calculated in this work. Some studies have pointed out that although the relative power is greatly affected by noise, it still shows better performance than the absolute power. The relative power is measured as the ratio of each band's power to the total band power. The descriptive results for the relative power are shown in Table 1. After training, the relative power of the θ and α_1 frequency bands was increased, while the other frequency bands remained unchanged. There was no significant change in the control group before and after training. The spectral topographic map shows the relative power of six frequency bands. In Figure 7, the data for each band in the two groups are normalized, and the color bar displayed below the figure is standardized within the range of 0 to 1, where red indicates higher activation. In the training group, the activation of the α_1, α_2 bands was more obvious; there was no significant difference in the control group.

Table 1. Average of each frequency band before and after training.

Frequency Band	Training Group		Control Group	
	Pre-Training	Post-Training	Pre-Training	Post-Training
δ	0.737 ± 0.070	0.699 ± 0.059	0.739 ± 0.065	0.745 ± 0.075
θ	0.113 ± 0.023	0.159 ± 0.025	0.121 ± 0.034	0.118 ± 0.033
α_1	0.046 ± 0.017	0.051 ± 0.012	0.047 ± 0.018	0.043 ± 0.012
α_2	0.059 ± 0.019	0.051 ± 0.015	0.056 ± 0.022	0.051 ± 0.017
β_1	0.024 ± 0.008	0.023 ± 0.007	0.022 ± 0.006	0.023 ± 0.009
β_2	0.021 ± 0.040	0.018 ± 0.035	0.018 ± 0.034	0.019 ± 0.038

A 2 (training group, control group) × 2 (pre-test, post-test) repeated-measures analysis of variance was conducted on the relative power of each frequency band, and the analysis results are shown in Table 2. The main effect of the θ band grouping was not significant, but the measurement time and interaction were significant. Further simple effects analysis showed that the θ relative power of the training group was significantly increased after training compared with that before training [$F(1, 19) = 15.458$, $p < 0.001$, $\eta^2 = 0.449$], while the control group had no significant difference before and after training [$F(1, 19) = 0.077$, $p = 0.785$, $\eta^2 = 0.004$]. The main effect edge of the α_1 band was significant, the measurement time edge was significant and the interaction was not significant. The main effect of the α_2 band group was significant, and the measurement time and interaction were significant. Further simple effects analysis showed that the relative power of α2 in the training group was significantly increased compared with that before training [$F(1, 19) = 12.963$, $p = 0.002 < 0.001$, $\eta^2 = 0.406$], while there was no significant difference in the control group before and after training [$F(1, 19) = 12.963$, $\eta^2 = 0.406$]. [$F(1, 19) = 0.033$, $p = 0.858$, $\eta^2 = 0.002$]. The main effect of the β1 band group was significant, and the measurement time and interaction were significant. Simple effects analysis showed that the relative power of β1 in the training group was significantly increased after training compared with that before training [$F(1, 19) = 21.142$, $p < 0.001$, $\eta^2 = 0.527$], while there was no significant difference in the control group before and after training [$F(1, 19) = 0.304$. $p = 0.588$, $\eta^2 = 0.016$]. The main effects of other frequency bands were not significant.

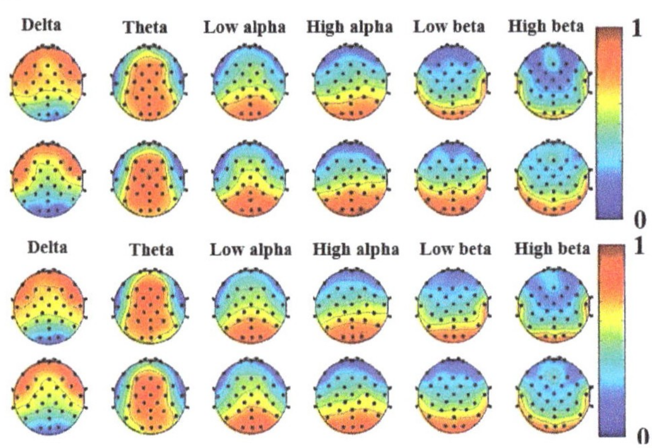

Figure 7. Comparison of brain topographic maps before and after training. (**a**) Working memory training group; (**b**) control group.

Table 2. ANOVA for each frequency band.

Norm		F	p	η^2
	group	2.376	0.140	0.111
δ	measure time	0.426	0.522	0.022
	group × measure time	0.781	0.388	0.039
	group	2.699	0.117	0.124
θ	measure time	7.047	0.016	0.271
	group × measure time	9.220	0.007	0.327
	group	3.050	0.097	0.138
α_1	measure time	3.293	0.085	0.148
	group × measure time	2.519	0.129	0.117
	group	4.232	0.054	0.182
α_2	measure time	7.459	0.013	0.282
	group × measure time	6.152	0.023	0.245
	group	7.476	0.013	0.282
β_1	measure time	5.287	0.033	0.218
	grouping × measure time	6.785	0.017	0.263
	group	0.360	0.555	0.019
β_2	measure time	0.106	0.749	0.006
	group × measure time	0.425	0.522	0.022

4. Discussion

In this study, the effectiveness of the working memory training process was investigated through the performance of the double-dimensional N-back training task. Similarly to previous research, it was found that the working memory capacity is malleable, and the most direct evidence of the plasticity of working memory comes from the direct training performance [20].

An improvement in working memory capacity is an important manifestation of enhanced working memory abilities, which is consistent with the results of previous, similar studies [12]. In terms of task performance, there was no improvement in the accuracy rate of the training group compared with the control group. The reason that the training effect did not transfer to the accuracy rate was probably the simplicity of the shearer monitoring task, and the accuracy rate had almost reached the maximum before training. In terms of the reaction time, working memory training effectively improves individuals' inhibitory control and attention transfer abilities, thereby reducing their reaction times, which has been shown in previous studies [21]. In terms of the SA scale, the subjective SART data showed that the interaction between the group and the measurement time was not significant, and the lack of consistency with previous results may be due to the fact that the participants could not accurately recognize their own states [22]. The objective SAGAT scale results showed that the SAGAT scores of the training group were significantly higher than those of the control group after training, indicating that the working memory training also had a positive impact on SA during the monitoring of the shearer. Previous studies have confirmed the strong relationship between working memory as measured by the SAGAT and SA [23]. In addition, EEG indicators showed the significant activation of the θ, α_2, and β_1 relative power after training. According to previous studies, the θ, α_2, and β_1 bands are related to situational awareness and have the potential to characterize SA [24].

This study also concludes that working memory training can prevent the failure of situational awareness, and several factors may be related to it. Endsley's information processing theoretical model emphasizes that attention and working memory are the key influencing factors that restrict individuals' access to SA [5]. The dual-dimensional N-back task involves multiple execution processes [25], including paying attention to the rational allocation of resources, the rapid updating and storage of memory information, monitoring the dual-task process at the same time, and inhibiting interference from irrelevant information. In addition, the brain has plasticity [26], which means that working memory training can prevent the failure of situational awareness and effectively improve the SA levels of individuals. In conclusion, the similar basic cognitive processing between working memory and SA and the adaptability of the working memory capacity are two important factors for the transfer of working memory training's effects to SA.

Nevertheless, there were some shortcomings in this study. First, due to the time limit during the training, the final time of the training effect is not clear at present. Secondly, there are many means of SA detection; each has its own advantages and disadvantages, and the emphasis of each method may be different. The methods selected in this study may not be complete, and more methods need to be selected to compensate for this. Third, the interval before and after the experiment was short, no long-term follow-up has been conducted, and the sustainability of the improvement in situational awareness needs further observation. These conclusions are for reference only.

5. Conclusions

This study preliminarily proved the effectiveness of working memory training in preventing the failure of situational awareness. The working memory capacity, operational performance of monitoring tasks and situational awareness level were all improved after training, and the θ, α_2 and β_1 bands emerged as sensitive indices to characterize SA. Based on this, the following suggestions are made: (1) operators should intentionally strengthen their memory training on a daily basis to enhance employees' thinking and thus develop their resilience in the face of complex decisions; (2) enterprises can design relevant situational simulations by incorporating accident cases and combining them with actual work scenarios in order to perceive the operating environment more efficiently and prevent situational awareness failures.

This study provides a scientific basis for the improvement of situational awareness in the operation of monitoring and control equipment, such as among coal miners, as well as new ideas and directions for the maintenance of good situational awareness in industries that require a high degree of vigilance and accurate judgment, such as mining, aviation, and transportation. In the future, we can consider using a variety of data for machine learning modeling to provide more theoretical support for SA measurement.

Author Contributions: Conceptualization, X.Y. and R.S.; methodology, X.Y. and R.S.; software, R.S.; validation, X.Y.; formal analysis, R.S.; investigation, R.S.; resources, X.Y. and L.S.; data curation, R.S.; writing—original draft preparation, R.S.; writing—review and editing, X.Y. and L.S.; visualization, R.S.; supervision, X.Y. and L.S.; project administration, X.Y.; funding acquisition, X.Y. and L.S. All authors have read and agreed to the published version of the manuscript.

Funding: This research was funded by the Social Science Foundation project of Shaanxi Province, grant number 2023R006, and by the Natural Science Foundation of Shaanxi Province, grant number 2023-JC-YB-624.

Institutional Review Board Statement: This study was conducted in accordance with the Declaration of Helsinki and approved by the Ethics Committee, Management, Xi'an University of Science and Technology (protocol code: 2024091501, date: 15 September 2024).

Informed Consent Statement: Informed consent was obtained from all subjects involved in the study.

Data Availability Statement: The raw data supporting the conclusions of this article will be made available by the authors on request.

Acknowledgments: The authors thank all of the participants who participated in the study.

Conflicts of Interest: The authors declare no conflicts of interest.

References

1. Kang, F.; Tian, R.; Zhang, J.; Wang, D.; Xie, H.; Pan, C.; Zho, T.; Deng, J. Study on microstructure evolution characteristics of coal in low temperature oxidation stage in goaf of Longfeng Coal Mine. *Combust. Sci. Technol.* **2024**, 1–20. [CrossRef]
2. Ma, L.; Zhang, P.Y.; Chen, X.K.; He, Y.P.; Wei, G.M.; Fan, J. Numerical investigation of coupling hazard zone of coal spontaneous combustion and gas in gob for high-gas mines. *Case Stud. Therm. Eng.* **2024**, *63*, 105341. [CrossRef]
3. Mahdinia, M.; Mohammadfam, I.; Mirzaei Aliabadi, M.; Aghaei, H.; Soltanian, A.R.; Soltanzadeh, A. The mediating effect of workers' situation awareness on the relationship between work-related factors and human error: A path analysis approach. *Int. J. Occup. Saf. Ergon.* **2022**, *28*, 1958–1966. [CrossRef] [PubMed]
4. Jiang, J.; Karran, A.J.; Coursaris, C.K.; Léger, P.M.; Beringer, J. A situation awareness perspective on human-AI interaction: Tensions and opportunities. *Int. J. Hum.–Comput. Interact.* **2023**, *39*, 1789–1806. [CrossRef]
5. Endsley, M.R.; Jones, D.G. Situation Awareness Oriented Design: Review and Future Directions. *Int. J. Hum.–Comput. Interact.* **2024**, *40*, 1487–1504. [CrossRef]
6. Cak, S.; Say, B.; Misirlisoy, M. Effects of working memory, attention, and expertise on pilots' situation awareness. *Cogn. Technol. Work* **2020**, *22*, 85–94. [CrossRef]
7. Chi, Y.; Nie, J.; Zhong, L.; Wang, Y.; Delahaye, D. A Review of Situational Awareness in Air Traffic Control. *IEEE Access* **2023**, *11*, 134040–134057. [CrossRef]
8. Broadbent, D.P.; D'Innocenzo, G.; Ellmers, T.J.; Parsler, J.; Szameitat, A.J.; Bishop, D.T. Cognitive load, working memory capacity and driving performance: A preliminary fNIRS and eye tracking study. *Transp. Res. Part F Traffic Psychol. Behav.* **2023**, *92*, 121–132. [CrossRef]
9. Buschman, T.J.; Miller, E.K. Working memory is complex and dynamic, like your thoughts. *J. Cogn. Neurosci.* **2022**, *35*, 17–23. [CrossRef]
10. Unsworth, N.; Robison, M.K. Working memory capacity and sustained attention: A cognitive-energetic perspective. *J. Exp. Psychol. Learn. Mem. Cogn.* **2020**, *46*, 77. [CrossRef]
11. Finc, K.; Bonna, K.; He, X.; Lydon-Staley, D.M.; Kühn, S.; Duch, W.; Bassett, D.S. Dynamic reconfiguration of functional brain networks during working memory training. *Nat. Commun.* **2020**, *11*, 2435. [CrossRef] [PubMed]
12. Fuchs, L.S.; Fuchs, D.; Sterba, S.K.; Barnes, M.A.; Seethaler, P.M.; Changas, P. Building word-problem solving and working memory capacity: A randomized controlled trial comparing three intervention approaches. *J. Educ. Psychol.* **2022**, *114*, 1633. [CrossRef]

13. Studer-Luethi, B.; Toermaenen, M.; Margelisch, K.; Hogrefe, A.B.; Perrig, W.J. Effects of Working Memory Training on Children's Memory and Academic Performance: The Role of Training Task Features and Trainee's Characteristics. *J. Cogn. Enhanc.* **2022**, *6*, 340–357. [CrossRef]
14. Wang, Y.F.; Bian, W.; Wei, J.; Hu, S. Anxiety-reducing effects of working memory training: A systematic review and meta-analysis. *J. Affect. Disord.* **2023**, *331*, 269–278. [CrossRef]
15. Wang, L.; Sheng, A.; Chang, L.; Zhou, R. Improving fluid intelligence of children through working memory training: The role of inhibition control. *Front. Psychol.* **2022**, *13*, 1025036. [CrossRef]
16. Braarud, P.Ø. Investigating the validity of subjective workload rating (NASA TLX) and subjective situation awareness rating (SART) for cognitively complex human–machine work. *Int. J. Ind. Ergon.* **2021**, *86*, 103233. [CrossRef]
17. Endsley, M.R. The divergence of objective and subjective situation awareness: A meta-analysis. *J. Cogn. Eng. Decis. Mak.* **2020**, *14*, 34–53. [CrossRef]
18. Li, Q.; Ng, K.K.; Simon, C.M.; Yiu, C.Y.; Lyu, M. Recognising situation awareness associated with different workloads using EEG and eye-tracking features in air traffic control tasks. *Knowl. Based Syst.* **2023**, *260*, 110179. [CrossRef]
19. Zhang, T.; Yang, J.; Liang, N.; Pitts, B.J.; Prakah-Asante, K.; Curry, R.; Duerstock, B.; Wachs, J.P.; Yu, D. Physiological measurements of situation awareness: A systematic review. *Hum. Factors* **2023**, *65*, 737–758. [CrossRef]
20. Könen, T.; Strobach, T.; Karbach, J. *Working memory training. Cognitive Training: An Overview of Features and Applications*; Springer: Cham, Switzerland, 2021; pp. 155–167.
21. Johann, V.; Könen, T.; Karbach, J. The unique contribution of working memory, inhibition, cognitive flexibility, and intelligence to reading comprehension and reading speed. *Child Neuropsychol.* **2020**, *26*, 324–344. [CrossRef]
22. Rizzo, R.; Knight, S.P.; Davis, J.R.; Newman, L.; Duggan, E.; Kenny, R.A.; Romero-Ortuno, R. SART and Individual Trial Mistake Thresholds: Predictive Model for Mobility Decline. *Geriatrics* **2021**, *6*, 85. [CrossRef] [PubMed]
23. Liu, Y.; Chen, N.; Rau, P.L.P. The effects of motivation and noise on situation awareness: A study based on SAGAT and EEG. *Int. J. Ind. Ergon.* **2023**, *97*, 103491. [CrossRef]
24. Kästle, J.L.; Anvari, B.; Krol, J.; Wurdemann, H.A. Correlation between Situational Awareness and EEG signals. *Neurocomputing* **2021**, *432*, 70–79. [CrossRef]
25. Li, W.; Zhang, Q.; Qiao, H.; Jin, D.; Ngetich, R.K.; Zhang, J.; Jin, Z.; Li, L. Dual n-back working memory training evinces superior transfer effects compared to the method of loci. *Sci. Rep.* **2021**, *11*, 3072. [CrossRef] [PubMed]
26. Nęcka, E.; Gruszka, A.; Hampshire, A.; Sarzyńska-Wawer, J.; Anicai, A.E.; Orzechowski, J.; Nowak, M.; Wójcik, N.; Sandrone, S.; Soreq, E. The effects of working memory training on brain activity. *Brain Sci.* **2021**, *11*, 155. [CrossRef]